ANIXTER

Anixter is a leading global supplier of communications products used to connect voice, video, data and security systems. Anixter is also a leading provider of electrical and electronic wire and cable, fasteners and other small components to build, repair and maintain a variety of systems and equipment. We bundle our products with our innovative Supply Chain Services to take cost out of our customers' business processes and ensure they get the right product, the first time.

TRADEMARKS

The following registered trademarks appear in this handbook:

Alumel® is a registered trademark of Concept Alloys, LLC
Chromel® is a registered trademark of Concept Alloys, LLC
Copperweld® is a registered trademark of Copperweld Steel Company
CSA® is a registered trademark of the Canadian Standards Association
Duofoil® is a registered trademark of Belden
Flamarrest® is a registered trademark of Belden
Halar® is a registered trademark of Solvay Solexis
Hypalon® is a registered trademark of E. I. DuPont de Nemours & Company
Hypot® is a registered trademark of Associated Research, Inc.
IBM® is a registered trademark of International Business Machines Corporation
Kapton® is a registered trademark of E. I. DuPont de Nemours & Company
Kevlar® is a registered trademark of E. I. DuPont de Nemours & Company
K FIBER® is a registered trademark of General Cable Corporation
Kynar® is a registered trademark of Arkema, Inc.
Loc-Trac® is a registered trademark of Alpha Wire
Megger® is a registered trademark of Megger Group Ltd.
Mylar® is a registered trademark of E. I. DuPont de Nemours & Company
NEC® is a registered trademark of the National Fire Protection Association
Nicrosil® is a registered trademark of Harrison Alloys, Inc.
Nisil® is a registered trademark of Harrison Alloys, Inc.
Nomex® is a registered trademark of E. I. DuPont de Nemours & Company
Polywater® is a registered trademark of American Polywater Corporation
Scotch® is a registered trademark of 3M
Scotchlok® is a registered trademark of 3M
Solef® is a registered trademark of Solvay Solexis
Teflon® is a registered trademark of E. I. DuPont de Nemours & Company
Tefzel® is a registered trademark of E. I. DuPont de Nemours & Company
Tyrin™ is a trademark of Dow Chemical Company
UL® is a registered trademark of Underwriters Laboratories, Inc.
UniBlend® is a registered trademark of General Cable Corporation
UniShield® is a registered trademark of General Cable Corporation
UniStrand® is a registered trademark of Belden Inc.
Valox® is a registered trademark of General Electric Company
Z-Fold® is a registered trademark of Belden
Zytel® is a registered trademark of E. I. DuPont de Nemours & Company

REFERENCE INFORMATION

Information in this handbook has been drawn from many publications of the leading wire and cable companies in the industry and authoritative sources in their latest available editions. Some of these include:

- American Society for Testing and Materials (ASTM)
- Canadian Standards Association (CSA)
- Electronic Industries Alliance (EIA)
- Institute of Electrical and Electronics Engineers (IEEE)
- Insulated Cable Engineers Association (ICEA)
- International Electrotechnical Commission (IEC)
- National Electrical Manufacturers Association (NEMA)
- National Fire Protection Association (NFPA)
- Naval Ship Engineering Center (NAVSEC)
- Telecommunications Industry Association (TIA)
- Underwriters Laboratories (UL).

Note: National Electrical Code (NEC) is a registered trademark of the National Fire Protection Association, Quincy, MA. The term, National Electrical Code, as used herein, means the triennial publication constituting the National Electrical Code and is used with permission of the National Fire Protection Association.

PREFACE

The Anixter Wire and Cable Technical Handbook is an easily accessible collection of engineering and technical information about electrical and electronic cable as well as their related products. Primarily intended for individuals who design, specify or troubleshoot wire and cable systems, the Anixter Wire and Cable Technical Handbook contains information about topics such as:

- Basic principles of electricity
- Conductor, insulation and jacket materials along with their electrical and mechanical properties
- Cable types, selection criteria and application guidelines for electrical and optical wire and cable
- Installation and testing guidelines and recommendations
- Application tips for cable accessories such as connectors, lugs and terminations
- Packaging, handling and shipping guidelines
- References to hundreds of key domestic and international wire and cable standards
- Conversion tables (e.g., AWG to mm^2) and basic engineering equations used in the industry.

The information contained in this handbook will assist engineers and individuals in designing and constructing safe, reliable, cost-effective and environmentally responsible electrical and communications networks.

Anixter wishes to acknowledge the contributions of the many individuals who assisted in the preparation of this edition of the handbook. Anixter would especially want to recognize the efforts of Deborah Altman, Dana Anderson, Milt Barber, Harmony Borris, Eric Bulington, Mark Fordham, Jeff Gronemeyer, Andy Jimenez, Jason Kreke, Al Landsberger, Jonathan Meyer, Nader Moubed and Bill Wilkens.

Anixter hopes it has succeeded in making this handbook the best in the industry and welcomes your comments and suggestions for improvements in future editions.

ABOUT ANIXTER

The Anixter Difference

Anixter is proud to serve more than 100,000 customers across 49 countries every day with world-class inventory, global capabilities, technical expertise and Supply Chain Services. Anixter's three specialized sales forces focus on Enterprise Cabling & Security Solutions, Electrical and Electronic Wire & Cable, and Fasteners.

Global Capabilities

Anixter is proud of the quality products it distributes, but the reason customers count on Anixter day in and day out isn't just about world-class, quality products.

Anixter's customers know they can count on Anixter to deliver consistent, superior service and support around the world. Anixter doesn't just ship products from the U.S. to various countries. By stocking inventory in each country, Anixter is a specialist at moving inventory across borders efficiently and cost effectively.

Being physically located in 49 countries around the globe means Anixter's in-country sales specialists are familiar with the trends and needs of the local markets, as well as the local currencies, standards and customs. Anixter speaks 30 different languages to serve its many customers' diverse needs.

Products

It seems simple enough: you need something, you call a distributor and buy it. Unfortunately, nothing is really that simple. In the real world, there are complicated systems, small parts, and constantly changing technical developments and requirements. Just determining what a project needs can be an all-consuming process, one that is only aggravated by hard to find or out of stock items. Missing a crucial component can significantly add costs and delays to a project.

Anixter takes the worry out of just-in-time product availability. By maintaining more than $1 billion in its global inventory, Anixter has the product expertise to make sure the right products arrive when and where they are needed.

Technical Expertise

Across the world, Anixter has more than 2,600 sales specialists who support its customers. These sales specialists have years of experience in specifying products and solutions for customers. Anixter's salespeople are well-trained to truly identify and understand any needs and requirements.

CONTENTS

1. BASIC PRINCIPLES OF ELECTRICITY

1.1 ELECTRICITY

Electricity, simply put, is the flow of electric current along a conductor. This electric current takes the form of free electrons that transfer from one atom to the next. Thus, the more free electrons a material has, the better it conducts. There are three parameters involved in the electrical equation: the volt, the ampere and the ohm.

1.2 THE VOLT

The pressure that is put on free electrons that causes them to flow is known as electromotive force (EMF). The volt is the unit of pressure, i.e., the volt is the amount of electromotive force required to push a current of one ampere through a conductor with a resistance of one ohm.

1.3 THE AMPERE

The ampere defines the flow rate of electric current. For instance, when one coulomb (or 6×10^{18} electrons) flows past a given point on a conductor in one second, it is defined as a current of one ampere.

1.4 THE OHM

The ohm is the unit of resistance in a conductor. Three things determine the amount of resistance in a conductor: its size, its material, e.g., copper or aluminum, and its temperature. A conductor's resistance increases as its length increases or diameter decreases. The more conductive the materials used, the lower the conductor resistance becomes. Conversely, a rise in temperature will generally increase resistance in a conductor.

1.5 OHM'S LAW

Ohm's Law defines the correlation between electric current (I), voltage (V), and resistance (R) in a conductor. Ohm's Law can be expressed as:

$$\mathbf{V = I \times R}$$

Where: V = volts
I = amps
R = ohms

1.6 AMPACITY

Ampacity is the amount of current a conductor can handle before its temperature exceeds accepted limits. These limits are given in the National Electrical Code (NEC), the Canadian Electrical Code and in other engineering documents such as those published by the Insulated Cable Engineers Association (ICEA). It is important to know that many external factors affect the ampacity of an electrical conductor and these factors should be taken into consideration before selecting the conductor size.

1.7 ELECTRICAL SYSTEMS

1.7.1 Medium Voltage

The most widely used medium voltage (2.4 to 35 kV) alternating current (AC) electrical distribution systems in North America are illustrated below:

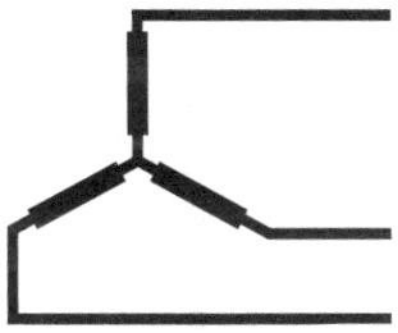

Figure 1.1 – Three phase wye (star), three wire

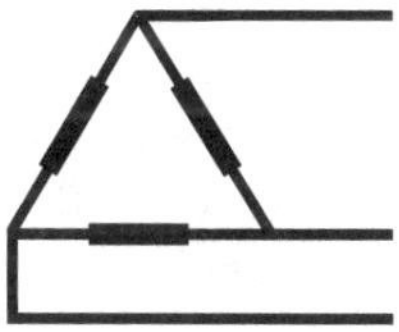

Figure 1.2 – Three phase delta, three wire

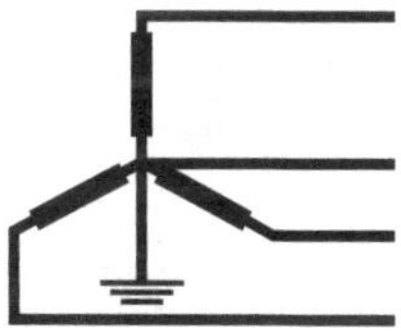

Figure 1.3 – Three phase star, four wire, grounded neutral

1.7.2 Low Voltage

Typical low-voltage systems (0 to 2,000 V) are illustrated below:

Figure 1.4 – Three phase wye (star), three wire, grounded neutral

Figure 1.5 – Three phase delta, four wire, grounded neutral

2. CONDUCTORS

CONDUCTORS

The conductor is the metallic component of cables through which electrical power or electrical signals are transmitted. Conductor size is usually specified by American Wire Gauge (AWG), circular mil area or in square millimeters.

AWG

The American Wire Gauge (sometimes called Brown and Sharpe or B. and S.) is used almost exclusively in the USA for copper and aluminum wire. The Birmingham Wire Gauge (BWG) is used for steel armor wire.

The diameters according to the AWG are defined as follows: The diameter of size 4/0 (sometimes written 0000) equals 0.4600 inch and that of size #36 equals 0.0050 inch; the intermediate sizes are found by geometric progression. That is, the ratio of the diameter of one size to that of the next smaller size (larger gauge number) is:

$$\sqrt[39]{\frac{0.4600}{0.0050}} = 1.122932$$

Circular Mil

Sizes larger than 4/0 are specified in terms of the total area of a cross-section of the copper in circular mils (cmil). A circular mil is a unit of area equal to the area of a circle one mil in diameter. It is $\pi/4$ (equal to 0.7854) of a square mil (one mil = 0.001 inch). The area of a circle in circular mils is therefore equal to the square of its diameter in mils. A solid wire one inch in diameter has an area of 1,000,000 cmils, whereas one square inch equals $4/\pi \times 1{,}000{,}000$ cmils (equal to 1,273,200 cmils).

Square Millimeters

Metric sizes are given in terms of square millimeters (mm^2).

Conductor Characteristics

Relative electrical and thermal conductivities of common metal conductors are as follows:

Table 2.1–Relative Electrical and Thermal Conductivities of Common Conductor Materials

Metal	Relative Electrical Conductivity at 20°C	Relative Thermal Conductivity at 20°C
Silver	106	108
Copper (annealed)	100	100
Copper (hard drawn)	97	–
Gold	72	76
Aluminum	62	56
Magnesium	39	41
Zinc	29	29
Nickel	25	15
Cadmium	23	24

Continued on next page >>

Table 2.1–Relative Electrical and Thermal Conductivities of Common Conductors (Continued)

Metal	Relative Electrical Conductivity at 20°C	Relative Thermal Conductivity at 20°C
Cobalt	18	17
Iron	17	17
Platinum	16	18
Tin	15	17
Steel	12	–
Lead	8	9

Additional electrical properties can be found in Section 7 of this handbook.

2.1 STRAND TYPES

2.1.1 Concentric Strand

A concentric stranded conductor consists of a central wire or core surrounded by one or more layers of helically laid wires. Each layer after the first has six more wires than the preceding layer. Except in compact stranding, each layer is usually applied in a direction opposite to that of the layer under it.

If the core is a single wire and if it and all of the outer strands have the same diameter, the first layer will contain six wires; the second, twelve; the third, eighteen; etc.

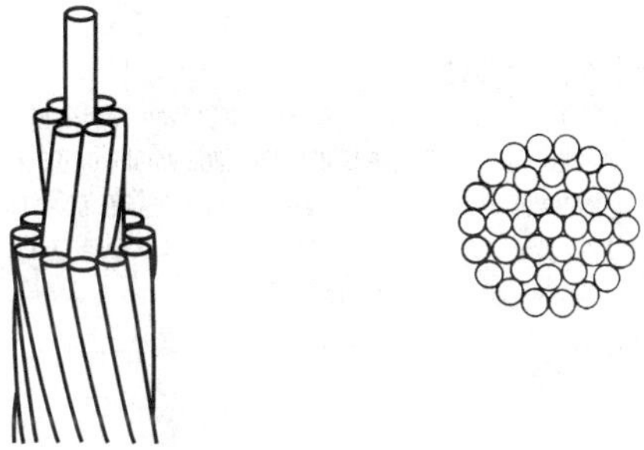

Figure 2.1–Concentric Strand

2.1.2 Bunch Strand

The term bunch stranding is applied to a collection of strands twisted together in the same direction without regard to the geometric arrangement.

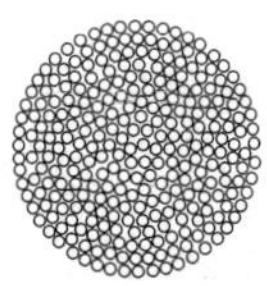

Figure 2.2–Bunch Strand

2.1.3 Rope Strand

A rope stranded conductor is a concentric stranded conductor each of whose component strands is itself stranded. A rope stranded conductor is described by giving the number of groups laid together to form the rope and the number of wires in each group.

Figure 2.3–Rope Strand

2.1.4 Sector Conductor

A sector conductor is a stranded conductor whose cross-section is approximately the shape of a sector of a circle. A multiple conductor insulated cable with sector conductors has a smaller diameter than the corresponding cable with round conductors.

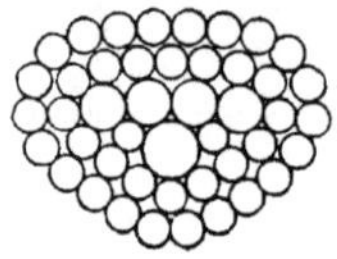

Figure 2.4–Sector Conductor

2.1.5 Segmental Conductor

A segmental conductor is a round, stranded conductor composed of three or four sectors slightly insulated from one another. This construction has the advantage of lower AC resistance (less skin effect).

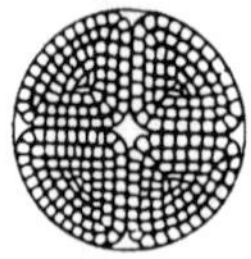

Figure 2.5–Segmental Conductor

2.1.6 Annular Conductor

An annular conductor is a round, stranded conductor whose strands are laid around a suitable core. The core is usually made wholly or mostly of nonconducting material. This construction has the advantage of lower total AC resistance for a given cross-sectional area of conducting material by eliminating the greater skin effect at the center.

Figure 2.6–Annular Conductor

2.1.7 Compact Strand

A compact stranded conductor is a round or sector conductor having all layers stranded in the same direction and rolled to a predetermined ideal shape. The finished conductor is smooth on the surface and contains practically no interstices or air spaces. This results in a smaller diameter.

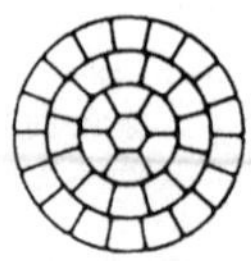

Figure 2.7–Compact Strand

2.1.8 Compressed Strand

Compressed conductors are intermediate in size between standard concentric conductors and compact conductors. A comparison is shown below:

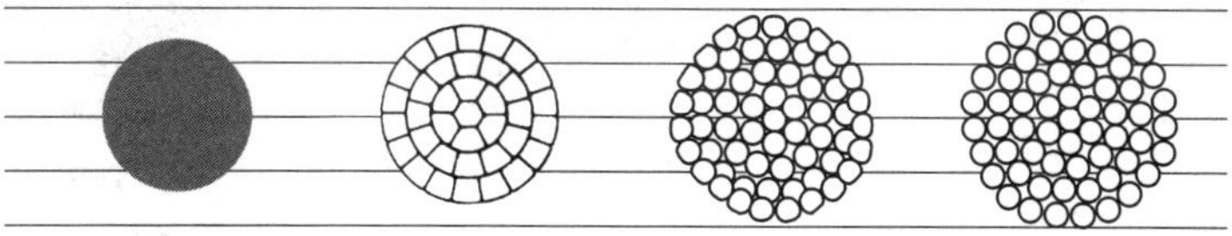

Figure 2.8–Comparative Sizes and Shapes of 1,000 kcmil Conductors

In a concentric stranded conductor, each individual wire is round and considerable space exists between wires. In a compressed conductor, the conductor has been put through a die that "squeezes out" some of the space between wires. In a compact conductor each wire is preformed into a trapezoidal shape before the wires are stranded together into a finished conductor. This results in even less space between wires. A compact conductor is, therefore, the smallest in diameter (except for a solid conductor, of course). Diameters for common conductor sizes are given in the table below.

Table 2.2–Diameters for Copper and Aluminum Conductors

Conductor Size		Nominal Diameters (in.)			
(AWG)	(kcmil)	Solid	Class B Compact	Class B Compressed	Class B Concentric
8	16.51	0.1285	0.134	0.141	0.146
6	26.24	0.1620	0.169	0.178	0.184
4	41.74	0.2043	0.213	0.225	0.232
3	52.62	0.2294	0.238	0.252	0.260
2	66.36	0.2576	0.268	0.283	0.292
1	83.69	0.2893	0.299	0.322	0.332
1/0	105.6	0.3249	0.336	0.361	0.373
2/0	133.1	0.3648	0.376	0.406	0.419
3/0	167.8	0.4096	0.423	0.456	0.470
4/0	211.6	0.4600	0.475	0.512	0.528
–	250	0.5000	0.520	0.558	0.575
–	300	0.5477	0.570	0.611	0.630
–	350	0.5916	0.616	0.661	0.681
–	400	0.6325	0.659	0.706	0.728
–	450	0.6708	0.700	0.749	0.772

Continued on next page >>

Table 2.2–Diameters for Copper and Aluminum Conductors (Continued)

Conductor Size		Nominal Diameters (in.)			
(AWG)	(kcmil)	Solid	Class B Compact	Class B Compressed	Class B Concentric
–	500	0.7071	0.736	0.789	0.813
–	550	0.7416	0.775	0.829	0.855
–	600	0.7746	0.813	0.866	0.893
–	650	0.8062	0.845	0.901	0.929
–	700	0.8367	0.877	0.935	0.964
–	750	0.8660	0.908	0.968	0.998
–	800	0.8944	0.938	1.000	1.031
–	900	0.9487	0.999	1.061	1.093
–	1,000	1.0000	1.060	1.117	1.152

Sources: ASTM B8 and B496
ICEA S-95-658 (NEMA WC-70)

2.2 COATINGS

There are three materials commonly used for coating a copper conductor. These are tin, silver and nickel.

Tin is the most common and is used for improved corrosion resistance and solderability.

Silver plated conductors are used in high-temperature environments (150°C–200°C). It is also used for high-frequency applications where silver's high conductivity (better than copper) and the "skin effect" work together to reduce attenuation at high frequencies.

Nickel coatings are used for conductors that operate between 200°C and 450°C. At these high temperatures, copper oxidizes rapidly if not nickel plated. One drawback of nickel, however, is its poor solderability.

2.3 TENSILE STRENGTH OF COPPER WIRE

Table 2.3–Tensile Strength of Copper Wire

Size	Soft or Annealed	Medium Hard Drawn	Hard Drawn
(AWG)	Max. Breaking Load (lb.)	Min. Breaking Load (lb.)	Min. Breaking Load (lb.)
4/0	6,000	6,970	8,140
3/0	4,750	5,660	6,720
2/0	3,765	4,600	5,530
1/0	2,985	3,730	4,520
1	2,435	3,020	3,690
2	1,930	2,450	3,010
3	1,535	1,990	2,440
4	1,215	1,580	1,970
6	765	1,010	1,280
8	480	645	825
10	315	410	530
12	200	262	335
14	125	167	215
16	78.5	106	135
18	49.5	67.6	85.5
20	31.0	43.2	54.2
22	19.4	27.3	34.1
24	12.7	17.5	21.7
26	7.94	11.1	13.7
28	4.99	7.02	8.64
30	3.14	4.48	5.47
32	2.01	2.90	3.53
34	1.25	1.82	2.20
36	0.79	1.16	1.40

2.4 COPPER STRAND PROPERTIES

2.4.1 Strand Classes

Table 2.4–Strand Classes

ASTM Standard	Construction	Class	Application
B8	Concentric lay	AA	For bare conductors – usually used in overhead lines.
		A	For bare conductors where greater flexibility than is afforded by Class AA is required.
		B	For conductors insulated with various materials such as EP, XLP, PVC, etc. This is the most common class.
		C	For conductors where greater flexibility is required than is provided by Class B.
		D	N/A
B173	Rope lay with concentric stranded members	G	Conductor constructions having a range of areas from 5,000,000 circular mils and employing 61 stranded members of 19 wires each down to No. 14 AWG containing seven stranded members stranded members of seven wires each. Typical uses are for portable (flexible) conductors and similar applications.
		H	Conductor constructions having a range of areas from 5,000,000 circular mils and employing 91 stranded members of 19 wires each down to No. 9 AWG containing 19 stranded members of seven wires each. Typical uses are for rubber-jacketed cords and conductors where flexibility is required, such as for use on take-up reels, over sheaves and apparatus conductors.

Continued on next page >>

Table 2.4–Strand Classes (Continued)

ASTM Standard	Construction	Class	Conductor Size	Individual Wire Size		Application
			(kcmil/AWG)	Diameter (in.)	(AWG)	
B172	Rope lay with with bunch stranded members	I	Up to 2,000	0.0201	24	Typical use is for special apparatus cable.
		K	Up to 2,000	0.0100	30	Typical use is for portable cord.
		M	Up to 1,000	0.0063	34	Typical use is for welding cable.
B174	Bunch stranded	I	7, 8, 9, 10	0.0201	24	Rubber-covered conductors.
		J	10, 12, 14, 16, 18, 20	0.0126	28	Fixture wire.
		K	10, 12, 14, 16, 18, 20	0.0100	30	Fixture wire, flexible cord and portable cord.
		L	10, 12, 14, 16, 18, 20	0.0080	32	Fixture wire and portable cord with greater flexibility than Class K.
		M	14, 16, 18, 20	0.0063	34	Heater cord and light portable cord.
		O	16, 18, 20	0.0050	36	Heater cord with greater flexibility than Class M.
		P	16, 18, 20	0.0040	38	More flexible conductors than provided in preceding classes.
		Q	18, 20	0.0031	40	Oscillating fan cord. Very good flexibility.

Source: Compiled from ASTM standards listed

2.4.2 Solid Copper

Table 2.5–Standard Nominal Diameters and Cross-sectional Areas of Solid Copper Wire

Size (AWG)	Diameter (mils)	Cross-sectional Area (kcmils)	Weight (lb./1,000 ft.)	Breaking Strength Soft or Annealed (lb.)
3/0	409.6	167.800	–	–
2/0	364.8	133.100	–	–
1/0	324.9	105.600	–	–
1	289.3	83.690	–	–
2	257.6	66.360	–	–
3	229.4	52.620	–	–
4	204.3	41.740	–	–
5	181.9	33.090	–	–
6	162.0	26.240	–	–
7	144.3	20.820	–	–
8	128.5	16.510	–	–
9	114.4	13.090	–	–
10	101.9	10.380	31.43	314.0
11	90.7	8.230	24.92	249.0
12	80.8	6.530	19.77	197.5
13	72.0	5.180	15.68	156.6
14	64.1	4.110	12.43	142.2
15	57.1	3.260	9.86	98.5
16	50.8	2.580	7.82	78.1
17	45.3	2.050	6.20	61.9
18	40.3	1.620	4.92	49.2
19	35.9	1.290	3.90	39.0
20	32.0	1.020	3.09	30.9
21	28.5	0.812	2.452	24.5
22	25.3	0.640	1.945	19.4
23	22.6	0.511	1.542	15.4
24	20.1	0.404	1.223	12.7
25	17.9	0.320	0.970	10.1
26	15.9	0.253	0.770	7.98
27	14.2	0.202	0.610	6.33
28	12.6	0.159	0.484	5.02
29	11.3	0.128	0.384	3.98

Continued on next page >>

Table 2.5–Standard Nominal Diameters and Cross-sectional Areas of Solid Copper Wire (Continued)

Size (AWG)	Diameter (mils)	Cross-sectional Area (kcmils)	Weight (lb./1,000 ft.)	Breaking Strength Soft or Annealed (lb.)
30	10.0	0.100	0.303	3.16
31	8.9	0.0792	0.241	2.50
32	8.0	0.0640	0.191	1.99
33	7.1	0.0504	0.152	1.58
34	6.3	0.0397	0.120	1.25
35	5.6	0.0314	0.095	0.990
36	5.0	0.0250	0.076	0.785
37	4.5	0.0202	0.060	0.623
38	4.0	0.0160	0.048	0.494
39	3.5	0.0122	–	–
40	3.1	0.00961	–	–
41	2.8	0.00784	–	–
42	2.5	0.00625	–	–
43	2.2	0.00484	–	–
44	2.0	0.00400	–	–
45	1.76	0.00310	–	–
46	1.57	0.00246	–	–

Source: ASTM B258, Specification for Standard Nominal Diameters and Cross-Sectional Areas of AWG Sizes of Solid Round Wires Used as Electrical Conductors

2.4.3 Class B, C and D Copper Strand

Table 2.6–Class B Concentric-Lay-Stranded Copper Conductors

Size	Number of Wires	Diameter of Each Strand	Weight	Nominal Overall Diameter
(AWG or kcmil)		(mils)	(lb./1,000 ft.)	(in.)
5,000	217	151.8	15,890	2.58
4,500	217	144	14,300	2.45
4,000	217	135.8	12,590	2.31
3,500	169	143.9	11,020	2.16
3,000	169	133.2	9,353	2.00
2,500	127	140.3	7,794	1.82
2,000	127	125.5	6,175	1.63
1,900	127	122.3	5,866	1.59
1,800	127	119.1	5,558	1.55
1,750	127	117.4	5,402	1.53
1,700	127	115.7	5,249	1.50
1,600	127	112.2	4,940	1.46
1,500	91	128.4	4,631	1.41
1,400	91	124.0	4,323	1.36
1,300	91	119.5	4,014	1.32
1,250	91	117.2	3,859	1.30
1,200	91	114.8	3,705	1.26
1,100	91	109.9	3,396	1.21
1,000	61	128.0	3,088	1.15
900	61	121.5	2,779	1.09
800	61	114.5	2,470	1.03
750	61	110.9	2,316	1.00
700	61	107.1	2,161	0.964
650	61	103.2	2,007	0.929
600	61	99.2	1,853	0.893
550	61	95.0	1,698	0.855
500	37	116.2	1,544	0.813
450	37	110.3	1,389	0.772
400	37	104.0	1,235	0.728
350	37	97.3	1,081	0.681
300	37	90.0	926.3	0.630
250	37	82.2	711.9	0.575
4/0	19	105.5	653.3	0.528

Continued on next page >>

Table 2.6–Class B Concentric-Lay-Stranded Copper Conductors (Continued)

Size	Number of Wires	Diameter of Each Strand	Weight	Nominal Overall Diameter
(AWG or kcmil)		(mils)	(lb./1,000 ft.)	(in.)
3/0	19	94.0	518.1	0.470
2/0	19	83.7	410.9	0.419
1/0	19	74.5	325.8	0.373
1	19	66.4	258.4	0.332
2	7	97.4	204.9	0.292
3	7	86.7	162.5	0.260
4	7	77.2	128.9	0.232
5	7	68.8	102.2	0.206
6	7	61.2	81.05	0.184
7	7	54.5	64.28	0.164
8	7	48.6	50.97	0.146
9	7	43.2	40.42	0.130
10	7	38.5	32.06	0.116
12	7	30.5	20.16	0.0915
14	7	24.2	12.68	0.0726
16	7	19.2	7.974	0.0576
18	7	15.2	5.015	0.0456
20	7	12.1	3.154	0.0363

Source: ASTM B8 Specification for Concentric-Lay-Stranded Copper Conductors, Hard, Medium-Hard, or Soft

Table 2.7–Copper Strand Diameters

Conductor Size		Stranding				
(AWG)	(kcmil)	Class B Compact (in.)	Class B Compressed (in.)	Class B Concentric (in.)	Class C Concentric (in.)	Class D Concentric (in.)
14	4.11	–	0.0704	0.0727	0.0735	0.0735
13	5.18	–	0.0792	0.0816	0.0825	0.0826
12	6.53	–	0.0888	0.0915	0.0925	0.0931
11	8.23	–	0.0998	0.103	0.104	0.104
10	10.39	–	0.113	0.116	0.117	0.117
9	13.09	–	0.126	0.130	0.131	0.132
8	16.51	0.134	0.141	0.146	0.148	0.148
7	20.82	–	0.158	0.164	0.166	0.166
6	26.24	0.169	0.178	0.184	0.186	0.186
5	33.09	–	0.200	0.206	0.208	0.209
4	41.74	0.213	0.225	0.232	0.234	0.235
3	52.62	0.238	0.252	0.260	0.263	0.264
2	66.36	0.268	0.283	0.292	0.296	0.297
1	83.69	0.299	0.322	0.332	0.333	0.333
1/0	105.6	0.336	0.362	0.372	0.374	0.374
2/0	133.1	0.376	0.406	0.418	0.420	0.420
3/0	167.8	0.423	0.456	0.470	0.471	0.472
4/0	211.6	0.475	0.512	0.528	0.529	0.530
–	250	0.520	0.558	0.575	0.576	0.576
–	300	0.570	0.611	0.630	0.631	0.631
–	350	0.616	0.661	0.681	0.681	0.682
–	400	0.659	0.706	0.728	0.729	0.729
–	450	0.700	0.749	0.772	0.773	0.773
–	500	0.736	0.789	0.813	0.814	0.815
–	550	0.775	0.829	0.855	0.855	0.855
–	600	0.813	0.866	0.893	0.893	0.893
–	650	0.845	0.901	0.929	0.930	0.930
–	700	0.877	0.935	0.964	0.965	0.965
–	750	0.908	0.968	0.999	0.999	0.998
–	800	0.938	1.000	1.030	1.032	1.032
–	900	0.999	1.061	1.094	1.093	1.095
–	1,000	1.060	1.117	1.152	1.153	1.153
–	1,100	–	1.173	1.209	1.210	1.211
–	1,200	–	1.225	1.263	1.264	1.264
–	1,250	–	1.251	1.289	1.290	1.290
–	1,300	–	1.275	1.314	1.316	1.316
–	1,400	–	1.323	1.365	1.365	1.365
–	1,500	–	1.370	1.412	1.413	1.413
–	1,600	–	1.415	1.459	1.460	1.460

Continued on next page >>

Table 2.7–Copper Strand Diameters (Continued)

Conductor Size		Stranding				
(AWG)	(kcmil)	Class B Compact (in.)	Class B Compressed (in.)	Class B Concentric (in.)	Class C Concentric (in.)	Class D Concentric (in.)
–	1,700	–	1.459	1.504	1.504	1.504
–	1,750	–	1.480	1.526	1.527	1.527
–	1,800	–	1.502	1.548	1.548	1.549
–	1,900	–	1.542	1.590	1.590	1.591
–	2,000	–	1.583	1.632	1.632	1.632
–	2,500	–	1.769	1.824	1.824	1.824
–	3,000	–	1.938	1.998	1.998	1.998

2.4.4 Class H Copper

Table 2.8–Class H Rope-Lay-Stranded Copper Conductors

Size (AWG or kcmil)	Number of Strands	Construction	Nominal Diameter of Each Strand (in.)	Nominal O.D. (in.)	Nominal Weight (lb./1,000 ft.)
8	133	19x7	0.0111	0.167	52
7	133	19x7	0.0125	0.188	65
6	133	19x7	0.0140	0.210	82
5	133	19x7	0.0158	0.237	105
4	133	19x7	0.0177	0.266	132
3	133	19x7	0.0199	0.299	167
2	133	19x7	0.0223	0.335	208
2	259	37x7	0.0160	0.336	210
1	259	37x7	0.0180	0.378	266
1/0	259	37x7	0.0202	0.424	334
2/0	259	37x7	0.0227	0.477	422
3/0	259	37x7	0.0255	0.536	533
3/0	427	61x7	0.0198	0.535	532
4/0	259	37x7	0.0286	0.601	670
4/0	427	61x7	0.0223	0.602	675
250	427	61x7	0.0242	0.653	795
300	427	61x7	0.0265	0.716	953
350	427	61x7	0.0286	0.772	1,110

Continued on next page >>

Table 2.8–Class H Rope-Lay-Stranded Copper Conductors (Continued)

Size (AWG or kcmil)	Number of Strands	Construction	Nominal Diameter of Each Strand (in.)	Nominal O.D. (in.)	Nominal Weight (lb./1,000 ft.)
400	427	61x7	0.0306	0.826	1,270
450	427	61x7	0.0325	0.878	1,435
500	427	61x7	0.0342	0.923	1,590
550	703	37x19	0.0280	0.980	1,770
600	703	37x19	0.0292	1.022	1,920
650	703	37x19	0.0304	1.064	2,085
700	703	37x19	0.0316	1.106	2,255
750	703	37x19	0.0327	1.145	2,410
800	703	37x19	0.0337	1.180	2,560
900	703	37x19	0.0358	1.253	2,895
1,000	703	37x19	0.0377	1.320	3,205
1,100	703	37x19	0.0396	1.386	3,535
1,200	703	37x19	0.0413	1.446	3,845
1,250	703	37x19	0.0422	1.477	4,015
1,300	703	37x19	0.0430	1.505	4,170
1,400	703	37x19	0.0446	1.561	4,485
1,500	703	37x19	0.0462	1.617	4,815
1,600	1,159	61x19	0.0372	1.674	5,145
1,700	1,159	61x19	0.0383	1.724	5,455
1,750	1,159	61x19	0.0389	1.751	5,625
1,800	1,159	61x19	0.0394	1.773	5,770
1,900	1,159	61x19	0.0405	1.823	6,100
2,000	1,159	61x19	0.0415	1.868	6,400

Source: ICEA S-95-658 (NEMA 70) Appendix K

2.4.5 Class I Copper

Table 2.9–Class I (24 AWG Strands) Rope-Lay-Stranded Copper Conductors

Size (AWG or kcmil)	Construction	Nominal Number of Strands	Nominal O.D. (in.)	Nominal Weight (lb./1,000 ft.)
10	1x26	26	0.125	32.5
9	1x33	33	0.138	41
8	1x41	41	0.156	51
7	1x52	52	0.185	65
6	7x9	63	0.207	80
5	7x12	84	0.235	105
4	7x15	105	0.263	134
3	7x19	133	0.291	169
2	7x23	161	0.319	205
1	7x30	210	0.367	267
1/0	19x14	266	0.441	342
2/0	19x18	342	0.500	439
3/0	19x22	418	0.549	537
4/0	19x28	532	0.613	683
250	7x7x13	637	0.682	825
300	7x7x15	735	0.737	955
350	7x7x18	882	0.800	1,145
400	7x7x20	980	0.831	1,270
450	7x7x23	1,127	0.894	1,460
500	7x7x25	1,225	0.941	1,590
550	7x7x28	1,372	0.980	1,780
600	7x7x30	1,470	1.027	1,905
650	19x7x12	1,596	1.152	2,090
700	19x7x13	1,729	1.194	2,260
750	19x7x14	1,862	1.235	2,435
800	19x7x15	1,995	1.290	2,610
900	19x7x17	2,261	1.372	2,965
1,000	19x7x19	2,527	1.427	3,305
1,100	19x7x21	2,793	1.495	3,655
1,200	19x7x22	2,926	1.537	3,830
1,250	19x7x23	3,059	1.564	4,000
1,300	19x7x24	3,192	1.605	4,175
1,400	19x7x26	3,458	1.674	4,560

Continued on next page >>

Table 2.9–Class I (24 AWG Strands) Rope-Lay-Stranded Copper Conductors (Continued)

Size (AWG or kcmil)	Construction	Nominal Number of Strands	Nominal O.D. (in.)	Nominal Weight (lb./1,000 ft.)
1,500	19x7x28	3,724	1.715	4,875
1,600	19x7x30	3,990	1.797	5,220
1,700	19x7x32	4,256	1.852	5,570
1,750	19x7x33	4.389	1.880	5,745
1,800	19x7x34	4,522	1.921	5,920
1,900	19x7x36	4,788	1.976	6,265
2,000	19x7x37	4,921	2.003	6,440

Source: ICEA S-75-381 (NEMA WC 58) Appendix K

2.4.6 Class K Copper

Table 2.10–Class K (30 AWG Strands) Rope-Lay-Stranded Copper Conductors

Size	Rope-Lay with Bunch Stranding		Bunch Stranding		Weight
(AWG or kcmil)	Nominal Number of Strands	Strand Construction	Nominal Number of Strands	Approx. O.D. (in.)	(lb./1,000 ft.)
1,000	10,101	37x7x39	10,101	1.419	3,270
900	9,065	37x7x35	9,065	1.323	2,935
800	7,980	19x7x60	7,980	1.305	2,585
750	7,581	19x7x57	7,581	1.276	2,455
700	6,916	19x7x52	6,916	1.207	2,240
650	6,517	19x7x49	6,517	1.166	2,110
600	5,985	19x7x45	5,985	1.125	1,940
550	5,453	19x7x41	5,453	1.056	1,765
500	5,054	19x7x38	5,054	0.988	1,635
450	4,522	19x7x34	4,522	0.933	1,465
400	3,990	1x7x30	3,990	0.878	1,290
350	3,458	19x7x26	3,458	0.809	1,120
300	2,989	7x7x61	2,989	0.768	960
250	2,499	7x7x51	2,499	0.682	802
4/0	2,107	7x7x43	2,107	0.627	676
3/0	1,666	7x7x34	1,666	0.533	535
2/0	1,323	7x7x27	1,323	0.470	425
1/0	1,064	19x56	1,064	0.451	338

Continued on next page >>

Table 2.10–Class K (30 AWG Strands) Rope-Lay-Stranded Copper Conductors (Continued)

Size	Rope-Lay with Bunch Stranding		Bunch Stranding		Weight
(AWG or kcmil)	Nominal Number of Strands	Strand Construction	Nominal Number of Strands	Approx. O.D. (in.)	(lb./1,000 ft.)
1	836	19x44	836	0.397	266
2	665	19x35	665	0.338	211
3	532	19x28	532	0.304	169
4	420	7x60	420	0.272	132
5	336	7x48	336	0.235	106
6	266	7x38	266	0.202	84
7	210	7x30	210	0.179	66
8	168	7x24	168	0.157	53
9	133	7x19	133	0.146	42
10	–	–	104	0.126	32.5
12	–	–	65	0.101	20.3
14	–	–	41	0.078	12.8
16	–	–	26	0.060	8.0
18	–	–	16	0.048	5.0
20	–	–	10	0.038	3.2

Sources: ASTM B172 Specification for Rope-Lay-Stranded Copper Conductors Having Bunch-Stranded Members and ICEA S-75-381 (NEMA WC58) Appendix K

2.4.7 Class M Copper

Table 2.11–Class M (34 AWG Strands) Rope-Lay-Stranded Copper Conductors

Size	Rope-Lay with Bunch Stranding		Bunch Stranding		Weight
(AWG or kcmil)	Nominal Number of Strands	Strand Construction	Nominal Number of Strands	Approx. O.D. (in.)	(lb./1,000 ft.)
1,000	25,193	61x7x59	25,193	1.404	3,240
900	22,631	61x7x53	22,631	1.331	2,910
800	20,069	61x7x47	20,069	1.256	2,580
750	18,788	61x7x44	18,788	1.207	2,415
700	17,507	61x7x41	17,507	1.183	2,250
650	16,226	61x7x38	16,226	1.133	2,085
600	14,945	61x7x35	14,945	1.084	1,920
550	13,664	61x7x32	13,664	1.035	1,755
500	12,691	37x7x49	12,691	0.997	1,630
450	11,396	37x7x44	11,396	0.940	1,465
400	10,101	37x7x39	10,101	0.901	1,300
350	8,806	37x7x34	8,806	0.825	1,130
300	7,581	19x7x57	7,581	0.768	975
250	6,384	19x7x48	6,384	0.713	821
4/0	5,320	19x7x40	5,320	0.645	684
3/0	4,256	19x7x32	4,256	0.576	547
2/0	3,325	19x7x25	3,325	0.508	427
1/0	2,646	7x7x54	2,646	0.423	337
1	2,107	7x7x43	2,107	0.376	268
2	1,666	7x7x34	1,666	0.337	212
3	1,323	7x7x27	1,323	0.305	169
4	1,064	19x56	1,064	0.269	134
5	836	19x44	836	0.240	105
6	665	19x35	665	0.215	84
7	532	19x28	532	0.196	67
8	420	7x60	420	0.162	53
9	336	7x48	336	0.146	42
10	259	7x37	259	0.126	32.5
12	168	7x24	168	0.101	21.0
14	–	–	104	0.078	12.8
16	–	–	65	0.060	8.0
18	–	–	41	0.048	5.0
20	–	–	26	0.038	3.2

Sources: ASTM B172 Specification for Rope-Lay-Stranded Copper Conductors Having Bunch-Stranded Members and ICEA S-75-381 (NEMA WC58)

2.5 ALUMINUM STRAND PROPERTIES

2.5.1 Solid Aluminum

Table 2.12–Aluminum 1350 Solid Round Wire

Size (AWG or kcmil)	Diameter (mils)	Cross-Sectional Area (kcmils)	Weight (lb./1,000 ft.)
4/0	460.0	211.600	194.4
3/0	409.6	167.800	154.2
2/0	364.8	133.100	122.3
1/0	324.9	105.600	97.0
1	289.3	83.690	76.91
2	257.6	66.360	60.98
3	229.4	52.620	48.36
4	204.3	41.740	38.35
5	181.9	33.090	30.40
6	162.0	26.240	24.12
7	144.3	20.820	19.13
8	128.5	16.510	15.17
9	114.4	13.090	12.03
10	101.9	10.380	9.542
11	90.7	8.230	7.559
12	80.8	6.530	5.999
13	72.0	5.180	4.764
14	64.1	4.110	3.776
15	57.1	3.260	2.996
16	50.8	2.580	2.371
17	45.3	2.050	1.886
18	40.3	1.620	1.492
19	35.9	1.290	1.184
20	32.0	1.020	0.9410
21	28.5	0.812	0.7464
22	25.3	0.640	0.5882
23	22.6	0.511	0.4693
24	20.1	0.404	0.3713
25	17.9	0.320	0.2944
26	15.9	0.253	0.2323

Continued on next page >>

Table 2.12–Aluminum 1350 Solid Round Wire (Continued)

Size (AWG or kcmil)	Diameter (mils)	Cross-Sectional Area (kcmils)	Weight (lb./1,000 ft.)
27	14.2	0.202	0.1853
28	12.6	0.159	0.1459
29	11.3	0.128	0.1173
30	10.0	0.100	0.09189

Source: ASTM B609 Specification for Aluminum 1350 Round Wire, Annealed and Intermediate Tempers

2.5.2 Class B Aluminum

Table 2.13–Class B Concentric-Lay-Stranded Compressed, Reverse-Lay Aluminum 1350 Conductors

Size (AWG or kcmil)	Number of Wires	Diameter of Each Wire (mils)	Nominal Overall Diameter (in.)
4,000	217	135.8	–
3,500	169	143.9	–
3,000	169	133.2	–
2,500	127	140.3	–
2,000	127	125.5	1.583
1,900	127	122.3	1.542
1,800	127	119.1	1.502
1,750	127	117.4	1.480
1,700	127	115.7	1.459
1,600	127	112.2	1.415
1,500	91	128.4	1.370
1,400	91	124.0	1.323
1,300	91	119.5	1.275
1,250	91	117.2	1.250
1,200	91	114.8	1.225
1,100	91	109.9	1.173
1,000	61	128.0	1.117
900	61	121.5	1.060
800	61	114.5	1.000
750	61	110.9	0.968
700	61	107.1	0.935

Continued on next page >>

ANIXTER

Table 2.13–Class B Concentric-Lay-Stranded Compressed, Reverse-Lay Aluminum 1350 Conductors (Continued)

Size (AWG or kcmil)	Number of Wires	Diameter of Each Wire (mils)	Nominal Overall Diameter (in.)
650	61	103.2	0.901
600	61	99.2	0.866
550	61	95.0	0.829
500	37	116.2	0.789
450	37	110.3	0.749
400	37	104.0	0.706
350	37	97.3	0.661
300	37	90.0	0.611
250	37	82.2	0.558
4/0	19	105.5	0.512
3/0	19	94.0	0.456
2/0	19	83.7	0.405
1/0	19	74.5	0.362
1	19	66.4	0.322
2	7	97.4	0.283
3	7	86.7	0.252
4	7	77.2	0.225
5	7	68.8	0.200
6	7	61.2	0.178
7	7	54.5	0.159
8	7	48.6	0.142
9	7	43.2	0.126
10	7	38.5	0.113
12	7	30.5	0.089
14	7	24.2	0.071
16	7	19.2	–
18	7	15.2	–
20	7	12.1	–

Source: ASTM B231 Concentric-Lay-Stranded Aluminum 1350 Conductors

2.5.3 ACSR

Table 2.14–Concentric-Lay-Stranded Aluminum Conductors, Coated-Steel Reinforced (ACSR)

Size	Stranding		Weight
(AWG or kcmil)	Aluminum Number/Diameter (in.)	Steel Number/Diameter (in.)	(lb./1,000 ft.)
2,156	84/0.1602	19/0.0961	2,511
1,780	84/0.1456	19/0.0874	2,074
1,590	54/0.1716	19/0.1030	2,044
1,590	45/0.1880	7/0.1253	1,792
1,431	54/0.1628	19/0.0977	1,840
1,431	45/0.1783	7/0.1189	1,613
1,272	54/0.1535	19/0.0921	1,635
1,272	45/0.1681	7/0.1121	1,434
1,113	54/0.1436	19/0.0862	1,431
1,113	45/0.1573	7/0.1049	1,255
954.0	54/0.1329	7/0.1329	1,229
954.0	45/0.1456	7/0.0971	1,075
795.0	45/0.1329	7/0.0886	896
795.0	26/0.1749	7/0.1360	1,094
795.0	24/0.1820	7/0.1213	1,023
636.0	26/0.1564	7/0.1216	875
636.0	24/0.1628	7/0.1085	819
636.0	18/0.1880	1/0.1880	690
556.5	26/0.1463	7/0.1138	766
556.5	24/0.1523	7/0.1015	717
556.5	18/0.1758	1/0.1758	604
477.0	30/0.1261	7/0.1261	747
477.0	26/0.1354	7/0.1053	657
477.0	24/0.1410	7/0.0940	615
477.0	18/0.1628	1/0.1628	518
397.5	26/0.1236	7/0.0961	547
397.5	24/0.1287	7/0.0858	512
397.5	18/0.1486	1/0.1486	432
336.4	30/0.1059	7/0.1059	527
336.4	26/0.1137	7/0.0884	463
336.4	18/0.1367	1/0.1367	365
266.8	26/0.1013	7/0.0788	367
266.8	18/0.1217	1/0.1217	290

Continued on next page >>

ANIXTER

Table 2.14–Concentric-Lay-Stranded Aluminum Conductors, Coated-Steel Reinforced (ACSR) (Continued)

Size	Stranding		Weight
(AWG or kcmil)	Aluminum Number/Diameter (in.)	Steel Number/Diameter (in.)	(lb./1,000 ft.)
4/0	6/0.1878	1/0.1878	291.1
211.3	12/0.1327	7/0.1327	527.5
203.2	16/0.1127	19/0.0977	676.8
190.8	12/0.1261	7/0.1261	476.3
176.9	12/0.1214	7/0.1214	441.4
3/0	6/0.1672	1/0.1672	230.8
159.0	12/0.1151	7/0.1151	396.8
134.6	12/0.1059	7/0.1059	336.0
2/0	6/0.1489	1/0.1489	183.1
110.8	12/0.0961	7/0.0961	276.6
1/0	6/0.1327	1/0.1327	145.2
101.8	12/0.0921	7/0.0921	254.1
80.0	8/0.1000	1/0.1670	149.0
2	7/0.0974	1/0.1299	106.7
2	6/0.1052	1/0.1052	91.3
4	7/0.0772	1/0.1029	67.0
4	6/0.0834	1/0.0834	57.4
6	6/0.0661	1/0.0661	36.1

Source: ASTM B232 Specification for Concentric-Lay-Stranded Aluminum Conductors, Coated-Steel Reinforced (ACSR)

2.6 ADDITIONAL CONDUCTOR PROPERTIES

2.6.1 Stranding, Diameter, Area and DC Resistance (32 Through 4/0 AWG)

Table 2.15–Stranding, Diameter, Area and DC Resistance

Size	Stranding	Diameter		Conductor Area		Copper DC Resistance at 20°C	
(AWG)	(No./AWG)	(in.)	(mm)	(cmils)	(mm^2)	(ohms/ft.)	(ohms/km)
32	Solid	0.008	0.203	63.0	0.032	176.1	577.8
	7/40	0.010	0.254	67.3	0.034	176.0	577.5
30	Solid	0.010	0.254	100.0	0.051	106.6	349.8
	7/38	0.012	0.305	112.0	0.057	96.2	315.6
28	Solid	0.013	0.320	159.0	0.081	70.8	232.3
	7/36	0.015	0.381	175.0	0.089	67.5	221.5
	19/40	0.016	0.406	182.6	0.093	58.9	193.3
26	Solid	0.016	0.406	253.0	0.128	44.5	146.0
	7/32	0.019	0.483	278.0	0.141	42.5	139.4
	19/38	0.021	0.533	304.0	0.154	38.9	127.6
24	Solid	0.020	0.511	404.0	0.205	27.2	89.2
	7/32	0.024	0.610	448.0	0.227	25.7	84.3
	10/34	0.023	0.584	397.0	0.201	28.8	94.5
	19/36	0.025	0.635	475.0	0.241	24.9	81.7
22	Solid	0.025	0.635	475.0	0.326	16.7	54.8
	7/30	0.031	0.787	700.0	0.355	16.6	54.5
	16/34	0.030	0.762	635.0	0.322	18.0	59.1
	19/34	0.032	0.813	754.0	0.382	15.5	50.9
20	Solid	0.032	0.813	1,020	0.517	10.5	34.5
	7/28	0.038	0.965	1,113	0.564	10.3	33.8
	10/30	0.036	0.914	1,000	0.507	11.4	37.4
	19/32	0.040	1.02	1,197	0.607	9.48	31.1
	26/34	0.037	0.940	1,032	0.523	11.3	37.1
18	Solid	0.040	1.02	1,620	0.821	6.77	22.2
	7/26	0.048	1.22	1,771	0.897	6.45	21.2
	16/30	0.046	1.17	1,600	0.811	6.15	23.5
	19/30	0.050	1.27	1,900	0.963	6.10	20.0
	41/34	0.047	1.19	1,627	0.824	7.08	23.2
16	Solid	0.051	1.29	2,580	1.31	4.47	14.7
	7/24	0.058	1.47	2,828	1.43	4.04	13.3
	19/29	0.057	1.45	2,426	1.23	4.82	15.8
	26/30	0.050	1.52	2,600	1.32	4.39	14.4
	65/34	0.060	1.52	2,580	1.32	4.47	14.7

Continued on next page >>

Table 2.15–Stranding, Diameter, Area and DC Resistance (Continued)

Size	Stranding	Diameter		Conductor Area		Copper DC Resistance at 20°C	
(AWG)	(No./AWG)	(in.)	(mm)	(cmils)	(mm²)	(ohms/ft.)	(ohms/km)
14	Solid	0.064	1.63	4,110	2.08	2.68	8.79
	19/27	0.071	1.80	3,838	1.94	3.05	10.0
	41/30	0.077	1.96	4,100	2.08	2.81	9.22
12	Solid	0.081	2.05	6,530	3.31	1.65	5.41
	19/25	0.090	2.29	6,080	3.08	1.87	6.14
	65/30	0.091	2.31	6,500	3.29	1.82	5.97
10	Solid	0.102	2.59	10,380	5.26	1.11	3.64
	37/36	0.112	2.84	9,361	4.74	1.38	4.53
	105/30	0.130	3.30	10,500	5.32	1.10	3.61
8	Solid	0.129	3.28	16,500	8.36	0.718	2.36
	133/29	0.166	4.22	17,024	8.63	0.710	2.33
	168/30	0.174	4.42	16,800	8.51	0.700	2.30
6	Solid	0.162	4.12	26,240	13.3	0.440	1.44
	133/27	0.210	5.33	26,866	13.6	0.430	1.41
	266/30	0.204	5.18	26,600	13.5	0.440	1.44
4	Solid	0.204	5.18	41,740	21.1	0.275	0.902
	133/25	0.257	6.53	42,560	21.6	0.290	0.951
	420/30	0.257	6.53	42,000	21.3	0.280	0.919
2	Solid	0.258	6.54	66,360	33.6	0.172	0.564
	665/30	0.338	8.59	66,500	33.7	0.180	0.591
1	Solid	0.289	7.34	83,690	42.4	0.142	0.466
	817/30	0.328	8.33	81,700	41.4	0.140	0.459
1/0	Solid	0.325	8.26	105,600	53.5	0.111	0.364
	1,045/30	0.410	10.4	104,500	53.0	0.120	0.394
2/0	Solid	0.365	9.27	133,100	67.4	0.088	0.289
	1,330/30	0.496	12.6	133,000	67.4	0.099	0.325
3/0	Solid	0.410	10.4	167,800	85.0	0.070	0.230
	1,661/30	0.464	11.8	166,100	84.2	0.068	0.223
4/0	Solid	0.460	11.7	211,600	107	0.055	0.180
	2,104/30	0.608	15.4	210,400	107	0.060	0.197

2.6.2 Stranding, Diameter, Area, DC Resistance and Weight (20 AWG Through 2,000 kcmil)

Table 2.16–Copper Conductor Stranding, Diameter, Area, Weight and DC Resistance

Nominal Area		Size	Number/Diameter of Individual Wires		Overall Diameter		Nominal Weight		DC Resistance at 20°C (68°F)	
(mm²)	(cmils)	(AWG)	(in.)	(mm)	(in.)	(mm)	(lb./1,000 ft.)	(kg/km)	(ohms/1,000 ft.)	(ohms/km)
0.50	987	–	1/0.032	1/0.613	0.032	0.81	3.100	4.613	10.13	32.33
–	1,020	20	7/0.0121	7/0.307	0.036	0.91	3.157	4.098	10.22	33.77
0.75	1,480	–	1/0.036	1/0.991	0.039	0.99	4.603	6.851	6.820	22.37
–	1,620	18	1/0.403	1/1.02	0.040	1.02	4.917	7.316	6.387	20.95
–	1,620	18	7/0.0152	7/0.386	0.046	1.16	4.980	7.410	6.523	21.40
1.0	1,970	–	1/0.045	1/1.14	0.045	1.14	6.130	9.122	5.127	16.80
1.0	1,970	–	7/0.017	7/0.432	0.051	1.30	6.293	9.266	5.213	17.11
–	2,580	16	1/0.0508	1/1.29	0.061	1.29	7.810	11.63	4.020	13.19
–	2,580	16	7/0.0192	7/0.488	0.058	1.46	7.877	11.82	4.087	13.41
1.5	2,960	–	1/0.055	1/1.40	0.055	1.40	9.157	13.63	3.430	11.25
1.5	2,960	–	7/0.021	7/0.533	0.063	1.60	8.837	14.14	3.417	11.21
–	4,110	14	1/0.641	1/1.63	0.064	1.63	12.44	18.51	2.524	8.281
–	4,110	14	7/0.0242	7/0.615	0.073	1.84	12.62	18.78	2.573	8.442
2.5	4,930	–	1/0.071	1/1.80	0.071	1.80	15.26	22.71	2.057	6.750
2.5	4,930	–	7/0.027	7/0.686	0.081	2.06	15.71	23.38	2.067	6.782
–	6,530	12	1/0.0808	1/2.05	0.081	2.05	19.76	29.41	1.589	5.212
–	6,530	12	7/0.0305	7/0.755	0.092	2.32	20.05	29.84	1.620	5.315
4	7,890	–	1/0.089	1/2.26	0.089	2.26	23.98	35.68	1.309	4.296
4	7,890	–	7/0.034	7/0.864	0.102	2.59	24.91	37.08	1.304	4.277
–	10,380	10	1/0.1019	1/2.59	0.102	2.59	31.43	46.77	0.999	3.277
–	10,380	10	7/0.0385	7/0.978	0.116	2.93	31.94	47.54	1.017	3.335
6	11,800	–	1/0.109	1/2.77	0.109	2.77	35.97	53.52	0.8730	2.864
6	11,800	–	7/0.042	7/0.107	0.126	3.21	38.00	56.55	0.8543	2.803
–	13,090	9	1/0.1144	7/2.91	0.1144	2.91	39.60	58.93	0.7923	2.600
–	13,090	9	7/0.0432	7/1.10	0.130	3.30	40.23	59.86	0.8073	2.649
–	16,510	8	1/0.1285	1/3.26	0.128	3.26	50.17	74.36	0.6147	2.061
–	16,510	8	7/0.0486	7/1.23	0.146	3.70	50.90	75.75	0.6380	2.093
10	19,700	–	1/0.141	1/3.58	0.141	3.58	60.17	89.54	0.5217	1.711
10	19,700	–	7/0.054	7/1.37	0.162	4.12	62.83	93.51	0.5167	1.695
–	20,820	7	1/0.1443	1/3.67	0.144	3.67	63.03	93.80	0.4980	1.634
–	20,820	7	7/0.0545	7/1.38	0.164	4.15	64.00	95.24	0.5073	1.664

Continued on next page >>

Table 2.16–Copper Conductor Stranding, Diameter, Area, Weight and DC Resistance (Continued)

Nominal Area		Size	Number/Diameter of Individual Wires		Overall Diameter		Nominal Weight		DC Resistance at 20°C (68°F)	
(mm^2)	(cmils)	(AWG)	(in.)	(mm)	(in.)	(mm)	(lb./1,000 ft.)	(kg/km)	(ohms/1,000 ft.)	(ohms/km)
–	26,240	6	1/0.162	1/4.11	0.162	4.11	79.43	118.2	0.3950	1.296
–	26,240	6	7/0.0612	7/1.55	0.184	4.66	80.73	120.1	0.4023	1.320
16	31,600	–	7/0.068	7/1.73	0.204	5.18	99.67	148.3	0.3259	1.069
–	33,090	6	7/0.0688	7/1.75	0.206	5.24	102.0	151.8	0.3183	1.044
–	41,740	4	7/0.0772	7/1.96	0.232	5.88	128.4	191.1	0.2528	0.8295
25	49,300	–	7/0.065	7/2.16	0.255	6.48	155.7	231.7	0.2176	0.6843
–	52,620	3	7/0.0867	7/2.20	0.260	6.61	162.0	241.1	0.2005	0.6577
35	69,100	–	7/0.100	7/2.54	0.300	7.62	215.5	320.7	0.1507	0.4944
35	69,100	–	19/0.061	19/1.55	0.305	7.75	218.1	324.5	0.1495	0.4909
–	83,690	1	19/0.0664	19/1.63	0.332	8.43	258.4	384.5	0.1261	0.4139
50	98,700	–	19/0.073	19/1.85	0.365	9.27	312.3	464.8	0.1044	0.3424
–	105,400	1/0	19/0.0745	19/1.89	0.373	9.46	325.3	484.1	0.10020	0.3288
–	133,100	2/0	19/0.0837	19/2.13	0.419	10.6	410.7	611.1	0.07940	0.2605
70	138,000	–	19/0.086	19/2.18	0.430	10.9	433.3	645.0	0.07520	0.2467
–	167,800	3/0	19/0.094	19/2.39	0.470	11.9	517.7	770.4	0.06293	0.2065
–	167,800	3/0	37/0.0673	37/1.71	0.471	12.0	517.0	769.4	0.06310	0.2070
95	187,000	–	19/0.101	19/2.57	0.505	12.8	597.7	889.4	0.05453	0.1789
–	211,600	4/0	19/0.1055	19/2.68	0.528	13.4	652.3	970.8	0.04997	0.1639
120	237,000	–	37/0.0811	37/0.0811	0.567	14.4	749.0	1,115	0.04357	0.1429
–	250,000	–	37/0.0822	37/0.0822	0.575	14.6	771.3	1,148	0.04230	0.1388
150	300,000	–	37/0.090	37/2.29	0.630	16.0	924.7	1,376	0.03527	0.1157
–	350,000	–	37/0.0973	37/2.47	0.681	17.3	1,081	1,609	0.03018	0.09903
185	365,000	–	37/0.100	37/2.54	0.700	17.8	1,142	1,699	0.02857	0.09375
–	400,000	–	37/0.104	37/2.64	0.728	18.5	1,235	1,838	0.02642	0.06668
240	474,000	–	37/0.114	37/2.90	0.798	20.3	1,484	2,206	0.02199	0.07214
240	474,000	–	61/0.089	61/2.26	0.801	20.3	1,491	2,219	0.02189	0.07181
–	500,000	–	37/0.1162	37/2.95	0.813	20.7	1,608	2,294	0.02116	0.06943
–	500,000	–	61/0.0905	61/2.30	0.814	20.7	1,549	2,295	0.02117	0.06944
300	592,000	–	61/0.099	61/2.51	0.891	22.6	1,842	2,746	0.02102	0.05803
–	600,000	–	61/0.0992	61/2.52	0.893	22.7	1,853	2,757	0.01762	0.05780
–	700,000	–	61/0.1071	61/2.72	0.964	24.5	2,160	3,214	0.01511	0.04959
–	750,000	–	61/0.1109	61/2.82	0.998	25.4	2,316	3,446	0.01410	0.04625
–	750,000	–	91/0.0908	91/2.31	0.999	25.4	2,316	3,447	0.01410	0.04625

Continued on next page >>

Table 2.16–Copper Conductor Stranding, Diameter, Area, Weight and DC Resistance (Continued)

Nominal Area		Size	Number/Diameter of Individual Wires		Overall Diameter		Nominal Weight		DC Resistance at 20°C (68°F)	
(mm^2)	(cmils)	(AWG)	(in.)	(mm)	(in.)	(mm)	(lb./1,000 ft.)	(kg/km)	(ohms/1,000 ft.)	(ohms/km)
400	789,000	–	61/0.114	61/2.90	1.026	26.1	2,447	3,642	0.01334	0.04377
–	800,000	–	61/0.1145	61/2.91	1.031	26.2	2,468	3,673	0.01322	0.04338
–	800,000	–	91/0.0938	91/2.38	1.032	26.2	2,471	3,678	0.01321	0.04334
500	1,000,000	–	61/0.1280	61/3.25	1.152	29.3	3,085	4,590	0.01058	0.03472
–	1,000,000	–	91/0.1048	91/2.66	1.153	29.3	3,085	4,591	0.01058	0.03472
625	1,234,000	–	91/0.117	91/2.97	1.287	32.7	3,845	5,722	0.00849	0.02786
–	1,250,000	–	91/0.1172	91/2.98	1.289	32.7	3,858	5,742	0.008460	0.02776
–	1,250,000	–	127/0.0992	127/2.52	1.290	32.8	3,858	6,741	0.008463	0.02777
–	1,500,000	–	91/0.1284	91/3.26	1.412	35.9	4,631	6,892	0.007050	0.02313
–	1,500,000	–	127/0.1087	127/2.76	1.413	35.9	4,632	6,894	0.007183	0.02312
800	1,580,000	–	91/0.132	91/3.35	1.452	36.9	4,894	7,284	0.006670	0.02188
1,000	1,970,000	–	91/0.147	91/3.73	1.617	41.1	6,070	9,033	0.005380	0.01765
–	2,000,000	–	127/0.1255	127/3.19	1.632	41.5	6,175	9,189	0.005287	0.01735
–	2,000,000	–	169/0.1088	169/2.76	1.632	41.5	6,176	9,191	0.005287	0.01735

Based on British (BSA), Canadian (CSA), American (ASTM and ICEA) and German (VDE) Standards

2.6.3 IEC Stranding

Table 2.17–Typical IEC Stranding

Cross Section	Ordinary Stranding (Class 2)	Multi-wire Stranding	Fine Wire Stranding (Class 5)	Extra-fine Wire Stranding (Class 6)			
(mm²)	No./Dia. (mm)	No./Dia. (mm)	No./Dia. (mm)	No./Dia. (mm)			
0.05	–	–	–	–	–	–	25/0.05
0.08	–	–	–	–	–	–	41/0.05
0.14	–	–	–	18/0.10	18/0.1	36/0.07	72/0.05
0.25	–	–	14/0.16	32/0.10	32/0.1	65/0.07	128/0.05
0.34	–	7/0.25	19/0.16	42/0.10	42/0.1	88/0.07	174/0.05
0.38	–	7/0.27	12/0.21	21/0.16	48/0.1	100/0.07	194/0.05
0.5	7/0.30	7/0.30	16/0.21	28/0.16	64/0.1	131/0.07	256/0.05
0.75	7/0.37	7/0.37	24/0.21	42/0.16	69/0.1	195/0.07	384/0.05
1.0	7/0.43	7/0.43	32/0.21	56/0.16	128/0.1	260/0.07	512/0.05
1.5	7/0.52	7/0.52	30/0.26	84/0.16	192/0.1	392/0.07	768/0.05
2.5	7/0.67	19/0.41	50/0.26	140/0.16	320/0.1	651/0.07	1,280/0.05
4	7/0.85	19/0.52	56/0.31	224/0.16	512/0.1	1,040/0.07	–
6	7/1.05	19/0.64	84/0.31	192/0.21	768/0.1	1,560/0.07	–
10	7/1.35	49/0.51	80/0.41	320/0.21	1,280/0.1	2,600/0.07	–
16	7/1.70	49/0.65	128/0.41	512/0.21	2,048/0.1	–	–
25	7/2.13	84/0.62	200/0.41	800/0.21	3,200/0.1	–	–
35	7/2.52	133/0.58	280/0.41	1,120/0.21	–	–	–
50	19/1.83	133/0.69	400/0.41	705/0.31	–	–	–
70	19/2.17	189/0.69	356/0.51	990/0.31	–	–	–
95	19/2.52	259/0.69	485/0.51	1,340/0.31	–	–	–
120	37/2.03	336/0.67	614/0.51	1,690/0.31	–	–	–
150	37/2.27	392/0.69	765/0.51	2,123/0.31	–	–	–
185	37/2.52	494/0.69	944/0.51	1,470/0.41	–	–	–
240	61/2.24	627/0.70	1,225/0.51	1,905/0.41	–	–	–
300	61/2.50	790/0.70	1,530x0.51	2,385x0.41	–	–	–
400	61/2.89	–	2.035x0.51	–	–	–	–
500	61/3.23	–	1,768x0.61	–	–	–	–

Note: Additional information is available in IEC 60228.

3. INSULATION AND JACKET MATERIALS

3.1 PURPOSE

Conductors need to be electrically isolated from other conductors and from the environment to prevent short circuits. Insulation is applied around a conductor to provide this isolation. Most wire and cable insulations consist of polymers (plastics), which have a high resistance to the flow of electric current. A jacket is the outermost layer of a cable whose primary function is to protect the insulation and conductor core from external physical forces and chemical deterioration.

3.2 TYPES AND APPLICATIONS

3.2.1 Thermoplastics

Chlorinated Polyethylene (CPE)
CPE is one of the few polymers available in both thermoplastic and thermoset (cross-linked) versions. As a rule, thermoset formulations have better high-temperature properties than thermoplastics but are also higher in cost. Thermoplastic CPE is more common than thermoset CPE. Properties of both thermoplastic and thermoset CPE are given in Section 3.4.

Polyvinyl Chloride (PVC)
Sometimes referred to simply as "vinyl," PVC does not usually exhibit extremely high- and low-temperature properties in one formulation. Certain formulations may have a –55°C to 105°C rating, while other common vinyls may have a –20°C to 60°C rating. The many varieties of PVC also differ in pliability and electrical properties. The price range can vary accordingly. Typical dielectric constant values range from 3.5 to 6.5.

When properly formulated, thermoplastic jackets of PVC provide cables with the ability to resist oils, acids, alkalis, sunlight, heat, weathering and abrasion. This range of properties makes PVC a suitable outer covering for such cable types as underground feeders (Type UF), control, aerial, street lighting and cables for direct burial.

PVC is frequently used as an impervious jacket over and/or under metal armor where the installation requires PVC's protective characteristics. Flamarrest is a plenum grade, PVC-based jacketing material with low smoke and low flame spread properties. Plenum-rated cables jacketed with Flamarrest meet NFPA 262 (formerly UL Standard 910).

Fluoropolymers
Fluoropolymers, with the exception of PTFE Teflon (sometimes called TFE), are extrudable thermoplastics used in a variety of low-voltage insulating situations. Fluoropolymers contain fluorine in their molecular composition, which contributes to their excellent thermal, chemical, mechanical and electrical characteristics. The most commonly used fluoropolymers are Teflon (PTFE, FEP and PFA), Tefzel (ETFE), Halar (ECTFE) and Kynar or Solef (PVDF).

Teflon has excellent electrical properties, temperature range and chemical resistance. It is not suitable where subjected to nuclear radiation and does not have good high-voltage characteristics. FEP Teflon is extrudable in a manner similar to PVC and polyethylene. This means that long wire and cable lengths are available. PTFE Teflon is extrudable in a hydraulic ram type process. Lengths are limited due to the amount of material in the ram, thickness of the insulation and preform size. PTFE must be extruded over a silver- or nickel-coated wire. The nickel- and silver-coated designs are rated 260°C and 200°C maximum, respectively. The cost of Teflon is approximately 8 to 10 times more per pound than PVC compounds.

Teflon PTFE is the original Teflon resin invented by DuPont in 1938. It is an opaque, white material, although some forms are translucent in thin sections. It does not melt in the usual sense. To coat wire for insulating purposes, Teflon PTFE is extruded around the conductor as a paste, then sintered. Conductors can also be wrapped with tape of Teflon PTFE. Maximum continuous service temperature of Teflon PTFE is 260°C (500°F).

Specific advantages of wire insulated with Teflon PTFE include:

- Nonflammability
- Extremely high insulation resistance
- Very low dielectric constant
- Small size compared to elastomer insulated wires
- Excellent lubricity for easier installation
- Chemical inertness.

Teflon FEP was also invented by DuPont and became commercially available in 1960. It has a glossy surface and is transparent in thin sections. Teflon FEP is a true thermoplastic. Wire insulated with Teflon FEP can be melt extruded by conventional methods. Maximum continuous service temperature is 400°F (205°C). Teflon FEP is an excellent nonflammable jacketing material for multiconductor cables.

Specific advantages of wire insulated with Teflon FEP include:

- High current carrying ability (ampacity)
- Easy color coding
- Smallest diameter of any high-temperature wire
- Nonflammability
- Very low moisture absorption.

Teflon PFA is a perfluoroalkoxy copolymer resin supplied by DuPont. Wire insulated with PFA is rated up to 250°C (482°F) and has excellent high-temperature creep resistance, low-temperature toughness and flame resistance.

Tefzel (ETFE) is commonly used in computer backplane wiring and has the highest abrasion and cut-through resistance of any fluoropolymer. Tefzel is a thermoplastic material having excellent electrical properties, heat resistance, chemical resistance, toughness, radiation resistance and flame resistance. Tefzel's temperature rating is –65°C to 150°C.

Halar (ECTFE) is similar to Tefzel and is also used in wirewrap applications, but because it is less expensive than Tefzel, it is often used as insulation on multipair plenum telephone cables. It has a maximum operating temperature of 125°C (UL). Halar has excellent chemical resistance, electrical properties, thermal characteristics and impact resistance. Halar's temperature rating is –70°C to 150°C.

Kynar (PVDF) is one of the least expensive fluoropolymers and is frequently used as a jacketing material on plenum cables. Because of its high dielectric constant, however, it tends to be a poor insulator. PVDF has a temperature maximum of 135°C (UL).

Polyolefins (PO)

Polyolefin is the name given to a family of polymers. The most common polyolefins used in wire and cable include polyethylene (PE), polypropylene (PP) and ethylene vinyl acetate (EVA).

Polyethylene (PE)

Polyethylene has excellent electrical properties. It has a low dielectric constant, a stable dielectric constant over a wide frequency range, and very high insulation resistance. However, polyethylene is stiff and very hard, depending on molecular weight and density. Low density PE (LDPE) is the most flexible, with high-density, high-molecular weight formulations being least flexible. Moisture resistance is excellent. Properly formulated PE has excellent weather resistance. The dielectric constant is 2.3 for solid and 1.6 for cellular (foamed) insulation. Flame retardant formulations are available, but they tend to have poorer electrical properties.

Polypropylene (PP)
Similar in electrical properties to polyethylene, this material is primarily used as an insulation material. Typically, it is harder than polyethylene. This makes it suitable for thin wall insulations. The UL maximum temperature rating may be 60°C or 80°C, but most UL styles call for 60°C maximum. The dielectric constant is typically 2.25 for solid and 1.55 for cellular designs.

Thermoplastic Elastomer (TPE)
TPE, sometimes called TPR (thermoplastic rubber), has excellent cold temperature characteristics, making it an excellent insulating and jacketing compound in cold climates. It is resistant to aging from sunlight, oxidation and atmospheric ozone. It retains most of its physical and electrical properties in the face of many severe environmental conditions such as a salt water environment. TPE compounds can be rated as high as 125°C (257°F).

TPE has good chemical resistance to all substances except hydrocarbons. It has a tendency to swell in a hydrocarbon environment, causing the material to degrade. It has good abrasion resistance. It will resist wear, cutting and impact. These properties make TPE jackets an excellent choice for use in control cables that are dragged around or frequently moved.

TPE compounds are used as insulating materials up to a 600-volt rating. The most common cables using TPE insulation are portable control cables such as SEO and SJEO.

Polyurethane (PUR)
Polyurethane is used primarily as a cable jacket material. It has excellent oxidation, oil and ozone resistance. Some formulations also have good flame resistance. It has excellent abrasion resistance. It has outstanding "memory" properties, making it an ideal jacket material for retractile cords.

3.2.2 Thermosets

Chlorinated Polyethylene (CPE)
Cross-linked chlorinated polyethylene is a material with outstanding physical and electrical properties for many cable jacket applications. It is highly resistant to cold flow (compression set) and other forms of external loading as well as heat, light and chemical attack. CPE is also often supplied in a thermoplastic (non-cross-linked) version.

CPE compares favorably with most other synthetic elastomers currently used for cable jacketing. It is resistant to ozone and ultraviolet (sunlight) degradation. Properly compounded, CPE will withstand prolonged immersion in water. It will not support combustion, but under the right conditions of excessive heat, oxygen supply and flame source, it will burn slowly. Removal of the ignition source, will extinguish the flame. CPE jacketed cables pass the IEEE 383, UL, CSA and ICEA flame tests.

CPE maintains its flexibility at –18°C (0°F) and does not become brittle unless temperatures are below –40°C (–40°F). Its low temperature impact resistance is excellent. CPE jackets are suited to 105°C (221°F) and intermittently to higher temperatures. They will maintain adequate flexibility after repeated aging at elevated temperatures. They are known for abrasion resistance and long life in mining cable applications. CPE does not support the growth of mold, mildew or fungus.

CPE is resistant to most strong acids and bases and many solvents except for chlorinated organics. It is particularly well-suited to chemical plant use where both above ground (ultraviolet and flame retardancy) and below ground (water and chemical resistance) properties are desired. CPE's resistance to oils and fuels is good. CPE can be conveniently colored over a wide range and will maintain color upon aging.

Neoprene (CP)

Neoprene is a vulcanized synthetic rubber also referred to as chloroprene. It provides a resilient jacket that resists permanent deformation under heat and load, and does not embrittle at low temperatures. It is highly resistant to aging from sunlight and oxidation, and is virtually immune to atmospheric ozone.

Samples of neoprene-jacketed cable, tested outdoors under constant exposure for 40 years, have remained tough, resilient, uncracked and completely serviceable. Neoprene jackets are "flame resistant," i.e., not combustible without directly applied heat and flame. Neoprene will burn slowly as long as an outside source of flame is applied, but is self-extinguishing as soon as the flame is removed. Neoprene-jacketed power cable can be flexed without damage to the jacket at –40°C (–40°F) and will pass a mandrel wrap test down to about –45°C (–49°F). Neoprene jackets resist degradation for prolonged periods at temperatures up to 121°C (250°F). Satisfactory performance at even higher temperatures is possible if the exposures are brief or intermittent.

Neoprene jackets have excellent resistance to soil acids and alkalis. Mildew, fungus and other biological agents do not deteriorate properly compounded neoprene. These jackets perform well in many chemical plants. They are tough, strong, resilient and have excellent resistance to abrasive wear, impact, crushing and chipping. Because of these properties, neoprene is the jacketing material frequently used for mine trailing cables and dredge cables.

Cross-linked Polyethylene (XLP or XLPE)

Cross-linked polyethylene is a frequently used polymer in wire and cable. It is most often used as the insulation of 600 volt building wire (e.g., Type XHHW), as the insulation in 5 to 69 kV and higher rated power cables and as the insulation in many control cables.

XLP has very high insulation resistance (IR), high dielectric strength and low dielectric constant (2.3). It also is a very tough material at temperatures below 100°C, so it is resistant to cutting, impact and other mechanical forces. Its low-temperature performance is also very good: down to –40°C and below. XLP's fire resistance, however, is poor unless flame retardants are added. XLP is lower in cost than EPR.

Ethylene Propylene Rubber (EP, EPR, or EPDM)

Ethylene propylene rubber is a common synthetic rubber polymer used as an insulation in electrical wire and cable. EPR is used as the insulation in 600 volt through 69 kV power cables, as an integral insulation/jacket on welding cables and as an insulation in many cords, portable mining cables and control/instrumentation cables.

Because of its rubber-like characteristics, EPR is used in many highly flexible cables. Its dielectric strength is good but not as high as that of PE or XLP. Dielectric constant ranges from 2.8 to 3.2 depending on the specific EPR formulation. EPR is abrasion resistant and is suitable for use at temperatures down to –60°C. It is fairly flame retardant and can be made even more flame retardant by careful formulation. Flame retardant versions are often referred to as "FREP" or "flame retardant EP." EPR's high-temperature characteristics are very good. Some formulations can withstand continuous temperatures as high as 150°C.

Hypalon (CSP)

Hypalon is a thermosetting, cross-linked, chlorosulfonated polyethylene made by DuPont with many excellent physical and electrical properties. It is inherently resistant to cold flow (compression set) resulting from clamping pressures and other forms of external loading; it is immune to attack by ozone; and it is highly resistant to aging from sunlight and oxidation. Water absorption of properly compounded Hypalon cable sheathing is extremely low.

Hypalon sheathing will not support combustion. It will burn slowly as long as an outside source of flame is applied but is self-extinguishing as soon as the flame is removed. It remains flexible at –18°C (0°F) and will not become brittle at –40°C (–40°F). Hypalon jacketed constructions pass both the Underwriters Laboratories' vertical flame test and the U.S. Bureau of Mines' flame test for mining cable.

At high temperatures, Hypalon will perform satisfactorily after short-term exposure at up to 148°C (300°F) – even higher if compounded for maximum heat resistance. It is well-known for its resistance to chemicals, oils, greases and fuels. It is particularly useful as a cable sheathing in plant processing areas, where airborne chemicals attack ordinary jacketing materials and metal conduit.

Hypalon surpasses most elastomers in resistance to abrasion. It is highly resistant to attack by hydrocarbon oils and fuels. It is especially useful in contact with oils at elevated temperatures. Sheathing of Hypalon provides high resistance to impact, crushing and chipping. Hypalon's electrical properties make it appropriate as insulation for low-voltage applications (up to 600 volts) and as jacketing for any type of wire and cable.

Silicone

Silicone is a soft, rubbery insulation that has a temperature range from –80°C to 200°C. It has excellent electrical properties plus ozone resistance, low moisture absorption, weather resistance, and radiation resistance. It typically has low mechanical strength and poor scuff resistance.

3.2.3 Fibrous Coverings

Fibrous coverings are commonly used on high-temperature cables due to their excellent heat resistance. They are normally constructed of a textile braid (e.g., fiberglass or K-fiber) impregnated with a flame and heat-resistant finish.

K-fiber insulating materials are a blend of polyaramid, polyamid, phenolic-based and fiberglass fibers. They are available as roving and yarn for insulating applications and as rope for use as fillers. They provide a non-asbestos, abrasion-, moisture-, flame- and temperature-resistant, non-melting insulating material for all applications requiring a 250°C (482°F) temperature rating, which would have previously utilized asbestos.

3.2.4 Additional Information

Additional information on the selection of cable jackets is available in IEEE 535 "Guide for Selecting and Testing Jackets for Power, Instrumentation and Control Cables."

3.3 COLOR CODING

3.3.1 Power, Control, Instrumentation and Thermocouple

ICEA standard S-73-532 (NEMA WC57) contains several methods for providing color coding in multiconductor power and control cables. Methods 1, 3 and 4 are the most widely used.

Method 1 – Colored compounds with tracers

Method 2 – Neutral colored compounds with tracers

Method 3 – Neutral or single-color compounds with surface printing of numbers and color designations

Method 4 – Neutral or single-color compounds with surface printing of numbers

Method 5 – Individual color coding with braids

Method 6 – Layer identification

Method 7 – Paired conductors

Historically, ICEA has established the sequence of colors used for Method 1 color coding, which consists of six basic colors, then a repeat of the colors with a colored band or tracer. This sequence of colors is referred to as K-1 color coding because it was formerly found in Table K-1 of many ICEA standards. (See Tables 3.1 through 3.5.) In the latest ICEA standard the color sequences are located in Tables E-1 through E-7.

The National Electrical Code (NEC) specifies that a conductor colored white can only be used as a grounded (neutral) conductor and that a conductor colored green can only be used as an equipment grounding conductor. The use of Table E-1 (formerly K-1) color coding would therefore be in violation of the Code in a cable having more than six conductors if conductors #7 (white/black), #9 (green/black), #14 (green/white), etc. are energized.

To address this issue, a different color coding sequence was developed by ICEA for cables that are used in accordance with the NEC. Table E-2 (formerly K-2) of the ICEA standard provides this color sequence. The ICEA standard provides further guidance stating that if a white conductor is required, this color may be introduced into Table E-2 as the second conductor in the sequence. If a green insulated conductor is required, it likewise can be introduced into the table. However, the white and green colors may only appear once.

The most popular multiconductor control cables in sizes 14 AWG–10 AWG have Method 1, Table E-2 color coding. The cables do not contain a white or green conductor. The most popular control cables used in sizes 8 AWG and larger are three conductor cables having black insulation surface ink printed with the numbers 1, 2 and 3. This is Method 4 color coding in the ICEA standards.

The electric utility industry often specifies control cables with the E-1 color coding sequence.

For applications where the NEC is applicable, such as in industrial and commercial applications, the E-2 color sequence is normally used.

ICEA standard S-82-552 (NEMA WC55) contains methods and color sequence tables for instrumentation and thermocouple cables.

Table 3.1–E-1 (Formerly K-1) Color Sequence for Control Cables

Conductor Number	Background or Base Color	First Tracer Color	Second Tracer Color	Conductor Number	Background or Base Color	First Tracer Color	Second Tracer Color
1	Black	–	–	31	Green	Black	Orange
2	White	–	–	32	Orange	Black	Green
3	Red	–	–	33	Blue	White	Orange
4	Green	–	–	34	Black	White	Orange
5	Orange	–	–	35	White	Red	Orange
6	Blue	–	–	36	Orange	White	Blue
7	White	Black	–	37	White	Red	Blue
8	Red	Black	–	38	Black	White	Green
9	Green	Black	–	39	White	Black	Green
10	Orange	Black	–	40	Red	White	Green
11	Blue	Black	–	41	Green	White	Blue
12	Black	White	–	42	Orange	Red	Green
13	Red	White	–	43	Blue	Red	Green
14	Green	White	–	44	Black	White	Blue
15	Blue	White	–	45	White	Black	Blue
16	Black	Red	–	46	Red	White	Blue
17	White	Red	–	47	Green	Orange	Red
18	Orange	Red	–	48	Orange	Red	Blue
19	Blue	Red	–	49	Blue	Red	Orange
20	Red	Green	–	50	Black	Orange	Red
21	Orange	Green	–	51	White	Black	Orange
22	Black	White	Red	52	Red	Orange	Black
23	White	Black	Red	53	Green	Red	Blue
24	Red	Black	White	54	Orange	Black	Blue
25	Green	Black	White	55	Blue	Black	Orange
26	Orange	Black	White	56	Black	Orange	Green
27	Blue	Black	White	57	White	Orange	Green
28	Black	Red	Green	58	Red	Orange	Green
29	White	Red	Green	59	Green	Black	Blue
30	Red	Black	Green	60	Orange	Green	Blue

Note: The former K-1 color sequence was the same as E-1 through conductor number 21. K-1 then repeated. The above table is only applicable to control cables. The color sequence for instrumentation cables can be found in ICEA S-82-552 (NEMA WC 55).

Table 3.2–E-2 (Formerly K-2) Color Sequence for Control Cables

Conductor Number	Background or Base Color	Tracer Color
1	Black	–
2	Red	–
3	Blue	–
4	Orange	–
5	Yellow	–
6	Brown	–
7	Red	Black
8	Blue	Black
9	Orange	Black
10	Yellow	Black
11	Brown	Black
12	Black	Red
13	Blue	Red
14	Orange	Red
15	Yellow	Red
16	Brown	Red
17	Black	Blue
18	Red	Blue
19	Orange	Blue
20	Yellow	Blue
21	Brown	Blue
22	Black	Orange
23	Red	Orange
24	Blue	Orange
25	Yellow	Orange
26	Brown	Orange
27	Black	Yellow
28	Red	Yellow
29	Blue	Yellow
30	Orange	Yellow
31	Brown	Yellow
32	Black	Brown
33	Red	Brown
34	Blue	Brown
35	Orange	Brown
36	Yellow	Brown

Note: The above table is only applicable to control cables. The color sequence for instrumentation cables can be found in ICEA S-82-552 (NEMA WC 55).

Table 3.3–E-3 (Formerly K-3) Color Sequence for Control Cables

Conductor Number	First Tracer Color (e.g., Wide Tracer)	Second Tracer Color (e.g., Narrow Tracer)
1	Black	–
2	White	–
3	Red	–
4	Green	–
5	Orange	–
6	Blue	–
7	White	Black
8	Red	Black
9	Green	Black
10	Orange	Black
11	Blue	Black
12	Black	White
13	Red	White
14	Green	White
15	Blue	White
16	Black	Red
17	White	Red
18	Orange	Red
19	Blue	Red
20	Red	Green
21	Orange	Green

Note: The above table is only applicable to control cables. The color sequence for instrumentation cables can be found in ICEA S-82-552 (NEMA WC 55).

Table 3.4–E-4 (Formerly K-4) Color Sequence for Control Cables

Conductor Number	First Tracer Color (e.g., Wide Tracer)	Second Tracer Color (e.g., Narrow Tracer)
1	Black	–
2	Red	–
3	Blue	–
4	Orange	–
5	Yellow	–
6	Brown	–
7	Red	Black
8	Blue	Black
9	Orange	Black
10	Yellow	Black
11	Brown	Black
12	Black	Red
13	Blue	Red
14	Orange	Red
15	Yellow	Red
16	Brown	Red
17	Black	Blue
18	Red	Blue
19	Orange	Blue
20	Yellow	Blue
21	Brown	Blue
22	Black	Orange
23	Red	Orange
24	Blue	Orange
25	Yellow	Orange
26	Brown	Orange
27	Black	Yellow
28	Red	Yellow
29	Blue	Yellow
30	Orange	Yellow
31	Brown	Yellow
32	Black	Brown
33	Red	Brown
34	Blue	Brown
35	Orange	Brown
36	Yellow	Brown

Note: The above table is only applicable to control cables. The color sequence for instrumentation cables can be found in ICEA S-82-552 (NEMA WC 55).

Table 3.5–E-5 (Formerly K-5) Color Sequence for Control Cables

Conductor Number	Background or Base Color	First Tracer Color	Second Tracer Color
1	Black	–	–
2	White	–	–
3	Red	–	–
4	Green	–	–
5	Orange	–	–
6	Blue	–	–
7	White	Black	–
8	Red	Black	–
9	Green	Black	–
10	Orange	Black	–
11	Blue	Black	–
12	Black	White	–
13	Red	White	–
14	Green	White	–
15	Blue	White	–
16	Black	Red	–
17	White	Red	–
18	Orange	Red	–
19	Blue	Red	–
20	Red	Green	–
21	Orange	Green	–
22	Black	White	Red
23	White	Black	Red
24	Red	Black	White
25	Green	Black	White
26	Orange	Black	White
27	Blue	Black	White
28	Black	Red	Green
29	White	Red	Green
30	Red	Black	Green
31	Green	Black	Orange
32	Orange	Black	Green
33	Blue	White	Orange

Continued on next page >>

Table 3.5–E-5 (Formerly K-5) Color Sequence for Control Cables (Continued)

Conductor Number	Background or Base Color	First Tracer Color	Second Tracer Color
34	Black	White	Orange
35	White	Red	Orange
36	Orange	White	Blue
37	White	Red	Blue

Note: The above table is only applicable to control cables. The color sequence for instrumentation cables can be found in ICEA S-82-552 (NEMA WC 55).

3.3.2 Belden Electronic Color Code

Table 3.6–Common Multiconductor Color Code (Belden Standard)

Conductor	Color
1	Black
2	White
3	Red
4	Green
5	Brown
6	Blue
7	Orange
8	Yellow
9	Purple
10	Gray
11	Pink
12	Tan

Table 3.7–Common Multipair Color Code (Belden Standard)

Pair No.	Color Combination	Pair No.	Color Combination
1	Black and Red	20	White and Yellow
2	Black and White	21	White and Brown
3	Black and Green	22	White and Orange
4	Black and Blue	23	Blue and Yellow
5	Black and Yellow	24	Blue and Brown
6	Black and Brown	25	Blue and Orange
7	Black and Orange	26	Brown and Yellow
8	Red and White	27	Brown and Orange
9	Red and Green	28	Orange and Yellow
10	Red and Blue	29	Purple and Orange
11	Red and Yellow	30	Purple and Red
12	Red and Brown	31	Purple and White
13	Red and Orange	32	Purple and Dark Green
14	Green and White	33	Purple and Light Blue
15	Green and Blue	34	Purple and Yellow
16	Green and Yellow	35	Purple and Brown
17	Green and Brown	36	Purple and Black
18	Green and Orange	37	Gray and White
19	White and Blue		

3.3.3 Telecommunication Color Codes

Individual telecommunication cable conductors are color-coded with solid colors (Table 3.8) or by applying a colored band of contrasting color to solid colored wires (Table 3.9). Bandmarking is used on inside wiring cable, plenum cable and switchboard cable. The color combinations are such that each wire is banded with the color of its mate. For example, in a blue and white pair, the blue wire has a white band and the white wire a blue band. Telephone wires (e.g., inside-outside station wire and distribution frame and jumper wire) that do not have paired constructions have solid color wires.

All colors must be readily distinguishable and lie within the Munsell color standard.

Large Pair Count Cables

In cables having more than 25 pairs, the pairs are arranged in groups, each containing a maximum of 25 pairs and wrapped with distinctively colored binder threads to permit distinction between groups.

Table 3.8–Telecommunication Cable Color Code (Solid Colors)

Pair No.	Tip	Ring
1	White	Blue
2	White	Orange
3	White	Green
4	White	Brown
5	White	Slate
6	Red	Blue
7	Red	Orange
8	Red	Green
9	Red	Brown
10	Red	Slate
11	Black	Blue
12	Black	Orange
13	Black	Green
14	Black	Brown
15	Black	Slate
16	Yellow	Blue
17	Yellow	Orange
18	Yellow	Green
19	Yellow	Brown
20	Yellow	Slate
21	Violet	Blue
22	Violet	Orange
23	Violet	Green
24	Violet	Brown
25	Violet	Slate

Table 3.9–Telecommunication Cable Color Code (Band Marked)

Pair No.	Tip	Ring
1	White-Blue	Blue-White
2	White-Orange	Orange-White
3	White-Green	Green-White
4	White-Brown	Brown-White
5	White-Slate	Slate-White
6	Red-Blue	Blue-Red
7	Red-Orange	Orange-Red
8	Red-Green	Green-Red
9	Red-Brown	Brown-Red
10	Red-Slate	Slate-Red
11	Black-Blue	Blue-Black
12	Black-Orange	Orange-Black
13	Black-Green	Green-Black
14	Black-Brown	Brown-Black
15	Black-Slate	Slate-Black
16	Yellow-Blue	Blue-Yellow
17	Yellow-Orange	Orange-Yellow
18	Yellow-Green	Green-Yellow
19	Yellow-Brown	Brown-Yellow
20	Yellow-Slate	Slate-Yellow
21	Violet-Blue	Blue-Violet
22	Violet-Orange	Orange-Violet
23	Violet-Green	Green-Violet
24	Violet-Brown	Brown-Violet
25	Violet-Slate	Slate-Violet
26	Red-White	White-Red

3.4 PROPERTIES

3.4.1 Thermoplastic

Table 3.10–Properties of Thermoplastic Insulation and Jacket Materials

	PVC	Low-Density Polyethylene	Cellular Polyethylene	High-Density Polyethylene	Polypropylene
Oxidation resistance	E	E	E	E	E
Heat resistance	G-E	G	G	E	E
Oil resistance	F	G-E	G	G-E	F
Low-temperature flexibility	P-G	E	E	E	P
Weather, sun resistance	G-E	E	E	E	E
Ozone resistance	E	E	E	E	E
Abrasion resistance	F-G	G	F	E	F-G
Electrical properties	F-G	E	E	E	E
Flame resistance	E	P	P	P	P
Nuclear radiation resistance	F	G-E	G	G-E	F
Water resistance	F-G	E	E	E	E
Acid resistance	G-E	G-E	G-E	E	E
Alkali resistance	G-E	G-E	G-E	E	E
Gasoline, kerosene, etc. (aliphatic hydrocarbons) resistance	P	G-E	G	G-E	P-F
Benzol, toluol, etc. (aromatic hydrocarbons) resistance	P-F	P	P	P	P-F
Degreaser solvents (halogenated hydrocarbons) resistance	P-F	G	G	G	P
Alcohol resistance	G-E	E	E	E	E
Underground burial	P-G	G	F	E	E

P = Poor, F = Fair, G = Good, E = Excellent, O = Outstanding
These ratings are based on average performance of general purpose compounds. Any given property can usually be improved by the use of selective compounding.
Source: Belden

Continued on next page >>

Table 3.10–Properties of Thermoplastic Insulation and Jacket Materials (Continued)

	Cellular Polypropylene	Polyurethane	Nylon	CPE	Plenum PVC
Oxidation resistance	E	E	E	E	E
Heat resistance	E	G	E	E	G-E
Oil resistance	F	E	E	E	F
Low-temperature flexibility	P	G	G	E	P-G
Weather, sun resistance	E	G	E	E	G
Ozone resistance	E	E	E	E	E
Abrasion resistance	F-G	O	E	E-O	F-G
Electrical properties	E	P	P	E	G
Flame resistance	P	P	P	E	E
Nuclear radiation resistance	F	G	F-G	O	F
Water resistance	E	P-G	P-F	O	F
Acid resistance	E	F	P-E	E	G
Alkali resistance	E	F	E	E	G
Gasoline, kerosene, etc. (aliphatic hydrocarbons) resistance	P	P-G	G	E	P
Benzol, toluol, etc. (aromatic hydrocarbons) resistance	P	P-G	G	G-E	P-F
Degreaser solvents (halogenated hydrocarbons) resistance	P	P-G	G	E	P-F
Alcohol resistance	E	P-G	P	E	G
Underground burial	F	G	P	E-O	P

P = Poor, F = Fair, G = Good, E = Excellent, O = Outstanding
These ratings are based on average performance of general purpose compounds. Any given property can usually be improved by the use of selective compounding.
Source: Belden

Continued on next page >>

Table 3.10–Properties of Thermoplastic Insulation and Jacket Materials (Continued)

	FEP Teflon	Tefzel (ETFE)	PTFE (TFE) Teflon	Solef/Kynar (PVDF)/PVF	Halar (ECTFE)
Oxidation resistance	O	E	O	O	O
Heat resistance	O	E	O	O	O
Oil resistance	O	E	E-O	E	O
Low-temperature flexibility	O	E	O	O	O
Weather, sun resistance	O	E	O	E-O	O
Ozone resistance	E	E	O	E	E
Abrasion resistance	E	E	O	E	E
Electrical properties	E	E	E	G-E	E
Flame resistance	O	G	E	E	E-O
Nuclear radiation resistance	P-G	E	P	E	E
Water resistance	E	E	E	E	E
Acid resistance	E	E	E	G-E	E
Alkali resistance	E	E	E	E	E
Gasoline, kerosene, etc. (aliphatic hydrocarbons) resistance	E	E	E	E	E
Benzol, toluol, etc. (aromatic hydrocarbons) resistance	E	E	E	G-E	E
Degreaser solvents (halogenated hydrocarbons) resistance	E	E	E	G	E
Alcohol resistance	E	E	E	E	E
Underground burial	E	E	E	E	E

P = Poor, F = Fair, G = Good, E = Excellent, O = Outstanding
These ratings are based on average performance of general purpose compounds. Any given property can usually be improved by the use of selective compounding.
Source: Belden

3.4.2 Thermoset

Table 3.11–Properties of Thermoset Insulation and Jacket Materials

	Neoprene	Hypalon (Chloro-Sulfonated Polyethylene)	EPR (Ethylene Propylene Rubber)	XLPE	CPE	Silicone Rubber
Oxidation resistance	G	E	E	E	E	E
Heat resistance	G	E	E	G	E	0
Oil resistance	G	G	P	G	G-E	F-G
Low-temperature flexibility	F-G	F	G-E	0	F	0
Weather, sun resistance	G	E	E	G	E	0
Ozone resistance	G	E	E	G	G-E	0
Abrasion resistance	G-E	G	G	F-G	G-E	P
Electrical properties	P	G	E	E	F-G	G
Flame resistance	G	G	P	P	G	F-G
Nuclear radiation resistance	F-G	E	G	E	G	E
Water resistance	E	E	G-E	G-E	G-E	G-E
Acid resistance	G	E	G-E	G-E	E	F-G
Alkali resistance	G	E	G-E	G-E	E	F-G
Gasoline, kerosene, etc. (aliphatic hydrocarbons) resistance	G	F	P	F	F	P-F
Benzol, toluol, etc. (aromatic hydrocarbons) resistance	P-F	F	F	F	F	P
Degreaser solvents (halogenated hydrocarbons) resistance	P	P-F	P	F	P	P-G
Alcohol resistance	F	G	P	E	G-E	G
Underground burial	G-E	E	E	E	E	G

P = Poor, F = Fair, G = Good, E = Excellent, 0 = Outstanding
These ratings are based on average performance of general purpose compounds. Any given property can usually be improved by the use of selective compounding.
Source: Belden

3.4.3 EPR Versus XLPE

Table 3.12–Properties of EPR Compared with Those of XLPE

Cross-linked Polyethylene (XLPE)	Ethylene Propylene Rubber (EPR)
Less deformation below 100°C	Less deformation above 100°C
Lower in cost	More heat resistance
Lower dissipation factor	Less shrinkback
Lower dielectric constant	Less thermal expansion
Higher dielectric strength	More corona resistant
Physically tougher	More flexible
More resistant to chemicals	More tree retardant
More oil resistant	More sunlight resistant

3.4.4 Thermal Characteristics

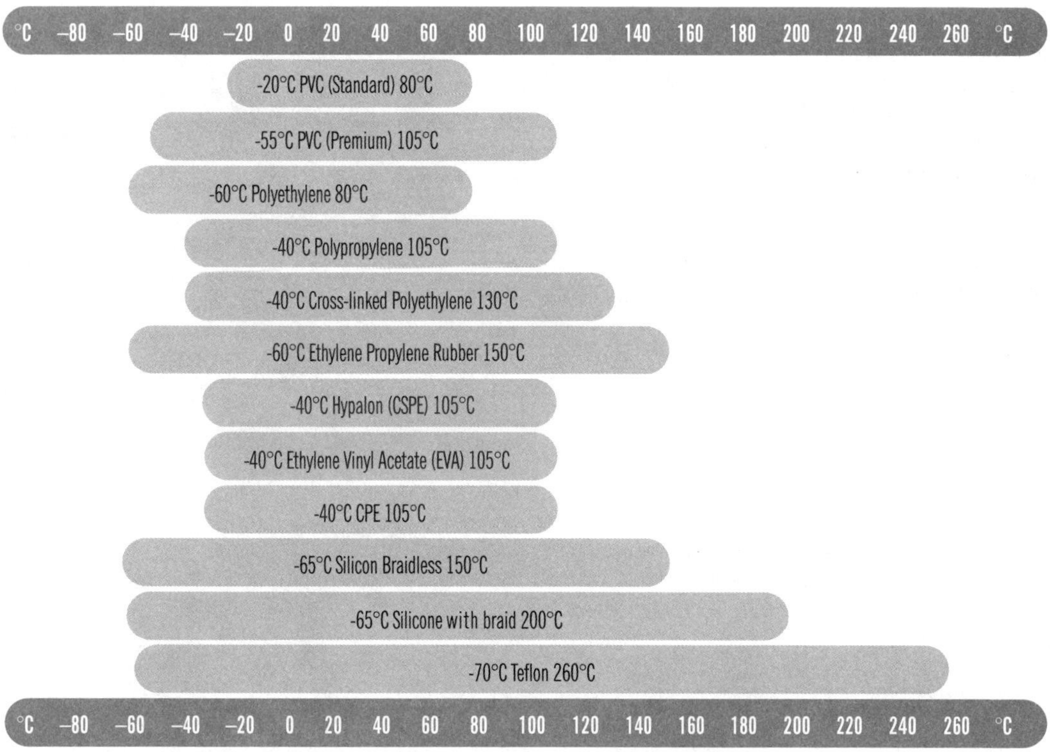

Figure 3.1–Nominal Temperature Range of Cable Polymers

3.4.5 Halogen Content

Table 3.13–Halogen Content in Typical Insulation and Jacket Materials

Material	Typical Halogen Content Percent by weight
PE insulation or jacket	<0.02
XLP insulation 600 V (6 AWG and larger)	<0.02
XLP insulation 5-35 kV	<0.02
EPR insulation 5-35 kV	<0.02
Polyurethane jacket	<0.02
EVA jacket	<0.02
XLP insulation 600 V (14-8 AWG)	7–13
FR-EPR insulation	9–14
Hypalon (insulation grade)	13–16
FR-XLP insulation	11–17
Hypalon jacket (heavy duty)	16–26
Neoprene jacket	16–18
CPE jacket	14–28
Hypalon jacket (extra heavy duty)	18–33
PVC jacket	22–29

NOTE: Halogen content can vary from manufacturer to manufacturer. The above values should be used for general comparisons only.

3.4.6 Limiting Oxygen Index (LOI)

LOI values are used to determine the relative flammability of polymers. Tests are usually conducted in accordance with ASTM D2863, which finds the percent oxygen required to sustain combustion. Typical values are shown below. The oxygen content of air is 20.9 percent.

Table 3.14–LOI of Common Wire and Cable Materials

Material	Percent Oxygen	Material	Percent Oxygen
Teflon	93	Neoprene	32
PVDF (Kynar)	43–85	Tefzel	30–32
Halar	55	PVC	28–32
Plenum grade PVC	38–42	Kevlar	29
FR-EP	30–40	NBR PVC	28
FR-XLP	30–40	XLP (Unfilled)	20–23
CPE	28–36	PE (Unfilled)	20–23
Ethylene Vinyl Acetate (EVA)	32–35		
Hypalon	34		

3.4.7 Dielectric Constant

Table 3.15–Dielectric Constant of Common Wire and Cable Materials

Material	Dielectric Constant	Material	Dielectric Constant
Teflon (FEP, PFA or TFE)	2.1	Polyester (Mylar)	3.3–3.8
Polypropylene	2.2–2.3	Silicone	3–4
Cross-linked Polyethylene	2.3	Nylon	3.2–5
Polyethylene	2.3	Mica	6.9
TPE	2.3–2.5	PVC	3.5–8
Halar (ECTFE)	2.6	Hypalon	8–10
Tefzel (ETFE)	2.6	Neoprene	9–10
EPR	2.8–3.5	Kynar (PVDF)	6–12
Ethylene Vinyl Acetate (EVA)	3.8		

4. SHIELDS

A shield is a metallic covering enclosing an insulated conductor or group of conductors. Though sometimes similar in appearance, shields for electronic and power cables perform very different functions. Electronic cable shields serve to both minimize the effect of external electromagnetic signals on the conductors in the cable and to reduce the radiated signal from the cable to an acceptable level. Power cable shields, on the other hand, help protect the user from shock hazards and increase cable reliability by preventing partial discharges (corona) in cables rated above 2,400 volts.

4.1 POWER CABLE

The use of shields in power cables rated above 2,400 volts reduces electrical shock hazard to people and provides uniform distribution of electrical stresses throughout the insulation. A uniform distribution of electrical stress extends the life of the cable by eliminating partial discharges. The various components of a power cable shield are discussed below.

4.1.1 Conductor Shield (Strand Shield)

The nonround geometry of stranded conductors permits air gaps between the outer surface of the conductor and the inner surface of the insulation. Without a stress control layer, high electric fields cause partial discharges within these gaps, which can harm the insulation. Energetic ions bombard the insulation, break molecular bonds and degrade the insulation. Microscopic channels called "trees" may form and ultimately cause premature failure of the insulation. Thus, the primary purpose of the conductor shield is to provide a smooth, continuous and void-free interface between the conductor and insulation.

There are two basic types of conductor shields – "conductive" and "emission" shields. An emission shield uses a material with a high dielectric constant to do its job. The most popular type, however, is the conductive shield. It is a material (either an extruded carbon black loaded polymer or carbon black impregnated fabric tape) with electrical conductivity midway between that of a metallic conductor such as copper and that of an insulation such as XLP. Such a material is commonly referred to as a "semiconductive" shield (not to be confused with semiconductors, i.e., transistors, used in the electronics industry). AEIC document CS8 and ICEA publication T-25-425 contain detailed specifications on the electrical and physical performance of the conductor shield.

Semiconductive shields must be as smooth, cylindrical and clean as possible to avoid electrical stress concentrations that can lead to insulation damage.

4.1.2 Outer Shield (Insulation Shield)

The insulation shield plays much the same role as the conductor shield in protecting the insulation from the damaging effects of corona, but at the outside of the cable's insulation. It too must remain in intimate contact with the insulation and be free of voids and defects. The insulation shield material is either electrically conductive or made of a high dielectric constant material and provides a uniform electrical field within the insulation. The insulation shield also provides an important safety function at terminations and splices where the metallic part of the shield may not completely cover the cable insulation surface. Volume resistivity of the insulation shield is normally less than 500 ohm-meters.

Copper Tape Shields

The copper tape used in power cable shields is usually 5 mils thick and 1- to 1 1/2- inches wide. It is generally helically applied over a semiconducting polymer insulation shield.

Power cables rated 5 to 35 kV and up frequently utilize copper tape as the metallic component of the metal/polymer shielding system. In combination with the extruded insulation shield, a copper tape shield increases insulation life by maintaining uniform electrical stress throughout the cable insulation and provides low end-to-end resistance of the shield system.

Figure 4.1–Typical Copper Tape Shielded Power Cable

Table 4.1–Power Cable Shielding

Advantages	Disadvantages
• When properly grounded, provides protection from electrical shock • Increases life of the cable insulation • Reduces electromagnetic interference (EMI)	• Must be terminated with a medium voltage termination to control electrical stresses • Higher cost

Wire Shields

Metallic wire shields on power cables come in two basic types: helically applied copper wires and UniShield.

Helically applied copper wire shields are sometimes used on 5 through 35 kV and higher rated power cables. They are sometimes used in combination with copper tape to provide additional shield fault current capacity.

UniShield cables have six corrugated copper wires longitudinally imbedded in a conducting CPE jacket. The wires can be used as “ripcords” to reduce termination time during installation.

4.2 ELECTRONIC CABLE

Electronic cable shielding provides an efficient way to manage electromagnetic interference (EMI).

When a shielded cable is present in an ambient electromagnetic field, an interference current is induced in the shield. The incident energy is partially reflected from the shield and partially absorbed by the shield and a small amount penetrates through the shield into the cable. The small amount of energy that makes it all the way through the shield generates an interference voltage in the signal carrying conductors of the cable. The smaller the interference voltage, the better the shield.

In addition to shielding effectiveness, electronic cable shields must satisfy a long list of electrical, mechanical, chemical and cost requirements. As a result, a diversified line of shield designs has evolved in the wire and cable industry.

4.2.1 Foil Shield

Foil shields are usually constructed of aluminum foil with a 1/2-mil thick polyester backing. This backing provides mechanical strength. The shield can be overlapped (Fig. 4.2) with the foil facing in or the foil facing out. This overlap creates a slot where signal leakage through the shield can occur. The "Z" fold (Fig. 4.3) construction provides the best electrical isolation between shields of adjacent pairs as well as 100 percent coverage. A tinned copper drain wire is placed in contact with the foil side of the shield to provide easier grounding of the shield at the cable terminations.

Foil shields are most common in electronic and coaxial cables. Foil shields provide excellent protection from electromagnetic interference, especially at high frequencies.

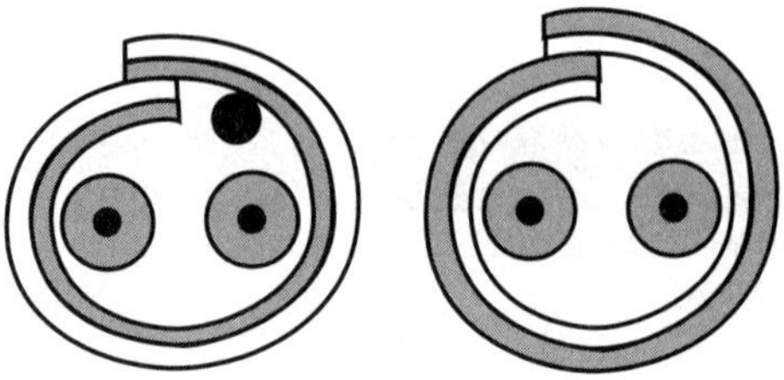

Figure 4.2–Foil Shield

Figure 4.3–Z-Fold Foil Shield

Table 4.2–Foil Shielding

Advantages	Disadvantages
• 100% coverage	• Poor mechanical strength
• Low cost	• Short flex life
• Ease of termination	• Less effective at low frequencies
• Good flexibility	
• Excellent shielding at high frequencies	

4.2.2 Copper Braid Shield

A braid shield typically consists of copper wire ranging in size from 32 to 40 AWG braided into a mesh around the cable core. The tightness of the braid determines the percent coverage. Typical coverage ranges from 60 percent to 90 percent. Generally, the higher the coverage the better the shield.

Braid shields are typically used on coaxial cables and on low-speed communication cables. Braid shields are most effective at low frequencies. Braid shields are also commonly used on cables where increased flex life and mechanical strength are required.

Figure 4.4–Dual Braid Shield Construction on a Multipair Cable

Figure 4.5–Copper Braid Construction on a Coaxial Cable

Table 4.3–Copper braid shields

Advantages	Disadvantages
• Best at low frequencies • Good mechanical strength • Increased flex life	• Increased cost • More difficult to terminate

4.2.3 Spiral (Serve) Shield

Spiral or serve shields, as they are sometimes called, are typically constructed with bare or tinned copper wires from 32 to 40 AWG in size that are helically applied in a flat or ribbon configuration (Fig. 4.6). Spiral shields range in coverage from 80 percent to about 97 percent.

Spiral shields are used primarily in audio, microphone and retractile cord cables where extreme flexibility and a long flex life are required. Spiral shields perform best when used at low (audio) frequencies.

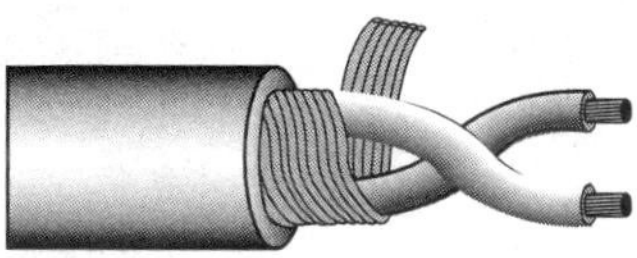

Figure 4.6–Spiral or Serve Shield

Table 4.4–Spiral Shields

Advantages	Disadvantages
• Excellent flexibility • Long flex life	• Poor electrical performance at high frequencies

5. ARMOR

Cables often need to be placed in areas where they are subjected to harsh mechanical stresses. These stresses could damage the insulated conductors or the optical fibers in the cable if they are not properly protected. Armor (usually a metal) is frequently applied over the cable core to provide this protection. The armor extends the life, while improving the reliability, safety and performance of the cable core. The following are some frequently used armor types.

5.1 INTERLOCKED ARMOR

Interlocked armor typically uses galvanized steel or aluminum. However, other metals are sometimes used for specialized applications. The interlocking construction protects the cable from damage during and after installation. The armor may be applied directly over the insulation or over an inner jacket. Materials and construction generally comply with the requirements of UL, CSA and/or ICEA.

Table 5.1–ICEA Recommended Thickness of Interlocked Armor

Diameter of Cable (in.)	Nominal Thickness (mils) Steel or Bronze	Aluminum
0 to 1.500	20	25
1.501 and larger	25	30

5.2 CONTINUOUSLY CORRUGATED AND WELDED (CCW)

CCW armor is made by forming an aluminum strip into a circle along its length and then welding it at the seam. This smooth tube is then rolled or crimped to form ridges to prevent kinking while bending (see Fig. 5.1). This type of sheath provides an impervious seal against moisture and other chemicals as well as physical protection.

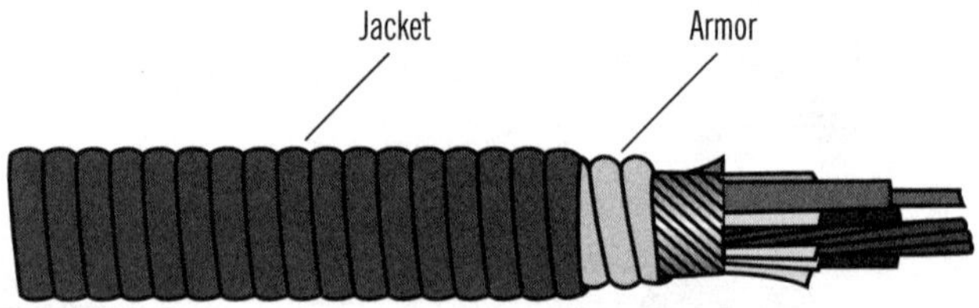

Figure 5.1–Continuously Corrugated and Welded (CCW) Armor

5.3 BASKET-WEAVE

Basket-weave armor is constructed of metal wires forming a braided outer covering. The wires may be of galvanized steel, aluminum or bronze. This armor is generally used on shipboard cables because it provides the mechanical protection of an armored cable, yet is much lighter in weight than other types of armored coverings. Materials and construction generally comply with the requirements of IEEE Standard 45 and various military specifications. This type of armor is referred to as GSWB (galvanized steel wire braid) in some international standards.

5.4 LEAD SHEATH

For underground installations in conduits, ducts and raceways, a lead sheath may be used to protect insulated cables from moisture. In locations where corrosive conditions may be encountered, a jacket over the lead is recommended.

Commercially pure lead is used on some lead-covered cables, which conforms to the requirements of ASTM B29 and ICEA S-93-639 (NEMA WC74). Lead alloy sheaths, containing added tin or antimony, are used where a harder sheath is desired or where vibration may be encountered.

5.5 WIRE SERVE

Wire serve armor is most commonly found on submarine cable because it provides excellent physical protection from boat anchors, sharp rocks, sharks, etc. This type of armor normally consists of 1/8- to 1/4- inch diameter solid steel wires, which are laid helically around the circumference of the cable. Tar or asphalt (bitumen) is placed over and around the steel wires to reduce the effects of corrosion. This type of armor is referred to as SWA (steel wire armor) in some international standards.

6. CABLE TYPES AND SELECTION CRITERIA

6.1 PORTABLE POWER AND CONTROL

6.1.1 Flexible Cords

Flexible cords come in a number of UL and CSA types including SO, SOW, SOOW, SJ, SJO, SJOW, STO and SJTO. In portable cord terminology, each letter of the cable type indicates the construction of the cable. For example: S = service, O = oil-resistant jacket, J = junior service (300 volts), W = weather resistant, T = thermoplastic, and OO = oil-resistant insulation and jacket.

The temperature rating of these cables can range from -50°C to +105°C for SOOW and -37°C to +90°C for other thermoset cords. Thermoplastic cords typically have temperature ratings that range from -20°C to +60°C. Thermoset portable cords have excellent cold bend characteristics and are extremely durable.

Table 6.1–Flexible Cord Type Designations

TST	Tinsel Service Thermoplastic
SPT-1	Service Parallel Thermoplastic – 1/64" Insulation
SPT-2	Service Parallel Thermoplastic – 2/64" Insulation
SPT-3	Service Parallel Thermoplastic – 3/64" Insulation
SPE-1	Service Parallel Elastomer – 1/64" Insulation
SPE-2	Service Parallel Elastomer – 2/64" Insulation
SPE-3	Service Parallel Elastomer – 3/64" Insulation
SV	Service Vacuum
SVO	Service Vacuum Oil-Resistant Jacket
SVOO	SVO with Oil-Resistant Insulation
SVT	Service Vacuum Thermoplastic
SVTO	SVT with Oil-Resistant Jacket
SVTOO	SVTO with Oil-Resistant Insulation
SVE	Service Vacuum Elastomer
SVEO	SVE with Oil-Resistant Jacket
SVEOO	SVEO with Oil-Resistant Insulation
SJ	Service Junior
SJO	SJ with Oil-Resistant Jacket
SJOO	SJO with Oil-Resistant Insulation
SJOOW	Weather Resistant SJOO
SJT	Service Junior Thermoplastic
SJTO	SJT with Oil-Resistant Jacket
SJTOO	SJTO with Oil-Resistant Insulation
SJTOOW	Weather-Resistant SJTOO
SJE	Service Junior Elastomer
SJEO	SJE with Oil-Resistant Jacket
SJEOO	SJEO with Oil-Resistant Insulation
SJEOOW	Weather Resistant SJEOO
S	Service
SO	Service with Oil-Resistant Jacket

Continued on next page >>

Table 6.1–Flexible Cord Type Designations (Continued)

SOO	SO with Oil-Resistant Insulation
SOOW	Weather-Resistant SOO
ST	Service Thermoplastic
STO	ST with Oil-Resistant Jacket
STOO	STO with Oil-Resistant Insulation
STOOW	Weather-Resistant STOO
SE	Service Elastomer
SEO	SE with Oil-Resistant Jacket
SEOO	SEO with Oil-Resistant Insulation
SEOOW	Weather-Resistant SEOO
HPN	Heater Parallel Neoprene
HSJ	Heater Service Junior
HSJO	HSJ with Oil-Resistant Jacket

6.1.2 Mining Cable

Mine power cables are generally designed to be used as flexible feeder cables for circuits between the main power source and mine load centers or as equipment trailing cables.

Mine power feeder (MPF) cables typically have voltage ratings of 5, 8, 15 or 25 kV and are available with or without a ground check conductor. A ground check (GC) conductor is a separate insulated ground wire that is used to monitor the "health" of the normal ground wire. MPF cables are flexible but are designed for only limited or occasional movement.

Shovel (SHD) cables are generally used to power heavy duty mobile mining equipment. SHD cables are unique in that they not only carry voltage ratings up to 25 kV but also have great flexibility and incredible physical toughness. Like mine power cables, SHD cables are generally available with or without a ground check conductor.

For low-voltage applications, there are a number of portable cables used by the mining industry. Among the most common are Type W and Type G. Both cables are a heavy-duty construction, can withstand frequent flexing and carry a voltage rating of up to 2 kV.

6.2 CONSTRUCTION AND BUILDING WIRE

Construction and building wire encompasses a wide variety of 300- and 600-volt wire and cable including UL Types THW, THW-2, THWN, THWN-2, THHN, TFFN, TFN, RHH, RHW, RHW-2, USE, USE-2, thermostat wire, SER, SE-U, XHHW, XHHW-2 and others. This category of wire is typically used as the permanent wiring in residential, commercial and industrial facilities. UL types with a "-2" suffix are rated 90°C in both dry and wet locations. In building wire terminology, each letter of the wire type indicates something about the construction. For example:

THHN – Thermoplastic, High Heat resistant, Nylon jacket

THWN-2 – Thermoplastic, Heat resistant, Wet and dry locations (-2 means 90°C wet), Nylon jacket

XHHW-2 – Cross-linked (X) insulation, High Heat resistant, Wet and dry locations (-2 means 90°C wet)

RHHW-2 – Rubber insulation, High Heat resistant, Wet and dry locations (-2 means 90°C wet)

USE-2 – Underground Service Entrance wire (-2 means 90°C wet)

6.3 CONTROL, INSTRUMENTATION AND THERMOCOUPLE

6.3.1 Control Cable

Control cables differ from power cables in that they are used to carry intermittent control signals, which generally require little power. Therefore, current loading is rarely a deciding factor in the choice of control cable. Primary criteria that are applied to the selection of control cable are voltage level and environmental conditions. The voltage level for control circuits may range anywhere from millivolts up to several hundred volts.

Environmental Conditions

Control cables are generally subject to rather severe environmental conditions. For this reason an examination of these conditions is at least as important as electrical considerations. High ambient temperature conditions (such as near boilers and steam lines), along with possible exposure to oils, solvents and other chemicals (in chemical, petroleum, steel, pulp and paper and cement plants), are vital considerations.

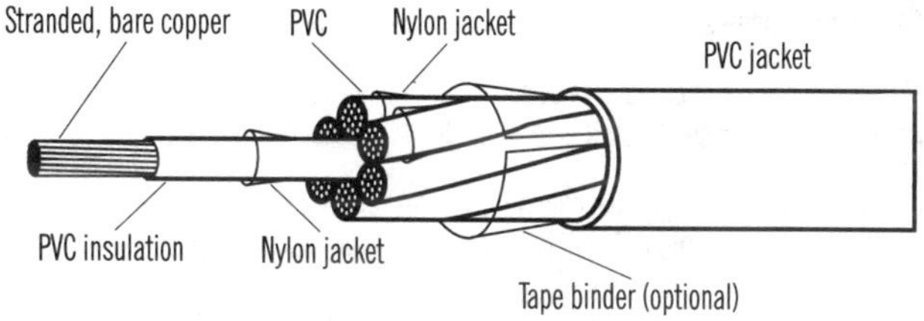

Figure 6.1–A Typical 600 V Control Cable

6.3.2 Instrumentation Cable

Instrumentation cable is generally used to transmit a low-power signal from a transducer (measuring for example, pressure, temperature, voltage, flow, etc.) to a PLC or DCS process control computer or to a manually operated control panel. It is normally available in 300- or 600-volt constructions with a single overall shield, or with individual shields over each pair (or triad) and an overall shield.

Figure 6.2–Control Cable with Overall Shield

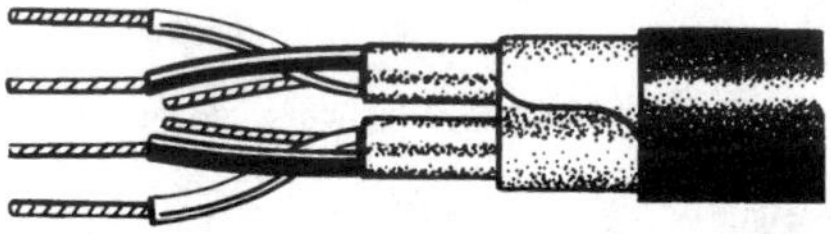

Figure 6.3–Control Cable with Individually Shielded Pairs and An Overall Shield

6.3.3 Thermocouple Wire

A thermocouple is a temperature measuring device consisting of two conductors of dissimilar metals or alloys that are connected together at one end. At this thermocouple junction, as it is called, a small voltage is produced. Electronic equipment senses this voltage and converts it to temperature. Thermocouple wire is available in either thermocouple grade or extension grade. Extension grade wire is normally lower in cost and is recommended for use in connecting thermocouples to the sensing or control equipment. The conditions of measurement determine the type of thermocouple wire and insulation to be used. Temperature range, environment, insulation requirements, response and service life should be considered. Note that thermocouple wire color codes can vary around the world.

Thermocouple Types

Type J (Iron vs Constantan) is used in vacuum, oxidizing, inert or reducing atmospheres. Iron oxidizes rapidly at temperatures exceeding 538°C (1,000°F), and therefore heavier gauge wire is recommended for longer life at these temperatures.

Type K (Chromel vs Alumel) is used in oxidizing, inert or dry reducing atmospheres. Exposure to a vacuum should be limited to short time periods. Must be protected from sulfurous and marginally oxidizing atmospheres. Reliable and accurate at high temperatures.

Type T (Copper vs Constantan) is used for service in oxidizing, inert or reducing atmospheres or in a vacuum. It is highly resistant to corrosion from atmospheric moisture and condensation and exhibits high stability at low temperatures; it is the only type with limits of error guaranteed for cryogenic temperatures.

Type E (Chromel vs Constantan) may be used in oxidizing, inert or dry reducing atmospheres, or for short periods of time under vacuum. Must be protected from sulfurous and marginally oxidizing atmospheres. Produces the highest EMF per degree of any standardized thermocouple.

Type R and S (Platinum vs Rhodium) are used in oxidizing or inert atmospheres. Must be protected from contamination. Reliable and accurate at high temperatures.

Thermocouple wire can be fabricated into an accurate and dependable thermocouple by joining the thermoelements at the sensing end.

Thermocouple wire or thermocouple extension wire of the same type must be used to extend thermocouples to indicating or control instrumentation. Red color code is negative throughout circuit.

Hook up red color-coded wire to negative terminal of instrument.

Temperature limit of the thermocouple depends on the thermocouple wire: wire size; wire insulation; and environmental factors.

Use thermocouple connectors if required. They are made of the same alloys and have the same color codes as extension wire.

Source: PMC Corporation

Figure 6.4–A Typical Thermocouple Circuit

Table 6.2–Color Code for Thermocouple Wire Per ANSI/ISA MC96.1

Thermocouple Type		Color Code	
Wire Alloys	ANSI Symbol	+/– Individual	Jacket
*Iron (+) vs Constantan (–)	J	White/Red	Brown
Chromel (+) vs *Alumel (–)	K	Yellow/Red	Brown
Copper (+) vs Constantan (–)	T	Blue/Red	Brown
Chromel (+) vs Constantan (–)	E	Purple/Red	Brown
Platinum (+) vs 13% Rhodium (–)	R	–	–
Platinum (+) vs 10% Rhodium (–)	S	–	–

*Magnetic

Table 6.3–Color Code for Thermocouple Extension Wire Per ANSI/ISA MC96.1

Thermocouple Type		Color Code	
Wire Alloys	ANSI Symbol	+/– Individual	Jacket
*Iron vs Constantan	JX	White/Red	Black
Chromel vs *Alumel	KX	Yellow/Red	Yellow
Copper vs Constantan	TX	Blue/Red	Blue
Chromel vs Constantan	EX	Purple/Red	Purple
Platinum vs 13% Rhodium (–)	RX	Black/Red	Green
Platinum vs 10% Rhodium (–)	SX	Black/Red	Green

*Magnetic

6.4 HIGH TEMPERATURE

High temperature generally refers to wire or cable with a temperature rating of 125°C (257°F) or higher. The table below lists some of the most common high-temperature wire and cable types along with their temperature rating.

Table 6.4–High-temperature Wire and Cable

°C	°F	Type
538	1,000	MG (Non-UL)
450	842	MG (UL Style 5107)
250	482	TGGT (UL Styles 5196 and 5214), TKGT (UL Style 5214) TMMG, TCGT (UL Style 5288)
200	392	SRG (UL Styles 3071, 3074, 3075, 3125, 3172 and 3231), SRK, SRGK and UL Types SF-2 and SFF-2
150	302	SRG, TGS and UL Styles 3212, 3213 and 3214
125	257	UL Style 3284 and CSA CL1254

6.5 POWER

Below are some of the key considerations when selecting a power cable:

- System voltage
- Current loading (ampacity)
- External thermal conditions such as ambient temperature, proximity of other cables, adjacent sources of heat, thermal conductivity of soil, etc.
- Voltage drop
- Special conditions, such as the presence of corrosive agents, flexibility and flame resistance

6.5.1 Voltage Rating

The system voltage on which the cable is to operate determines the required cable voltage rating. Cables rated 5 kV and above are separated into two classifications: grounded systems (100 percent insulation level) and ungrounded systems (133 percent insulation level). In case of a phase-to-ground fault in a three-phase system, it is possible to operate ungrounded systems for up to one hour with one conductor at ground potential. This condition results in full line-to-line voltage stress across the insulation of each of the other two conductors. For this reason each conductor of such a circuit must have additional insulation. Cables designed for use on grounded systems take advantage of the absence of this full line-to-line voltage stress across the insulation and use thinner insulation. The direct result of such a design is lower cost, as well as reduced cable diameter.

A recent change in the NEC now requires all cables operating above 2,400 volts in the U.S. to be shielded.

6.5.2 Conductor Size

Conductor size is based principally on three considerations:

- Current-carrying capacity (ampacity)
- Short-circuit current
- Voltage drop

The current-carrying capacity of a cable is affected primarily by the permissible operating temperature of its insulation. The higher the operating temperature of the insulation, the higher the current-carrying capacity of a given conductor size. The temperature at which a particular cable will operate is affected by the ability of the surrounding material to conduct away the heat. Therefore, the current-carrying capacity is materially affected by the ambient temperature as well as by the installation conditions. For example, a cable installed in a 40°C ambient temperature has an ampacity that is only about 90 percent of the ampacity in a 30°C ambient.

Running a single-conductor cable through a magnetic conduit will increase the apparent resistance of the cable and will also result in a lower current-carrying capacity due to the additional resistance and magnetic losses. Similarly, when a cable is run close to other cables, the presence of the other cables effectively increases the ambient temperature, which decreases the ability of the cable to dissipate its heat. It is apparent from the above that many conditions must be known before an accurate current-carrying capacity can be determined for a particular cable installation.

Occasionally, emergency overload conditions are also involved and may affect conductor size.

6.5.3 Short Circuit Current

A second consideration in selection of conductor size is that of the short circuit current, which the cable must be able to carry in an emergency. From a thermal standpoint there is a limit to the amount of short-circuit current that a cable can handle without damage.

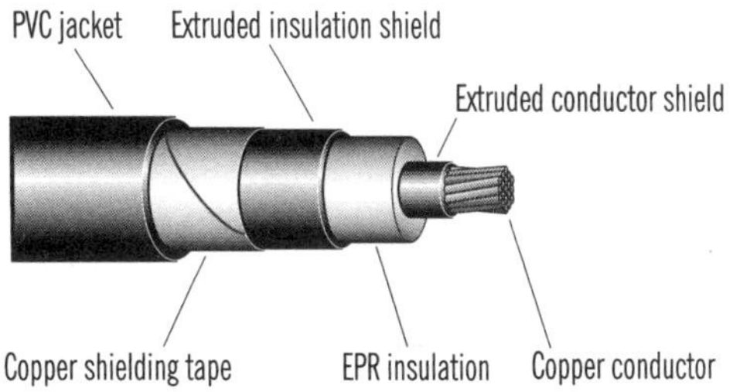

Figure 6.5–Typical Tape Shielded 15 kV Power Cable

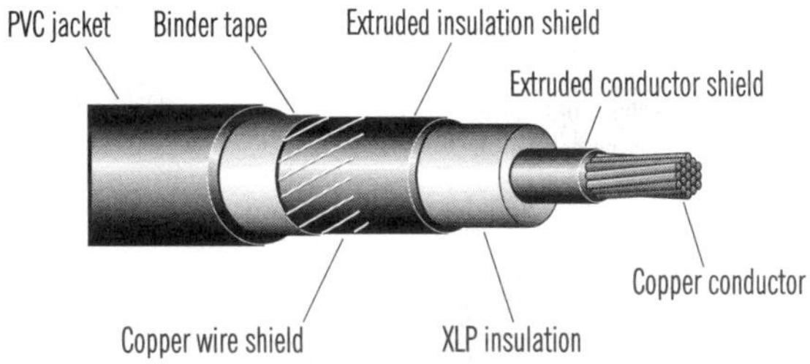

Figure 6.6–Typical Wire Shielded 15 kV Power Cable

6.5.4 Voltage Drop Considerations

Cable conductor size is sometimes governed by voltage drop rather than by heating. Generally, conductor size on long, low-voltage lines is governed by voltage drop; on short, high-voltage lines by heating. Due to voltage drop considerations, it might be necessary to increase conductor size, even though the current load is adequately handled by a smaller size conductor.

6.5.5 Special Conditions

The following are only a few of the many special conditions that may affect cable selection:

- The presence of large sources of heat (boilers, steam lines, etc.)
- The effect of magnetic materials such as pipes or structural members close to large cables carrying heavy current loads
- The presence of corrosive chemicals in the soil or other locations in which the cable is installed
- The interference that may occur in telecommunication circuits because of adjacent power cables
- Flame and radiation resistance
- Mechanical toughness
- Moisture resistance
- Overload and fault current requirements

All special conditions should be carefully investigated, and the advice of competent engineers obtained, before proceeding with an important cable installation.

6.6 ARMORED POWER AND CONTROL

Armored cables comprise a group of cables that are designed to withstand severe mechanical and chemical environments. For information on the various types and their applications, see Section 5 on armor.

6.7 ELECTRONIC CABLE

This category of wire and cable covers thousands of small gauge single-conductor wire types along with many types of multiconductor cables. These basic types come in various combinations of stranding, insulation material, conductor count, jacket material, etc. Some common types and key characteristics are described below.

6.7.1 Coaxial Cable

A coaxial cable consists of four basic parts:

- Inner conductor (center conductor)
- Outer conductor (shield)
- Dielectric, which separates the inner and outer conductors
- Jacket, which is the outer polymer layer protecting the parts inside

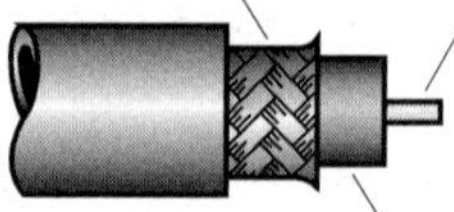

Figure 6.7–Typical Coaxial Cable

Characteristic Impedance

The characteristic impedance of a coaxial cable is a function of its geometry and materials. Characteristic impedance is independent of length and typically ranges from 35 to 185 ohms. The most common values are 50, 75 and 93 ohms. The characteristic impedance of a cable should not be confused with the impedance of the conductors in a cable, which is dependent on length.

The most efficient transfer of energy from a source to a load occurs when all parts of the system have the same characteristic impedance. For example, a transmitter, interconnecting cable and receiver should all have the same impedance. This need for impedance matching is especially critical at higher frequencies, where the consequences of mismatches are more severe.

VSWR

The Voltage Standing-Wave Ratio (VSWR) is a measure of the standing waves that result from reflections. It expresses the uniformity or quality of a cable's characteristic impedance. Uniformity is also measured as structural return loss (SRL).

Velocity of Propagation

Velocity of propagation is the speed at which electromagnetic energy travels along the cable. In free space or air, electromagnetic energy travels at the speed of light, which is 186,000 miles per second. In other materials, however, the energy travels slower, depending on the dielectric constant of the material. Velocity of propagation is expressed as a percentage of the speed of light. For example, a velocity of 65 percent means that the energy travels at 120,900 miles per second – or 35 percent slower than in free space.

The dielectric (insulation) separating the two conductors determines the velocity of propagation. Although the electromagnetic energy travels in the dielectric, the current associated with the energy travels primarily on the outside of the center conductor and the inside of the outer conductor (shield). The two conductors bind the energy within the cable. Consequently, the quality of the dielectric is important to efficient, speedy transfer of energy. Speed is important to engineers who must know the transit time of signals for digital transmission.

Voltage Rating

This is the maximum voltage the cable is designed to handle.

Operating Temperature Range

This is the minimum and maximum temperatures at which the cable can operate.

Coaxial Types

The following paragraphs describe four common types of coaxial cable.

- **Flexible Coax**
  The most common type, flexible coax has a braided outer conductor (shield) of extremely fine wires. While the braid makes the cable flexible, it does not provide complete shielding – energy (RF signals) can leak through the shield via minute gaps in the braid. To combat this, many cables have several layers in the outer conductor. In addition, thin foils are sometimes used to supplement the braid to provide better coverage for greater shielding effectiveness. The greater the coverage, the better the shield.
- **Semirigid Coax**
  Semirigid coax has a solid, tubular metallic outer conductor, similar to a pipe. This construction gives the cable a very uniform characteristic impedance (low VSWR) and excellent shielding, but at the expense of flexibility.
- **Triaxial Cable (Triax)**
  This coax has two outer conductors (shields) separated by a dielectric layer. One outer conductor (shield) serves as a signal ground, while the other serves as earth ground, providing better noise immunity and shielding. One caution: Do not confuse a flexible cable having a multilayer outer shield with triaxial cable.
- **Dual Coax**
  This cable contains two individual coaxial cables surrounded by a common outer jacket.

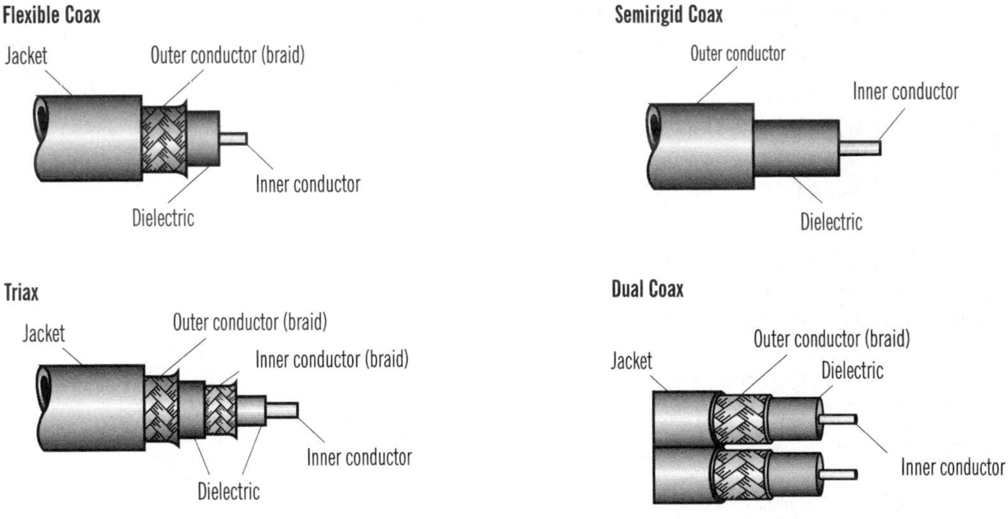

Figure 6.8–Common Types of Coaxial Cable

6.7.2 Twinaxial Cable (Twinax)

Twinax has a pair of insulated conductors encased in a common outer conductor (shield). The center conductors may be either twisted or run parallel to one another. In appearance, the cable is often similar to a shielded twisted pair, but it is held to the tighter tolerances common to fixed-impedance coaxial cable. A common use of twinax is high-speed, balanced-mode multiplexed transmission in large computer systems. Balanced mode means that the signal is carried on both conductors, which provides greater noise immunity.

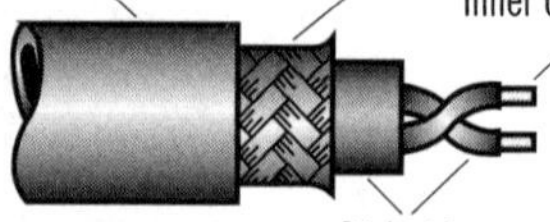

Figures 6.9–A Typical Twinaxial Cable

6.7.3 100 ohm Twisted Pair Cable

100 ohm unshielded twisted pair (UTP) and shielded twisted pair are low pair count cables (usually 4 pairs) that have been designed for use in local area networks such as Ethernet. Because of their relatively low cost these cable types are widely used and are available in several different performance categories (levels) – currently Categories 3, 5e, 6 and 6A. Insertion loss, crosstalk, impedance and other electrical parameters are specified in TIA/EIA-568-B.2 and its related addenda. A summary of their electrical requirements are shown below.

Table 6.5–Category 3 Performance (100 meters)

Frequency (MHz)	Insertion Loss (dB)	NEXT (dB)	PSNEXT (dB)
0.772	2.2	43.0	43
1.0	2.6	40.3	41
4.0	5.6	32.3	32
8.0	8.5	27.8	28
10.0	9.7	26.3	26
16.0	13.1	23.2	23

Maximum propagation delay: 545 ns/100 m at 10 MHz
Maximum delay skew: 45 ns/100 m at 16 MHz
Characteristic impedance: 100±15 ohms from 1 to 16 MHz

Table 6.6–Category 5e Performance (100 meters)

Frequency (MHz)	Insertion Loss (dB)	NEXT (dB)	PSNEXT (dB)	ELFEXT (dB)	PSELFEXT (dB)	Return Loss (dB)
0.772	1.8	67.0	64.0	–	–	19.4
1.0	2.0	65.3	62.3	63.8	60.8	20.0
4.0	4.1	56.3	53.3	51.8	48.8	23.0
8.0	5.8	51.8	48.8	45.7	42.7	24.5
10.0	6.5	50.3	47.3	43.8	40.8	25.0
16.0	8.2	47.2	44.2	39.7	36.7	25.0
20.0	9.3	45.8	42.8	37.8	34.8	25.0
25.0	10.4	44.3	41.3	35.8	32.8	24.3
31.25	11.7	42.9	39.9	33.9	30.9	23.6
62.5	17.0	38.4	35.4	27.9	24.9	21.5
100.0	22.0	35.3	32.3	23.8	20.8	20.1

Maximum propagation delay: 538 ns/100 m at 100 MHz
Maximum delay skew: 45 ns/100 m at 100 MHz

Table 6.7–Category 6 Performance (100 meters)

Frequency (MHz)	Insertion Loss (dB)	NEXT (dB)	PSNEXT (dB)	ELFEXT (dB)	PSELFEXT (dB)	Return Loss (dB)
0.772	1.8	76.0	74.0	70.0	67.0	19.4
1.0	2.0	74.3	72.3	67.8	64.8	20.0
4.0	3.8	65.3	63.3	55.8	52.8	23.0
8.0	5.3	60.8	58.8	49.7	46.7	24.5
10.0	6.0	59.3	57.3	47.8	44.8	25.0
16.0	7.6	56.2	54.2	43.7	40.7	25.0
20.0	8.5	54.8	52.8	41.8	38.8	25.0
25.0	9.5	53.3	51.3	39.8	36.8	24.3
31.25	10.7	51.9	49.9	37.9	34.9	23.6
62.5	15.4	47.4	45.4	31.9	28.9	21.5
100.0	19.8	44.3	42.3	27.6	24.8	20.1
200.0	29.0	39.8	37.8	21.8	18.8	18.0
250.0	32.8	38.3	36.3	19.8	16.8	17.3

Maximum propagation delay: 538 ns/100 m at 100 MHz (536 at 250 MHz)
Maximum delay skew: 45 ns/100 m at all frequencies

Table 6.8–Category 6A Performance (100 meters)

Frequency (MHz)	Insertion Loss (dB)	NEXT (dB)	PSNEXT (dB)	ACRF (dB)	PSACRF (dB)	Return Loss (dB)	PSANEXT (dB)	PSAACRF (dB)
1.0	2.1	74.3	72.3	67.8	64.8	20.0	67.0	67.0
4.0	3.8	65.3	63.3	55.8	52.8	23.0	67.0	66.2
8.0	5.2	60.8	58.8	49.7	46.7	24.5	67.0	60.1
10.0	5.9	59.3	57.3	47.8	44.8	25.0	67.0	58.2
16.0	7.5	56.2	54.2	43.7	40.7	25.0	67.0	54.1
20.0	8.4	54.8	52.8	41.8	38.8	25.0	67.0	52.2
25.0	9.4	53.3	51.3	39.8	36.8	24.3	67.0	50.2
31.25	10.5	51.9	49.9	37.9	34.9	23.6	67.0	48.3
62.5	15.0	47.4	45.4	31.9	28.9	21.5	67.0	42.3
100.0	19.1	44.3	42.3	27.8	24.8	20.1	67.0	38.2
150.0	23.7	41.7	39.7	24.3	21.3	18.9	67.0	34.7
200.0	27.6	39.8	37.8	21.8	18.8	18.0	64.5	32.2
250.0	31.1	38.3	36.3	19.8	16.8	17.3	62.5	30.2
300.0	34.3	37.1	35.1	18.3	15.3	16.8	61.0	28.7
350.0	37.2	36.1	34.1	16.9	13.9	16.3	59.6	27.3
400.0	40.1	35.3	33.3	15.8	12.8	15.9	58.5	26.2
450.0	42.7	34.5	32.5	14.7	11.7	15.5	57.4	25.1
500.0	45.3	33.8	31.8	13.8	10.8	15.2	56.5	24.2

Maximum propagation delay: 538 ns/100 m at 100 MHz
Maximum delay skew: 45 ns/100 m at all frequencies

100-ohm Unshielded Twisted Pair (UTP) vs Shielded Twisted Pair

There are two basic types of electromagnetic interference (EMI) that cable engineers worry about – EMI emissions and EMI immunity. Emissions refer to energy that is radiated by the cable that might affect the proper operation of a neighboring circuit or system. Immunity is the ability of the cable to reject outside signals that might interfere with the proper operation of the circuit or system to which the cable is attached.

Electromagnetic interference is present in all types of cabling to some degree. In local area networks (LANs), failure to properly manage EMI can have an adverse effect on the integrity of the transmitted information.

Shielded cables generally use an aluminum or copper shield to provide protection. When properly grounded (connected) to the associated electronic equipment, the shield acts as a barrier to incoming as well as outgoing EMI.

In an unshielded (UTP) cable, careful design of the cable and the associated electronic equipment results in a "balance" of the currents in the two conductors of a pair. That is, the currents in the two conductors are equal in magnitude but flowing in opposite directions. In a balanced system, there is very little radiation of EMI since the external field from one conductor is effectively canceled by the external field from the other conductor of the pair.

Generally, the more twists per foot in the conductor pairs of the cable, the better the cable is electrically balanced. For example, Category 5e cable has more twists per foot than Category 3 cable and, therefore, offers better protection from EMI problems.

6.7.4 IBM Cabling System

The IBM Cabling System is a structured building wiring system that is compatible with IEEE 802.5 (Token Ring) networks and equipment. Cable types consist of various combinations of shielded data grade media (DGM) and non-shielded voice grade media (VGM). Cable types include "Type 1," which is a 2-pair DGM cable, "Type 2," which contains two DGM pairs plus four VGM pairs and "Type 6," which is a 2-pair DGM cable with smaller conductors (26 AWG instead of 22 AWG).

6.8 TELEPHONE

Telephone cables play a major role in modern communications. In conjunction with microwave and satellite transmission, copper and optical fiber cables provide the communication links that have become essential to society.

With the advent of optical fiber cables in the early 1980s, telephone wire and cable has generally been grouped into three broad categories: 1) fiber, 2) copper and 3) hybrid (composite) cable with both fiber and copper components under one jacket.

Telephone cable is usually classified according to its location of use. Cable used outdoors between the telephone company's central office and the building being served is referred to as outside cable, or sometimes called black cable. Wire or cable used indoors, e.g., inside homes and commercial buildings, is referred to as premises distribution wiring or more simply as inside cable.

6.8.1 Outside Cables

Outside cables typically range in size from small (2 to 6 pair) constructions, which are usually referred to as "service drop" or "buried distribution" wire (the cable installed in many residential backyards), up to large 1,500 pair "exchange" cables, which are typically installed between central offices of the telephone company. Many high pair-count copper cables have been replaced by optical fiber cables.

Exchange cables, because they are often installed in underground ducts or directly buried in the earth, are designed with various combinations of polyethylene (PE) jackets and aluminum, copper or steel sheaths. The PE jacket and metal armoring isolate signal-carrying conductor pairs from moisture, mechanical damage and lightning induced voltages.

Exchange cables are manufactured in "filled" and "unfilled" (aircore) versions. With filled cables, the interstices between insulated conductors are filled with a waterproofing gel to prevent the ingress and longitudinal movement of water. Some aircore cable designs are kept dry by pressurizing the core of the cable with dry air or nitrogen. Water is the "Achilles' heel" of outdoor telephone cable because it increases capacitance (normally 0.083 μF per mile) between the "tip" and "ring" conductors and compromises crosstalk (pair-to-pair signal coupling) performance of the cable.

The terms tip and ring are carryovers from earlier days when each twisted pair was terminated with a 1/4-inch diameter plug at a manually operated switchboard. One conductor was attached to the tip, the other to the ring of the plug.

6.8.2 Indoor Cables

Inside wire and cable is usually divided into 1) station wire and 2) inside cable (sometimes called IC). Station wire is usually 2 to 4 pair, 22 or 24 AWG wire and is typically installed in residences.

While station wire is one type of inside wire, it is usually designed for both indoor and outdoor use because it often extends to the exterior of the building. True inside cable, on the other hand, is typically larger (25 to 200 pair) 22 or 24 AWG cable, which is installed exclusively indoors in larger public and commercial buildings. Station wire and inside cables are usually used in plenum, riser, and general purpose versions. The plenum version is a highly flame retardant construction that is capable of passing the Steiner Tunnel Flame Test (NFPA-262).

Article 800 of the National Electrical Code (NEC) requires that telephone wire and cable be plenum rated when installed indoors in plenums (air handling spaces) without conduit, i.e., it must carry the marking CMP (CM for communication and P for plenum). When installed in vertical risers in multistory buildings, a riser rating, i.e., Type CMR, is required. General purpose communication cables must be labeled Type CM. Cables installed in one- and two-family dwellings must be identified as Type CMX.

6.8.3 Insulation and Jacket Materials

Two thermoplastic polymers are generally used to insulate the conductors of outdoor telephone wire and cable: polypropylene (PP) or polyethylene (PE). These polymers are used primarily because of their low dielectric constant, high dielectric strength (to withstand lightning induced overvoltages), excellent moisture resistance, mechanical toughness, extrudability in thin walls and low cost. Indoor dielectrics include PP and PE but, in addition, include FEP (fluorinated ethylene-propylene or Teflon), ECTFE (ethylene-chlorotrifluoroethylene or Halar) and PVC (polyvinyl chloride). FEP and ECTFE are used in plenum cables to provide the necessary flame retardancy and are extruded on the wire in either solid or foamed (expanded) versions.

The most important telephone wire and cable electrical characteristics and their usual units of measurement include capacitance (microfarads per mile), conductor resistance (ohm per loop-mile), crosstalk (decibel isolation between pairs) and attenuation (decibels per mile). When used for high-speed digital applications, characteristic impedance (ohm) and structural return loss (decibels) also become important.

The mechanical and chemical characteristics of telephone cable insulation are as important as the electrical characteristics. Several important mechanical and chemical characteristics include compression cut resistance, low-temperature brittleness, resistance to the base oils used in filling gels, adequate tensile and elongation properties, and acceptable long-term aging characteristics.

6.9 MILITARY

The U.S. military has developed extensive specifications for many wire and cable types used in military applications. This includes hook-up and lead wire, airframe wire, control cable and coax. A MIL-Spec wire or cable must meet rigorous performance requirements. Tests that prove the wire or cable meets the specified requirements must be conducted by the manufacturer and must be carefully documented.

Following is a partial list of military wire and cable types.

Type	Description
MIL-C-5756	Cable and wire, portable power, rubber insulated (replaced by SAE-AS5756)
MIL-C-7078	Cable, aerospace vehicle (replaced by NEMA WC27500)
MIL-C-13294	Field wire (replaced by MIL-DTL-49104)
MIL-DTL-915	Shipboard cable (inactive for new design except outboard types)
MIL-DTL-3432	Power and special purpose cables used for ground support systems ("CO" types), 300 and 600 V
MIL-DTL-8777	Aircraft wire, silicone insulated, 600 V, 200°C
MIL-DTL-13486	Cable, special purpose, low tension, single and multiconductor, shielded and unshielded
MIL-DTL-16878	General purpose hook-up and lead wire
MIL-DTL-24640	Shipboard cable, lightweight
MIL-DTL-24643	Shipboard cable, low smoke
MIL-DTL-25038	Aircraft wire, inorganic fibrous/Teflon insulation, high temperature and fire resistant, engine zone wire
MIL-DTL-23053	Tubing, heat shrink (replaced by AMS-DTL-23053)
MIL-DTL-27072	Cable, power and special purpose, multiconductor and single shielded (replaced by NEMA WC27500)
MIL-DTL-27500	Aerospace and other general application wire

Continued on next page >>

(Continued)

Type	Description
MIL-DTL-49055	Cable, power, flat, unshielded
MIL-DTL-55021	Cable, shielded singles, twisted pairs and triples, internal hook-up
MIL-I-22129	Tubing, PTFE, nonshrink
MIL-W-76	General purpose hook-up wire
MIL-W-5845	Thermocouple wire, iron and Constantan
MIL-W-5846	Thermocouple wire, chromel and alumel
MIL-W-81822	Solderless wrap (wire wrap), insulated and uninsulated
MIL-W-47206	Cable, single conductor, twisted pairs; and multiconductor, high temperature (replaced by MIL-DTL-27500)

6.10 SHIPBOARD CABLES (MIL-DTL-24643, MIL-DTL-24640 AND MIL-DTL-915)

Due to concern about flammability, smoke and toxicity, the U.S. Navy introduced the MIL-DTL-24643 cable specification. Generally, this document provides low-smoke, fire-retardant cables that are approximately equivalent in size, weight and electricals to many of the older MIL-DTL-915 constructions.

In consideration of circuit density, weight and size, the U.S. Navy produced the MIL-DTL-24640 cable document. The cables covered by this specification are also low-smoke, fire-retardant constructions, but they are significantly lighter in weight and smaller in diameter. MIL-DTL-24640 cables are used to interconnect systems where weight and space savings are critical; however, they are not direct replacements. Because the overall diameters have been reduced and electrical characteristics may have been changed, they should not be used to replace existing MIL-DTL-915 or MIL-DTL-24643 constructions unless a comprehensive electrical and physical system evaluation or redesign has been completed.

For many years, most of the shipboard power and lighting cables for fixed installation had silicone-glass insulation, polyvinyl chloride jacket, and aluminum armor and were of watertight construction. It was determined that cables with all of these features were not necessary for many applications, especially for applications within watertight compartments and noncritical areas above the watertightness level. Therefore, for applications within watertight compartments and noncritical areas, a new family of non-watertight lower cost cables was designed. This family of cables is electrically and dimensionally interchangeable with silicone-glass insulated cables of equivalent sizes and is covered by Military Specification MIL-DTL-915.

6.11 OPTICAL FIBER CABLES

In all types of optical fiber cables, the individual optical fibers are the signal transmission media that act as individual optical wave guides. The fibers consist of a central transparent core region that propagates the optical radiation and an outer cladding layer that completes the guiding structure. The core and the cladding are typically made of pure silica glass, though other materials can be used. To achieve high signal bandwidth capabilities, the core region sometimes has a varying (or graded) refractive index.

6.11.1 Fiber Types

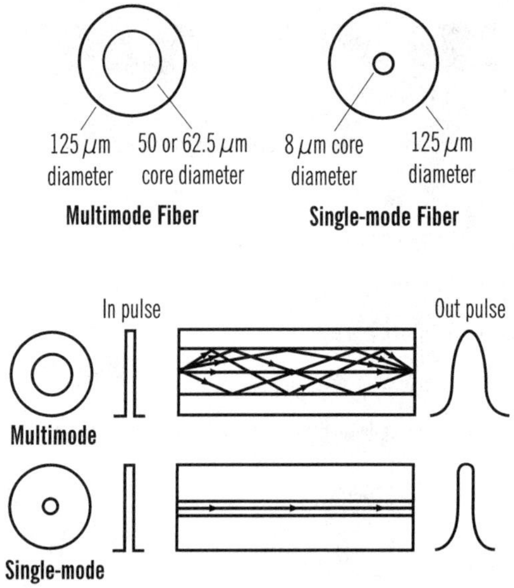

Figure 6.10—Optical Fiber Types

There are two basic fiber types – single-mode and multimode. Single-mode has a core diameter of 8 to 10 microns and is normally used for long distance requirements (e.g., interstate) and high-bandwidth (information carrying capacity) applications. Multimode, on the other hand, has a core diameter of 50 or 62.5 microns and is usually used intrabuilding.

Laser-optimized fibers are a fairly recent development in which 50-micron multimode fibers are optimized for 850 nm VCSEL (vertical cavity surface emitting laser) sources and can provide significantly increased bandwidth performance when compared with standard multimode fiber types. The added bandwidth of laser-optimized 50-micron fiber allows for distance support up to 550 meters for 10 Gigabit Ethernet networks as well as providing a lower overall system cost when compared with single-mode systems utilizing higher cost 1300 or 1550 laser sources. Laser-optimized fiber is referred to as "OM3" fiber in ISO/IEC-11801. OM3 fibers are also referenced by other industry standards, such as the TIA-568 wiring standards and Institute of Electrical and Electronics Engineers (IEEE). OM1 and OM2 designations are specified for standard 62.5 and 50 micron multimode fibers, respectively.

6.11.2 Fiber Selection

The three major fiber parameters used in selecting the proper fiber for an application are bandwidth, attenuation and core diameter.

Bandwidth

The bandwidth at a specified wavelength represents the highest sinusoidal light modulation frequency that can be transmitted through a length of fiber with an optical signal power loss equal to 50 percent (-3 dB) of the zero modulation frequency component. The bandwidth is expressed in megahertz over a kilometer length (MHz-km).

Attenuation

The optical attenuation denotes the amount of optical power lost due to absorption and scattering of optical radiation at a specified wavelength in a length of fiber. It is expressed as an attenuation in decibels of optical power per kilometer (dB/km).

The attenuation is determined by launching a narrow spectral band of light into the full length of fiber and measuring the transmitted intensity. This measure is then repeated for the first 1.5 to 2.5 meters of the same fiber cable without disturbing the input end of the fiber. The dB/km attenuation is then calculated and normalized to 1 km.

Core Diameter

The fiber core is the central region of an optical fiber whose refractive index is higher than that of the fiber cladding. Various core diameters are available to permit the most efficient coupling of light from commercially available light sources, such as LEDs or laser diodes.

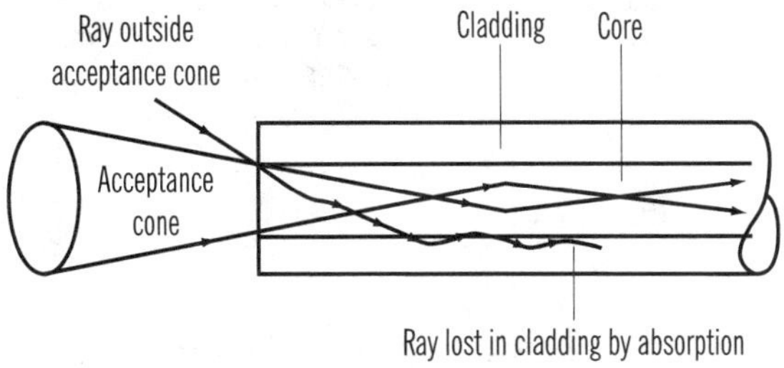

Figure 6.11–Optical Fiber Attenuation

6.11.3 Optical Fiber Cable Selection

Another important consideration when specifying optical fiber cable is the cable construction. Proper selection depends on the environment in which the cable will be installed. One of two different types of cable construction are generally employed to contain and protect the optical fibers.

Loose Buffer

The first is a loose buffer tube construction where the fiber is contained in a water-blocked polymer tube that has an inner diameter considerably larger than the fiber itself. This provides a high level of isolation for the fiber from external mechanical forces that might be present on the cable. For multifiber cables, a number of these tubes, each containing one or more fibers, are combined with the necessary longitudinal strength member. Loose buffer cables are typically used in outdoor applications and can accommodate the changes in external conditions (e.g., contraction in cold weather and elongation in warm weather).

Tight Buffer

The second cable construction is a tight buffer tube design. Here, a thick buffer coating is placed directly on the fiber.

Both constructions have inherent advantages. The loose buffer tube construction offers lower cable attenuation from a given fiber, plus a high level of isolation from external forces. This means more stable transmission characteristics under continuous mechanical stress. The tight buffer construction permits smaller, lighter weight designs and generally yields a more flexible cable. A comparison of these two cable constructions is shown below.

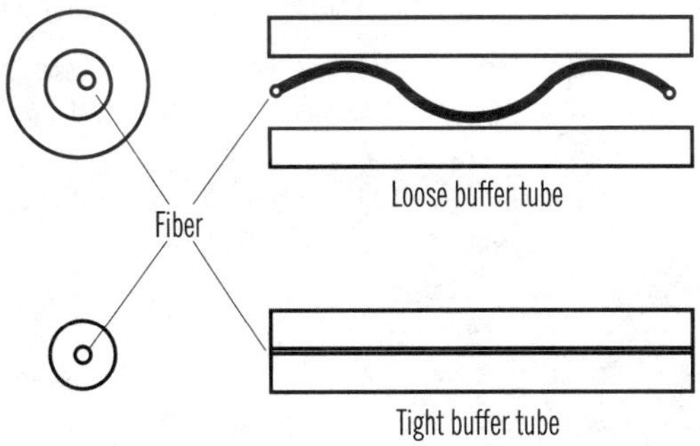

Figure 6.12–Optical Fiber Cable Designs

Table 6.9–A Comparison of Loose Tube and Tight Buffer Optical Fiber Cable

Cable Parameter	Cable Construction	
	Loose Tube	Tight Buffer
Bend radius	Larger	Smaller
Diameter	Larger	Smaller
Tensile strength, installation	Higher	Lower
Impact resistance	Higher	Lower
Crush resistance	Higher	Lower
Attenuation change at low temperatures	Lower	Higher

Strength Members

Once the optical fiber is surrounded with a buffer, either loose or tight, strength members are added to the cable structure to keep the fibers free from stress and to minimize elongation and contraction. Such strength members provide tensile load properties similar to electronic cables and, in some cases, are used as temperature stabilization elements.

Jacket

As with conventional metallic cables, the jacket protects the core from the external environment. With optical fibers, however, the selection of materials is influenced by the fact that the thermal coefficient of expansion of glass is significantly lower than that of the metal or plastic used in the cable structure.

Installation

Normal cable loads sustained during installation or environmental movements first stress the strength members without transferring the stress to the optical fibers. If the load is increased, the fiber may ultimately be placed in a tensile stress state. This level of stress may cause microbending losses that result in attenuation increase and possibly fatigue effects.

6.12 TRAY CABLES

Tray cables are a special class of cables designed to meet stringent flame test requirements. A tray cable rating is given to a cable if it can meet the UL or CSA Standard for the rating. To obtain the rating, a cable must pass the 70,000 BTU, UL 1685 Vertical Tray Flame test or the Vertical Flame Test described in CSA C22.2 No. 0.3 (See Section 11.2 Fire Safety Tests for additional information).

In effect, a cable does not have a tray cable rating unless it is so marked, for example “for CT use” or “Type TC.” Electrical inspectors will usually reject a cable even if it is capable of passing the tray cable fire test unless it is clearly marked on the cable as being a tray-rated cable.

A summary of applicable UL Standards, listings and markings is shown in Table 6.10. Note that, in some cases, the tray rating is an optional marking and is not an inherent part of the listing. Other UL and CSA Types that can be installed in tray in accordance with the NEC include CL2, CL2R, CL2P, CL3, CL3R, CL3P, CM, CMR, CMP, CMG, FPL, FPLR, FPLP, OFN, OFNR and OFNP.

Table 6.10–Tray Cable Listings and Markings

Standard	UL Listings (Types)	Optional Markings
UL 4	AC	For CT use
UL 13	PLTC	Direct burial Sunlight resistant ER (Exposed Run)
UL 44	XHHW-2 RHW-2, RHH, RH SIS, SA	For CT use Sunlight resistant Oil resistant Pump cable
UL 444	CM, CMR, CMP	Sunlight resistant
UL 1072	MV	For CT use Direct burial Sunlight resistant Oil resistant
UL 1277	TC	Direct burial Sunlight resistant Oil resistant ER (Exposed Run) LS (Limited Smoke)
UL 1424	FPL, FPLR, FPLP	Direct burial Sunlight resistant CI (Circuit Integrity) Limited combustible Wet location
UL 1425	NPLF, NPLFR, NPLFP	Direct burial Sunlight resistant CI (Circuit Integrity) Limited combustible Wet location
UL 2250	ITC	Direct burial Sunlight resistant Wet location

7. ELECTRICAL CHARACTERISTICS

For a wire or cable to perform its intended function reliably, safely and efficiently, the wire or cable must be selected so that its many electrical, physical, chemical and thermal properties match those of the application.

The following sections provide information on some of the most frequently requested electrical parameters.

7.1 DC RESISTANCE OF PLATED COPPER CONDUCTORS

Table 7.1–DC Resistance of Plated Copper Conductors

Wire Size	No. of Wires/Size	Strand	Nominal Area	Nominal DC Resistance ohms/1,000 ft. at 20°C (68°F)		
(AWG/kcmil)	(AWG or in.)	Class	(cmils)	Silver Plated	Nickel Plated	Tin Plated
777	1,952/24	AAR	788,728	–	–	0.0139
750	703/0.0327	H	751,711	–	–	0.0146
750	1,862/24	I	752,267	–	–	0.0146
750	7,448/30	K	744,800	–	–	0.0148
700	703/0.0316	H	701,988	–	–	0.0157
700	1,729/24	I	698,533	–	–	0.0158
700	6,916/30	K	691,600	–	–	0.0165
650	703/0.0304	H	649,684	–	–	0.0169
650	1,596/24	I	644,800	–	–	0.0171
650	6,517/30	K	651,700	–	–	0.0169
646	1,647/24	AAR	665,404	–	–	0.0165
600	703/0.0292	H	599,406	–	–	0.0183
600	1,470/24	I	593,895	–	–	0.0185
600	5,985/30	K	598,500	–	–	0.0184
550	703/0.028	H	551,152	–	–	0.0200
550	1,372/24	I	554,302	–	–	0.0200
550	5,453/30	K	545,300	–	–	0.0200
535	1,332/24	AAR	538,141	–	–	0.0204
500	427/0.0342	H	449,436	–	–	0.0220
500	1,125/24	I	494,912	–	–	0.0222
500	5,054/30	K	505,400	–	–	0.0218
450	427/0.0325	H	451,019	–	–	0.0244
450	1,127/24	I	455,319	–	–	0.0241
450	4,522/30	K	452,200	–	–	0.0243
444	1,110/24	AAR	448,451	–	–	0.025
400	427/0.0306	H	399,826	–	–	0.028
400	980/24	I	395,930	–	–	0.028
400	3,990/30	K	399,000	–	–	0.028
373	925/24	AAR	373,709	–	–	0.029
350	427/0.0286	H	349,269	–	–	0.031

Continued on next page >>

Table 7.1–DC Resistance of Plated Copper Conductors (Continued)

Wire Size	No. of Wires/Size	Strand	Nominal Area	Nominal DC Resistance ohms/1,000 ft. at 20°C (68°F)		
(AWG/kcmil)	(AWG or in.)	Class	(cmils)	Silver Plated	Nickel Plated	Tin Plated
350	882/24	I	356,337	–	–	0.031
350	3,458/30	K	345,800	–	–	0.032
313	777/24	AAR	313,916	–	–	0.035
300	427/0.0265	H	299,861	–	–	0.037
300	735/24	I	296,947	–	–	0.037
300	2,989/30	K	298,900	–	–	0.037
262	646/24	AAR	260,990	–	–	0.042
250	427/0.0242	H	250,068	–	–	0.043
250	637/24	I	257,354	–	–	0.043
250	2,499/30	K	249,900	–	–	0.044
4/0	2,109/30	K	210,900	0.052	0.053	0.052
4/0	427/0.0223	H	212,343	0.052	0.053	0.052
3/0	1,665/30	K	166,500	0.066	0.067	0.069
3/0	427/0.0198	H	167,401	0.066	0.067	0.066
2/0	1,330/30	K	133,000	0.083	0.084	0.088
2/0	427/0.0177	H	133,775	0.083	0.084	0.082
1/0	1,045/30	K	104,500	0.105	0.107	0.116
1/0	259/0.0202	H	105,682	0.105	0.107	0.103
1	817/30	K	81,700	0.134	0.137	0.144
1	259/0.018	H	83,916	0.134	0.137	0.129
2	665/30	K	66,500	0.165	0.168	0.177
2	259/0.016	H	66,304	0.165	0.168	0.164
2	133/0.0223	H	66,140	0.165	0.168	0.164
3	133/0.0199	H	52,669	0.165	0.168	0.205
4	133/25	H	42,615	0.249	0.259	0.264
5	133/0.0158	H	33,202	0.249	0.259	0.325
6	133/27	H	26,818	0.393	0.409	0.417
8	19/0.0295	C	16,535	0.628	0.689	0.640
8	37/0.0211	D	16,473	0.630	0.692	0.655
8	133/29	H	16,983	0.616	0.642	0.654
10	7/0.0385	B	10,376	1.00	1.10	1.02
10	19/0.0234	C	10,404	1.00	1.10	1.03
10	37/26	D	9,354	1.13	1.18	1.20
12	7/0.0305	B	6,512	1.59	1.75	1.65
12	19/25	C	6,088	1.71	1.78	1.81
12	19/0.0185	C	6,503	1.60	1.75	1.70
12	37/28	D	5,874	1.80	1.87	1.91
12	65/30	K	6,500	1.80	1.87	1.61
14	7/0.0242	B	4,099	2.53	2.69	2.63

Continued on next page >>

Table 7.1—DC Resistance of Plated Copper Conductors (Continued)

Wire Size	No. of Wires/Size	Strand	Nominal Area	Nominal DC Resistance ohms/1,000 ft. at 20°C (68°F)		
(AWG/kcmil)	(AWG or in.)	Class	(cmils)	Silver Plated	Nickel Plated	Tin Plated
14	19/27	C	3,831	2.70	2.81	2.86
14	19/0.0147	C	4,105	2.62	2.65	2.78
14	37/0.0105	D	4,079	2.62	2.65	2.59
14	41/30	K	4,100	2.62	2.65	2.58
16	7/0.0192	B	2,580	4.02	4.28	4.27
16	19/29	C	2,426	4.23	4.41	4.49
16	19/0.0117	C	2,600	4.14	4.20	4.39
16	26/30	K	2,600	4.14	4.20	4.07
18	7/0.0152	B	1,617	6.58	6.67	6.99
18	7/26	B	1,769	5.86	6.10	6.22
18	16/30	K	1,600	5.86	6.10	6.61
18	19/30	C	1,900	5.38	5.60	5.77
18	19/0.0092	C	1,608	6.69	6.82	7.18
20	7/.28	B	1,111	9.27	9.65	9.84
20	10/30	K	1,000	–	–	10.58
20	19/32	C	1,216	8.53	9.07	9.15
22	7/30	B	700	14.60	15.20	15.60
22	19/34	C	754	13.70	14.60	14.70
24	7/34	B	448	23.10	24.60	24.80
24	19/36	C	475	21.50	22.90	23.10
25	7/0.0067	B	314	33.00	34.80	36.40
26	7/34	B	277	37.10	39.50	39.80
26	19/38	C	304	33.30	36.60	35.70
28	7/36	B	175	58.40	62.10	66.50
28	19/40	C	182	54.60	60.00	58.60
30	7/38	B	112	90.30	99.20	96.20
30	19/42	C	118	82.70	94.00	88.80
32	7/40	B	67	148.0	163.0	159.0
34	7/42	B	43	225.0	256.0	241.5
36	7/44	B	28	244.0	391.0	369.2

Note: AAR – American Association of Railroads
Strand Classes B, C, D, H, I and K per ASTM

7.2 DC AND AC RESISTANCE OF COPPER CONDUCTORS

Table 7.2–DC and AC Resistance of Copper Conductors, Nominal Ohms Per 1,000 ft.

Size	20°C Conductor Temperature	60°C Conductor Temperature			75°C Conductor Temperature			90°C Conductor Temperature		
	DC	DC	60 Hz AC		DC	60 Hz AC		DC	60 Hz AC	
(AWG/kcmil)			*Single Cond.	†Multi-Cond.		*Single Cond.	†Multi-Cond.		*Single Cond.	†Multi-Cond.
40	1,080.0	1,250.0	–	–	1,314.0	–	–	1,378.0	–	–
38	648.6	750.7	–	–	789.1	–	–	827.3	–	–
36	414.8	480.1	–	–	504.6	–	–	529.1	–	–
34	260.9	302.0	–	–	317.4	–	–	332.8	–	–
32	164.1	189.9	–	–	199.6	–	–	209.3	–	–
30	103.2	119.4	–	–	125.5	–	–	131.6	–	–
28	64.9	75.1	–	–	79.0	–	–	82.8	–	–
26	40.8	47.2	–	–	49.7	–	–	52.1	–	–
24	26.1	30.2	–	–	31.8	–	–	33.3	–	–
22	16.4	19.0	–	–	20.0	–	–	20.9	–	–
20	10.3	11.9	–	–	12.5	–	–	13.1	–	–
18	6.54	7.57	–	–	7.96	–	–	8.34	–	–
16	4.1	4.75	–	–	4.99	–	–	5.23	–	–
14	2.57	2.98	2.98	2.98	3.14	3.14	3.14	3.29	3.29	3.29
12	1.62	1.88	1.88	1.88	1.97	1.97	1.97	2.07	2.07	2.07
10	1.17	1.18	1.18	1.18	1.24	1.24	1.24	1.31	1.31	1.31
8	0.638	0.744	0.744	0.744	0.783	0.783	0.783	0.822	0.822	0.822
6	0.403	0.466	0.466	0.466	0.491	0.491	0.491	0.515	0.515	0.515
4	0.253	0.295	0.295	0.295	0.310	0.310	0.31	0.325	0.325	0.325
2	0.159	0.184	0.184	0.185	0.195	0.194	0.196	0.203	0.203	0.205
1	0.126	0.147	0.147	0.148	0.154	0.154	0.155	0.162	0.162	0.163
1/0	0.1	0.116	0.116	0.118	0.122	0.122	0.124	0.128	0.128	0.13
2/0	0.0794	0.0923	0.0923	0.0950	0.0971	0.0971	0.1	0.102	0.102	0.105
3/0	0.0629	0.073	0.073	0.0759	0.0769	0.0769	0.0799	0.0807	0.0807	0.0839
4/0	0.05	0.0579	0.0579	0.0608	0.061	0.061	0.064	0.0639	0.0639	0.0671
250	0.0423	0.049	0.0492	0.0519	0.0516	0.0518	0.0547	0.0541	0.0543	0.0573
300	0.0353	0.0409	0.0411	0.0437	0.0431	0.0433	0.0461	0.0452	0.0454	0.0483

Continued on next page >>

Table 7.2–DC and AC Resistance of Class B Copper Conductors, ohms per 1,000 ft. (Continued)

Size	20°C Conductor Temperature	60°C Conductor Temperature			75°C Conductor Temperature			90°C Conductor Temperature		
	DC	DC	60 Hz AC		DC	60 Hz AC		DC	60 Hz AC	
(AWG/kcmil)			*Single Cond.	†Multi-Cond.		*Single Cond.	†Multi-Cond.		*Single Cond.	†Multi-Cond.
350	0.0302	0.035	0.0353	0.0378	0.0369	0.0372	0.0398	0.0387	0.0390	0.0418
400	0.0264	0.0307	0.031	0.0338	0.0323	0.0326	0.0355	0.0339	0.0342	0.0373
500	0.0212	0.0246	0.025	0.0278	0.0258	0.0262	0.0291	0.0271	0.0275	0.0306
600	0.0177	0.0205	0.021	0.0238	0.0215	0.022	0.0249	0.0226	0.0231	0.0262
700	0.0151	0.0175	0.0181	0.0208	0.0184	0.019	0.0219	0.0193	0.0199	0.0229
750	0.0141	0.0164	0.0170	0.0198	0.0172	0.0178	0.0208	0.0181	0.0188	0.0219
1,000	0.0106	0.0123	0.0131	0.016	0.0129	0.0137	0.0167	0.0135	0.0144	0.0175
1,250	0.00847	0.00982	0.0108	0.0138	0.0103	0.0113	0.0145	0.0108	0.0119	0.0152
1,500	0.00705	0.00818	0.00934	0.0125	0.00861	0.00983	0.0132	0.00904	0.01030	0.0138
1,750	0.00604	0.00701	0.0083	0.0117	0.00738	0.00874	0.0123	0.00774	0.00917	0.0129
2,000	0.00529	0.00613	0.00755	0.0111	0.00645	0.00795	0.0117	0.00677	0.00835	0.0123
2,500	0.00428	0.00496	–	–	0.00521	–	–	0.00546	–	–
3,000	0.00356	0.00412	–	–	0.00433	–	–	0.00454	–	–
5,000	0.00218	0.00252	–	–	0.00265	–	–	0.00278	–	–

Note: 40 AWG through 26 AWG values are for solid conductors, all others are for ASTM Class B stranded conductors.
*One single conductor in air, buried or in nonmetallic conduit.
†Multiconductor cable or two or three single conductors in one metallic conduit.

Table 7.3–Temperature Correction Factors for Copper DC Resistance

Temperature (°C)	Multiplying Factors for Correction To:	
	20°C	25°C
60	0.864	0.881
75	0.822	0.838
90	0.784	0.800

Example:
The DC resistance of a 500 kcmil copper conductor at 60°C is 0.0246 ohms per 1,000 ft. The resistance at 25°C would be $0.0246 \times 0.881 = 0.0217$ ohms per 1,000 ft.

7.3 DC AND AC RESISTANCE OF ALUMINUM CONDUCTORS

Table 7.4–DC and AC Resistance of Class B Aluminum Conductors, ohms Per 1,000 ft.

Size	60°C Conductor Temperature			75°C Conductor Temperature			90°C Conductor Temperature		
	DC	60 Hz AC		DC	60 Hz AC		DC	60 Hz AC	
(AWG/kcmil)		*Single Cond.	†Multi-Cond.		*Single Cond.	†Multi-Cond.		*Single Cond.	†Multi-Cond.
12	3.08	3.08	3.08	3.24	3.24	3.24	3.40	3.40	3.40
10	1.93	1.93	1.93	2.03	2.03	2.03	2.13	2.13	2.13
8	1.21	1.21	1.21	1.28	1.28	1.28	1.34	1.34	1.34
6	0.765	0.765	0.765	0.080	0.808	0.808	0.848	0.848	0.848
4	0.483	0.483	0.483	0.507	0.507	0.507	0.533	0.533	0.533
3	0.382	0.382	0.382	0.402	0.402	0.402	0.422	0.422	0.422
2	0.303	0.303	0.303	0.319	0.319	0.319	0.335	0.335	0.335
1	0.240	0.240	0.240	0.253	0.253	0.253	0.266	0.266	0.266
1/0	0.191	0.191	0.191	0.201	0.201	0.201	0.211	0.211	0.211
2/0	0.151	0.151	0.151	0.159	0.159	0.159	0.167	0.167	0.167
3/0	0.119	0.119	0.120	0.126	0.126	0.127	0.132	0.132	0.133
4/0	0.0953	0.0954	0.0963	0.101	0.101	0.102	0.106	0.106	0.107
250	0.0806	0.0808	0.0822	0.0848	0.0850	0.0865	0.0890	0.0892	0.0908
300	0.0672	0.0674	0.0686	0.0706	0.0708	0.0720	0.0741	0.0744	0.0756
350	0.0575	0.0578	0.0593	0.0605	0.0608	0.0623	0.0635	0.0638	0.0654
400	0.0504	0.0507	0.0525	0.0500	0.0533	0.0552	0.0557	0.0560	0.0580
500	0.0403	0.0406	0.0428	0.0424	0.0427	0.0450	0.0445	0.0448	0.0472
600	0.0336	0.0340	0.0370	0.0353	0.0357	0.0381	0.0370	0.0374	0.0400
700	0.0288	0.0292	0.0320	0.0303	0.0307	0.0337	0.0318	0.0322	0.0353
750	0.0269	0.0273	0.0302	0.0283	0.0288	0.0317	0.0297	0.0302	0.0333
1,000	0.0201	0.0207	0.0239	0.0212	0.0218	0.0253	0.0222	0.0228	0.0265
1,250	0.0162	0.0176	0.0215	0.0170	0.0177	0.0216	0.0179	0.0186	0.0228
1,500	0.0135	0.0143	0.0184	0.0142	0.0150	0.0193	0.0149	0.0158	0.0203
1,750	0.0115	0.0124	0.0168	0.0121	0.0131	0.0177	0.0127	0.0137	0.0186
2,000	0.0101	0.0111	0.0158	0.0106	0.0117	0.0165	0.0111	0.0122	0.0173

*One single conductor in air, buried or in nonmetallic conduit.
†Multiconductor cable or two or three single conductors in one metallic conduit.

Table 7.5–Temperature Correction Factors for Aluminum DC Resistance

Temperature (°C)	Multiplying Factors for Correction To:	
	20°C	25°C
60	0.861	0.878
75	0.818	0.835
90	0.780	0.796

Example:
The DC resistance of a 500 kcm aluminum conductor at 60°C is 0.0403 ohms per 1,000 ft. The resistance at 25°C would be $0.0403 \times 0.878 = 0.0354$ ohms per 1,000 ft.

7.4 REACTANCE AND IMPEDANCE AT 60 Hz

Table 7.6–Reactance and Impedance at 60 Hz for Single Copper Conductor Cables Installed in Air, Buried or in Separate Nonmetallic Conduits

Conductor Size	Approximate ohms per 1,000 ft. per Conductor at 25°C (77°F)							
	Distance Between Centers of Conductors							
	2 in.		4 in.		6 in.		8 in.	
(AWG/kcmil)	Reactance	Impedance	Reactance	Impedance	Reactance	Impedance	Reactance	Impedance
8	0.0816	0.659	0.0976	0.661	0.1070	0.662	0.1135	0.664
6	0.0764	0.417	0.0922	0.420	0.1016	0.422	0.1082	0.424
4	0.0710	0.255	0.0868	0.261	0.0962	0.264	0.1025	0.267
3	0.0682	0.216	0.0842	0.221	0.0934	0.225	0.1000	0.228
2	0.0656	0.175	0.0815	0.181	0.0908	0.186	0.0974	0.189
1	0.0627	0.143	0.0787	0.151	0.0880	0.156	0.0945	0.160
1/0	0.0600	0.118	0.0760	0.127	0.0853	0.133	0.0918	0.137
2/0	0.0598	0.0993	0.0732	0.109	0.0826	0.116	0.0892	0.121
3/0	0.0573	0.0884	0.0706	0.0954	0.0799	0.103	0.0866	0.108
4/0	0.0520	0.0728	0.0680	0.0850	0.0773	0.0926	0.0840	0.0982
250	0.0500	0.0661	0.0660	0.0789	0.0753	0.0869	0.0819	0.0926
300	0.0481	0.0602	0.0640	0.0734	0.0732	0.0816	0.0798	0.0876
350	0.0462	0.0557	0.0622	0.0695	0.0715	0.0779	0.0780	0.0840
400	0.0445	0.0522	0.0606	0.0664	0.0700	0.0750	0.0766	0.0814
500	0.0422	0.0476	0.0581	0.0621	0.0674	0.0709	0.0740	0.0772
600	0.0400	0.0441	0.0559	0.0588	0.0652	0.0678	0.0718	0.0741
700	0.0380	0.0412	0.0539	0.0561	0.0633	0.0652	0.0700	0.0718
750	0.0376	0.0404	0.0534	0.0554	0.0628	0.0645	0.0694	0.0710

Continued on next page >>

Table 7.6–Reactance and Impedance at 60 Hz for Single Copper Conductor Cables Installed in Air, Buried or in Separate Nonmetallic Conduits (Continued)

Conductor Size	Approximate ohms per 1,000 ft. per Conductor at 25°C (77°F)							
	Distance Between Centers of Conductors							
	2 in.		4 in.		6 in.		8 in.	
(AWG/kcmil)	Reactance	Impedance	Reactance	Impedance	Reactance	Impedance	Reactance	Impedance
800	0.0370	0.0396	0.0527	0.0546	0.0621	0.0636	0.0687	0.0701
900	0.0354	0.0376	0.0512	0.0527	0.0606	0.0619	0.0673	0.0685
1,000	0.0342	0.0360	0.0500	0.0512	0.0594	0.0605	0.0660	0.0670
1,250	0.0314	0.0328	0.0472	0.0481	0.0566	0.0574	0.0632	0.0639
1,500	0.0296	0.0307	0.0453	0.0460	0.0548	0.0554	0.0614	0.0619
1,750	0.0276	0.0285	0.0434	0.0440	0.0527	0.0532	0.0593	0.0597
2,000	0.0264	0.0272	0.0422	0.0427	0.0514	0.0518	0.0582	0.0585

For equations that can be used to calculate inductive reactance for other conductor spacings, see Section 18.10.

7.5 AC/DC RESISTANCE RATIO AT 60 Hz

Table 7.7–AC/DC Resistance Ratio at 60 Hz

To determine effective 60 Hz AC resistance, multiply DC resistance values corrected for proper temperature by the AC/DC resistance ratio given below.

Conductor Size (AWG/kcmils)	Single Copper Conductors in Air, or in Individual Nonmetallic Conduits	Multiple Copper Conductor Cable or Two or Three Single Conductor Cables in Same Metallic Conduit
Up to 3	1.00	1.00
2 and 1	1.00	1.01
1/0	1.00	1.02
2/0	1.00	1.03
3/0	1.00	1.04
4/0	1.00	1.05
250	1.005	1.06
300	1.006	1.07
350	1.009	1.08
400	1.011	1.10
500	1.018	1.13
600	1.025	1.16
700	1.034	1.19
750	1.039	1.21
800	1.044	–
1,000	1.067	–
1,250	1.102	–
1,500	1.142	–
1,750	1.185	–
2,000	1.233	–

Source: Underground Systems Reference Book, Edison Electric Institute, 1957.

The single conductor column in the table above covers single conductor nonshielded cable having spacing of six inches or more including all conditions of use except when two or more cables are pulled into the same metallic or nonmetallic conduit.

The multiple conductor column in the table above covers the following conditions:

(a) Single conductor cable; two or three cables in the same metallic conduit.
(b) Single conductor shielded cable; two or three cables in the same metallic or nonmetallic duct or conduit, but only with conductor sizes up to 250 kcmils. For larger conductor sizes the short-circuited sheath losses increase rapidly and the table above does not apply.
(c) Three conductor nonshielded cable; one cable in metal conduit.
(d) Three conductor shielded cable; all conditions of use in air, in ducts and in conduit.

The table represents maximum AC losses for the conditions outlined.

7.6 TEMPERATURE CORRECTION FACTORS FOR RESISTANCE

Table 7.8–Temperature Correction Factors for the Resistance of Copper Conductors

Temp°C	Multiplying Factor
25	1.000
40	1.058
50	1.096
55	1.116
60	1.135
65	1.154
70	1.173
75	1.193
80	1.212
85	1.231
90	1.250
100	1.289
105	1.308
125	1.385
130	1.404
150	1.482
200	1.674

The DC resistance of copper wire increases with increasing temperature in accordance with the formula:

$$R_t = R_0 [1 + a (T - T_0)]$$

Where:

R_t = Resistance at temperature T

R_0 = Resistance at temperature T_0

a = Temperature coefficient of resistance at T_0 [At 20°C (68°F) the temperature coefficient of copper is 0.00393 per degree Celsius

7.7 VOLTAGE DROP

The values in Tables 7.9 for copper conductors and 7.10 for aluminum conductors are calculated at 60°C, the estimated average temperature that may be anticipated in service. They may be used without significant error for conductor temperatures up to and including 75°C. For 90°C multiply by 1.102 for copper and by 1.105 for aluminum. To obtain values for other circuits, multiply by 1.155 for single-phase line-to-line and by 0.577 for single- or three-phase line-to-neutral.

$$\text{Voltage drop} = \frac{\text{Table value} \times \text{Current in amps} \times \text{Length of circuit in feet}}{100}$$

$$\text{Voltage drop in percent} = \frac{\text{Voltage drop in V} \times 100}{\text{Circuit voltage in V}}$$

Table 7.9–Phase-to-Phase Voltage Drop Per Amp Per 100 ft. of Circuit for a Three-Phase, 60 Hz System Operating at 60°C with Copper Conductors

Size	In Non-Magnetic Conduit			In Magnetic Conduit		
	Percent Power Factor			Percent Power Factor		
(AWG/kcmil)	80	90	100	80	90	100
12	0.2710	0.3030	0.3330	0.2720	0.3030	0.3320
10	0.1710	0.1910	0.2080	0.1720	0.1910	0.2080
8	0.1090	0.1200	0.1300	0.1100	0.1210	0.1300
6	0.0720	0.0790	0.0840	0.0730	0.0800	0.0840
4	0.0470	0.0510	0.0530	0.0480	0.0520	0.0530
2	0.0310	0.0330	0.0330	0.0320	0.0340	0.0340
1	0.0260	0.0270	0.0260	0.0260	0.0280	0.0260
1/0	0.0210	0.0220	0.0210	0.0220	0.0230	0.0210
2/0	0.0170	0.0180	0.0160	0.0190	0.0190	0.0170
3/0	0.0140	0.0150	0.0130	0.0160	0.0160	0.0140
4/0	0.0120	0.0120	0.0100	0.0140	0.0130	0.0110
250	0.0110	0.0110	0.0088	0.0120	0.0120	0.0093
300	0.0097	0.0095	0.0073	0.0110	0.0110	0.0078
350	0.0088	0.0085	0.0062	0.0100	0.0095	0.0067
400	0.0081	0.0076	0.0055	0.0095	0.0088	0.0061
500	0.0073	0.0067	0.0045	0.0085	0.0078	0.0050
600	0.0066	0.0059	0.0038	0.0080	0.0071	0.0042
700	0.0062	0.0055	0.0033	0.0074	0.0066	0.0037
750	0.0059	0.0054	0.0029	0.0073	0.0064	0.0035
1,000	0.0050	0.0043	0.0023	0.0066	0.0055	0.0023

Table 7.10–Phase-To-Phase Voltage Drop Per Amp Per 100 ft. of Circuit for a Three-Phase, 60 Hz System Operating at 60°C with Aluminum Conductors

Size	In Non-Magnetic Conduit			In Magnetic Conduit		
	Percent Power Factor			Percent Power Factor		
(AWG/kcmil)	80	90	100	80	90	100
12	0.4240	0.4750	0.5230	0.4260	0.4760	0.5230
10	0.2680	0.2990	0.3290	0.2690	0.2140	0.3290
8	0.1700	0.1890	0.2070	0.1720	0.1910	0.2070
6	0.1110	0.1230	0.1330	0.1120	0.1230	0.1320
4	0.0710	0.0780	0.0830	0.0730	0.0790	0.0840
2	0.0460	0.0500	0.0520	0.0470	0.0510	0.0520
1	0.0380	0.0400	0.0420	0.0390	0.0410	0.0420
1/0	0.0310	0.0330	0.0330	0.0320	0.0340	0.0330
2/0	0.0250	0.0260	0.0260	0.0260	0.0270	0.0260
3/0	0.0210	0.0220	0.0210	0.0220	0.0230	0.0210
4/0	0.0170	0.0180	0.0170	0.0180	0.0180	0.0170
250	0.0150	0.0150	0.0140	0.0160	0.0160	0.0140
300	0.0130	0.0130	0.0120	0.0140	0.0140	0.0120
350	0.0120	0.0120	0.0099	0.0130	0.0130	0.0100
400	0.0110	0.0110	0.0087	0.0120	0.0120	0.0091
500	0.0092	0.0089	0.0070	0.0100	0.0099	0.0074
600	0.0083	0.0079	0.0059	0.0095	0.0088	0.0062
700	0.0076	0.0071	0.0050	0.0088	0.0082	0.0055
750	0.0073	0.0068	0.0048	0.0085	0.0079	0.0052
1,000	0.0068	0.0063	0.0042	0.0077	0.0069	0.0042

7.8 MAXIMUM CONDUCTOR SHORT CIRCUIT CURRENT

Because of the high kilovolt-ampere (kVA) capacity of many power systems, possible short circuit currents must be considered in power system design. A cable's maximum short circuit current rating is the maximum allowable current that the cable can withstand without damage. The maximum allowable short circuit current for copper and aluminum conductors can be determined with the aid of Figures 7.1 and 7.2, respectively.

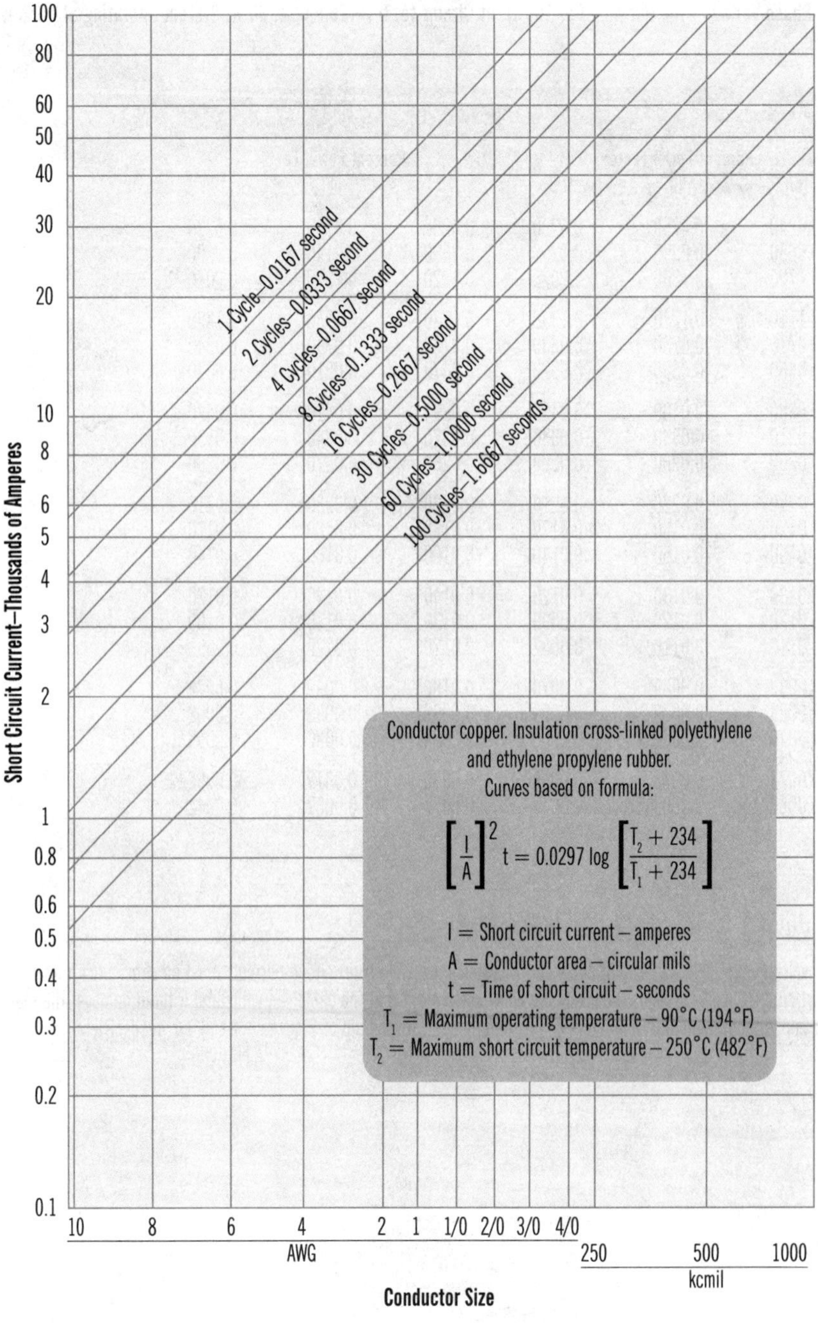

Source: ICEA P-32-382

Figure 7.1–Maximum Conductor Short Circuit Current for Copper Cables

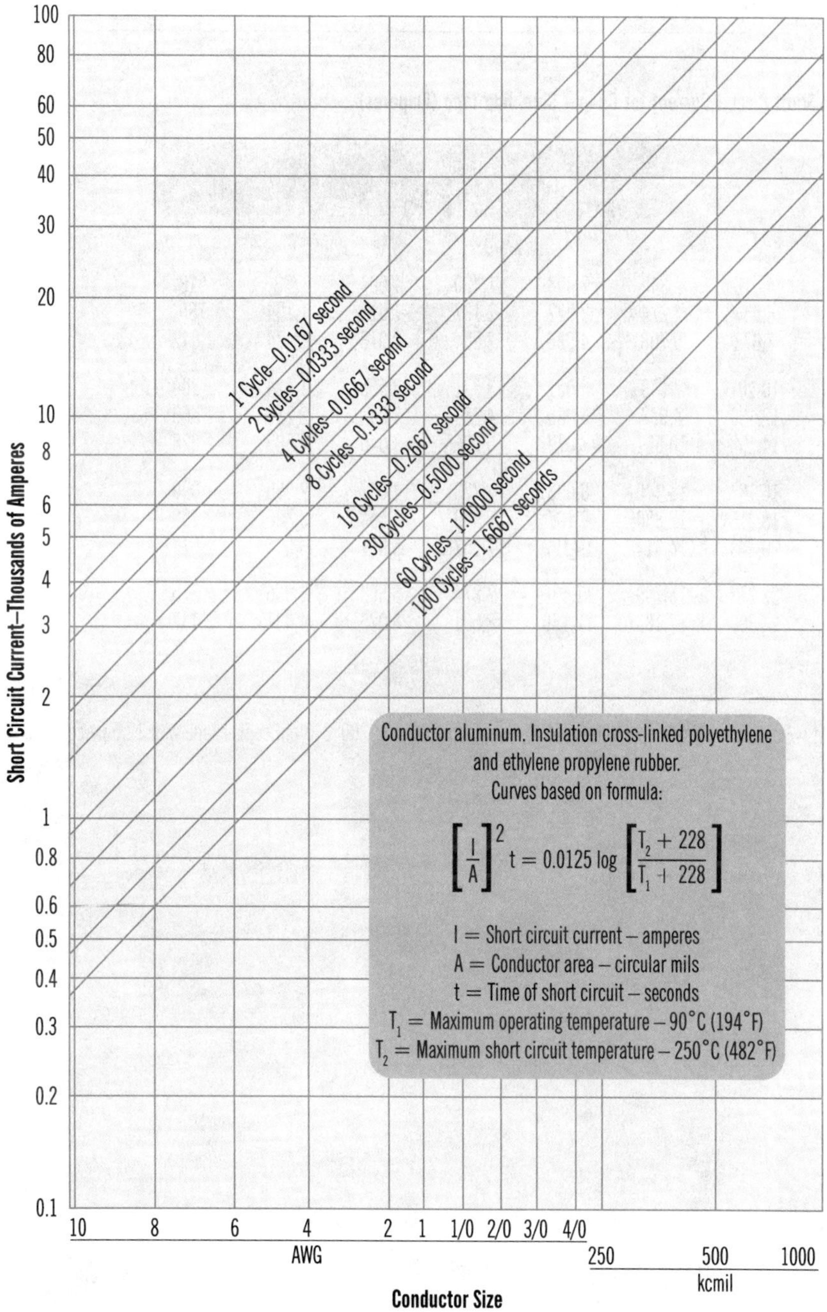

Source: ICEA P-32-382

Figure 7.2–Maximum Conductor Short Circuit Current for Aluminum Cables

7.9 MAXIMUM SHIELD SHORT CIRCUIT CURRENT

Table 7.11–Maximum Short Circuit Current for Copper Shielding Tape (Amperes)

Shield Dia.	Effective Shield Area	Short Circuit Time (Number of Cycles at 60 Hz)						
(in.)	(cmils)	1	2	4	8	16	30	60
1/2	7,484	4,016	2,840	2,008	1,420	1,004	733	518
3/4	11,264	6,044	4,274	3,022	2,137	1,511	1,104	780
1	15,044	8,073	5,708	4,036	2,854	2,018	1,474	1,042
1 1/4	18,824	10,101	7,143	5,051	3,571	2,525	1,844	1,304
1 1/2	22,604	12,130	8,577	6,065	4,289	3,032	2,215	1,566
1 3/4	26,384	14,158	10,011	7,079	5,006	3,540	2,585	1,828
2	30,164	16,187	11,446	8,093	5,723	4,047	2,955	2,090
2 1/4	33,944	18,215	12,880	9,107	6,440	4,554	3,326	2,352
2 1/2	37,724	20,243	14,314	10,122	7,157	5,061	3,696	2,613
2 3/4	41,504	22,272	15,749	11,136	7,874	5,568	4,066	2,875
3	45,284	24,300	17,183	12,150	8,591	6,075	4,437	3,137

Source: ICEA P-45-482

Information in this chart is based on initial temperature of 65°C, final temperature of 200°C, 5 mil copper tape with 12.5 percent overlap.

7.10 RESISTANCE AND AMPACITY AT 400 AND 800 HZ

Table 7.12–400 and 800 Hz Ampacity Factors for 600 V Copper Cables with Class B Strand, Installed with Minimum Triangular Spacing in Air or in Nonmetallic Conduit

				400 Hz		800 Hz	
Conductor Size (AWG/kcmil)	Conductor Diameter (in.)	Cable Diameter (in.)	DC Resistance 75°C (ohms/1,000 ft.)	AC/DC Resistance Ratio	Ampacity Derating Factor*	AC/DC Resistance Ratio	Ampacity Derating Factor*
14	0.073	0.21	3.14	1.00	1.00	1.00	1.00
12	0.092	0.23	1.97	1.00	1.00	1.00	1.00
10	0.116	0.25	1.24	1.00	1.00	1.00	1.00
8	0.146	0.32	0.780	1.00	1.00	1.00	1.00
6	0.184	0.39	0.490	1.00	1.00	1.00	1.00
4	0.232	0.44	0.310	1.00	1.00	1.05	0.98
2	0.292	0.50	0.194	1.03	0.98	1.12	0.94
1	0.332	0.61	0.154	1.05	0.98	1.16	0.93
1/0	0.372	0.65	0.122	1.08	0.96	1.25	0.89
2/0	0.418	0.69	0.097	1.15	0.93	1.40	0.84
3/0	0.470	0.75	0.0767	1.22	0.90	1.53	0.81
4/0	0.528	0.81	0.0608	1.33	0.87	1.70	0.77
250	0.575	0.92	0.0515	1.40	0.84	1.82	0.74
350	0.681	1.08	0.0368	1.56	0.80	2.05	0.70
500	0.813	1.16	0.0258	1.90	0.72	2.54	0.63
750	0.998	1.38	0.0172	2.30	0.66	3.06	0.57
1,000	1.152	1.54	0.0129	2.60	0.62	3.44	0.54

Source: ICEA P-43-457

* These derating factors do not apply to cables with metallic sheath or armor, nor to cables installed in conduit or adjacent to steel structures. Ampacity equals the 60 Hz ampacity multiplied by the derating factor.

7.11 CURRENT RATINGS FOR ELECTRONIC CABLES

The maximum continuous current rating for an electronic cable is limited by conductor size, number of conductors contained within the cable, maximum temperature rating of the cable and environmental conditions such as ambient temperature and airflow. To use the current capacity chart (Figure 7.3), first determine conductor gauge, temperature rating and number of conductors for the cable of interest.

Next, find the current value on the chart for the applicable temperature rise (temperature rating of cable minus ambient temperature) and conductor size. To calculate the maximum current rating per conductor, multiply the chart value by the appropriate conductor factor. The chart assumes the cable is surrounded by still air at an ambient temperature of 25°C. Current values are in RMS amperes and are valid for copper conductors only.

Note: Current ratings are intended as general guidelines for low power, electronic communications and control applications. Current ratings for power applications are published by codes and standards groups including NEC, UL, CSA, ICEA, NEMA, IEEE and IEC.

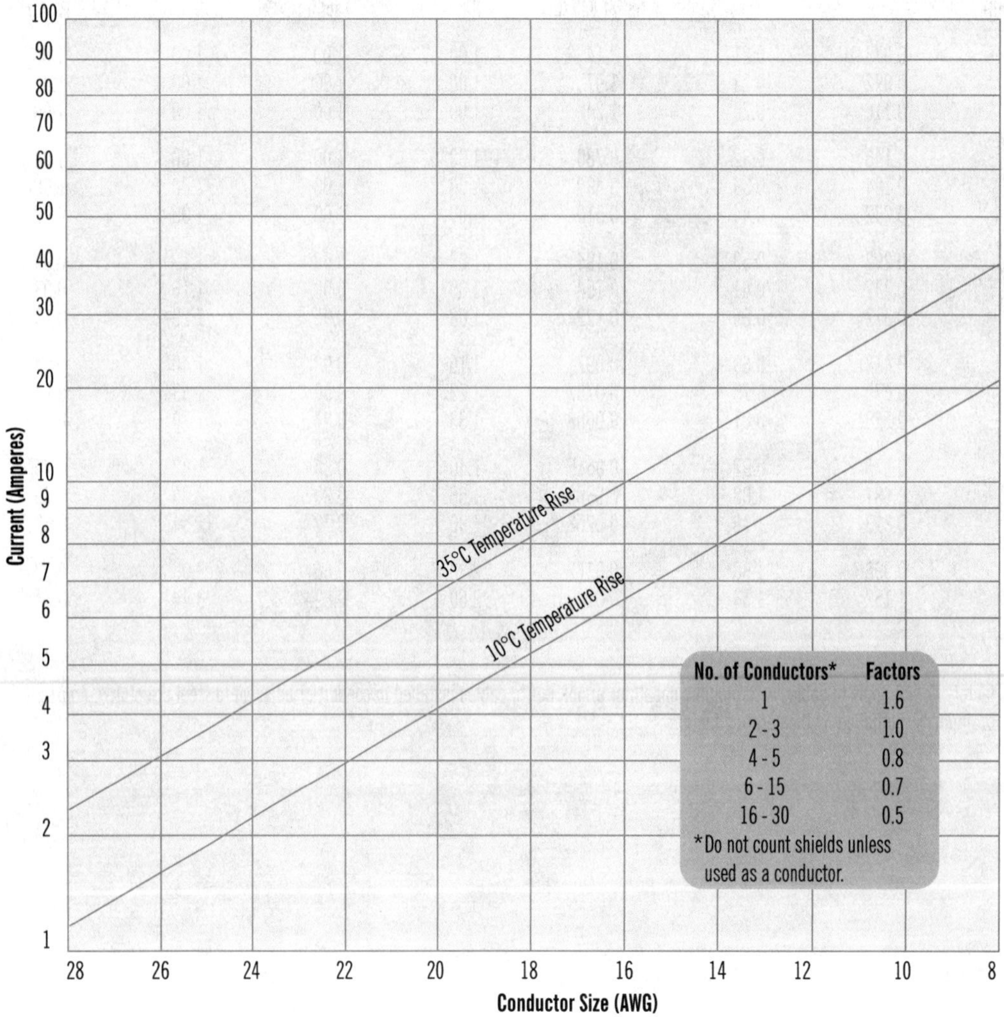

No. of Conductors*	Factors
1	1.6
2 - 3	1.0
4 - 5	0.8
6 - 15	0.7
16 - 30	0.5

*Do not count shields unless used as a conductor.

Figure 7.3–Current Ratings for Electronic Cables

7.12 AMPACITY OF POWER CABLES

The ampacity of a power cable depends primarily on its conductor size, conductor material (e.g., copper or aluminum), temperature rating, ambient temperature, installed cable configuration and other factors. Because so many external conditions affect ampacity, tables covering all situations are not possible. However, tables covering many common situations are available. Frequently used ampacity tables are contained in the following publications:

NFPA Standard 70 National Electrical Code
CSA Standard C22.1 Canadian Electrical Code
IEEE Standard 835 Power Cable Ampacity Tables
ICEA P-53-426 (NEMA WC 50) Ampacities Including Shield Losses for 15 Through 69 kV Cables
ICEA P-54-440 (NEMA WC 51) Ampacities of Cables in Open-Top Cable Trays
IEEE Standard 45 Recommended Practice for Installations on Shipboard

7.13 BASIC IMPULSE LEVEL (BIL) RATINGS

Electrical equipment, including wire and cable, is designed to withstand short-term, but very high-voltage pulses such as those sometimes caused by lightning and switching surges. These “spikes,” as they are sometimes called, typically have a risetime in the range of 1.5 microseconds and a falltime around 40 microseconds. The basic impulse level (BIL) is the maximum impulse voltage that a cable is designed to withstand. Common BIL ratings are shown below.

Table 7.13–Basic Impulse Level (BIL) Ratings

System Voltage Rating (kV)	Basic Impulse Level (kV)
2.5	60
5.0	75
8.0	95
15.0	110
25.0	150
35.0	200
69.0	350
138.0	650

Source: IEEE 82 Impulse Voltage Tests on Insulated Conductors

8. INSTALLATION AND TESTING

This section is intended as a guide for the installer's use in the field. The information has been obtained from many sources and covers some of the major considerations when installing and testing power, control, instrumentation, fiber and communication cable.

8.1 RECEIVING, HANDLING AND STORAGE

The following guidelines are recommended to prevent possible deterioration or damage of cable during handling or storage prior to installation.

8.1.1 Receiving

Before accepting any shipment, all reels should be visually inspected for both hidden and obvious damage. Be especially alert if:

- A reel is lying flat on its side
- Reels are poorly stacked
- Protective covering (packaging material) is removed or damaged
- Cable end seals are removed or damaged
- Reel flanges are broken
- A reel has been dropped
- Cable ties are loose
- Nails or staples have been driven into the reel flange.

8.1.2 Handling

Cable reels should always be rolled in the direction of the "roll this way" stenciled on the flanges. This prevents loosening of the cable turns, which may cause problems during installation. If the roll direction is not indicated, rotate the reel in the same direction it was rotated when the cable was wound onto the reel.

Cable reels should only be lifted by forklift trucks from the sides and only if the forks are long enough to cradle both flanges.

Steel lifting bars of a suitable diameter and length should be used when lifting cable reels by crane or other overhead lifting devices. With heavy reels or reels that may be unbalanced, the use of a lifting yoke is recommended to prevent reels from slipping or tipping during lifting.

8.1.3 Storage

Where possible, reels should be stored indoors on a hard, dry surface. If reels must be stored outside they should be supported off the ground and covered with a suitable weatherproof material.

- Each reel should be aligned flange to flange.
- Each reel should be chocked.
- Reels should be stored to allow easy access for lifting and moving.

When cable lengths are cut from a master cable reel, all exposed cable ends should be resealed with plastic weatherproof caps or tape to prevent the entrance of moisture.

8.2 CONDUIT FILL

Below is a table of the maximum number of conductors that can be installed in electrical metallic tubing (EMT). The table is based on Table 1, Chapter 9 of the National Electrical Code. For installation in other types of conduits, for other wire types or for installation of compact stranded conductors, refer to Tables C1 through C12 in Annex C of the 2008 NEC.

Table 8.1–Maximum Number of Conductors in Electrical Metallic Tubing (EMT)

Type Letters	Conductor Size	Conduit or Tubing Trade Size (in.)									
	(AWG/kcmil)	½	¾	1	1¼	1½	2	2½	3	3½	4
RHH, RHW RHW-2	14	4	7	11	20	27	46	80	120	157	201
	12	3	6	9	17	23	38	66	100	131	167
	10	2	5	8	13	18	30	53	81	105	135
	8	1	2	4	7	9	16	28	42	55	70
	6	1	1	3	5	8	13	22	34	44	56
	4	1	1	2	4	6	10	17	26	34	44
	3	1	1	1	4	5	9	15	23	30	38
	2	1	1	1	3	4	7	13	20	26	33
	1	0	1	1	1	3	5	9	13	17	22
	1/0	0	1	1	1	2	4	7	11	15	19
	2/0	0	1	1	1	2	4	6	10	13	17
	3/0	0	0	1	1	1	3	5	8	11	14
	4/0	0	0	1	1	1	3	5	7	9	12
	250	0	0	0	1	1	1	3	5	7	9
	300	0	0	0	1	1	1	3	5	6	8
	350	0	0	0	1	1	1	3	4	6	7
	400	0	0	0	1	1	1	2	4	5	7
	500	0	0	0	0	1	1	2	3	4	6
	600	0	0	0	0	1	1	1	3	4	5
	700	0	0	0	0	0	1	1	2	3	4
	750	0	0	0	0	0	1	1	2	3	4
	800	0	0	0	0	0	1	1	2	3	4
	900	0	0	0	0	0	1	1	1	3	3
	1,000	0	0	0	0	0	1	1	1	2	3
THHW, THW, THHW, THW THW-2	14	6	10	16	28	39	64	112	169	221	282
	12	4	8	13	23	31	51	90	136	177	227
	10	3	6	10	18	24	40	70	106	138	177
	8	1	4	6	10	14	24	42	63	83	106
	6	1	3	4	8	11	18	32	48	63	81
	4	1	1	3	6	8	13	24	36	47	60
	1/0	0	1	1	2	3	6	10	16	20	26
	2/0	0	1	1	1	3	5	9	13	17	22
	3/0	0	1	1	1	2	4	7	11	15	19

Continued on next page >>

Table 8.1–Maximum Number of Conductors in Electrical Metallic Tubing (EMT) (Continued)

Type Letters	Conductor Size	Conduit or Tubing Trade Size (in.)									
	(AWG/kcmil)	½	¾	1	1¼	1½	2	2½	3	3½	4
	4/0	0	0	1	1	1	3	6	9	12	16
	250	0	0	1	1	1	3	5	7	10	13
	300	0	0	1	1	1	2	4	6	8	11
	350	0	0	0	1	1	1	4	6	7	10
	400	0	0	0	1	1	1	3	5	7	9
	500	0	0	0	1	1	1	3	4	6	7
	600	0	0	0	1	1	1	2	3	4	6
	700	0	0	0	0	1	1	1	3	4	5
	750	0	0	0	0	1	1	1	3	4	5
	800	0	0	0	0	1	1	1	3	3	5
	900	0	0	0	0	0	1	1	2	3	4
	1,000	0	0	0	0	0	1	1	2	3	4
THHN, THWN, THWN-2	14	12	22	35	61	84	138	241	364	476	608
	12	9	16	26	45	61	101	176	266	347	443
	10	5	10	16	28	38	63	111	167	219	279
	8	3	6	9	16	22	36	64	96	126	161
	6	2	4	7	12	16	26	46	69	91	116
	4	1	2	4	7	10	16	28	43	56	71
	3	1	1	3	6	8	13	24	36	47	60
	2	1	1	3	5	7	11	20	30	40	51
	1	1	1	1	4	5	8	15	22	29	37
	1/0	1	1	1	3	4	7	12	19	25	32
	2/0	0	1	1	2	3	6	10	16	20	26
	3/0	0	1	1	1	3	5	8	13	17	22
	4/0	0	1	1	1	2	4	7	11	14	18
	250	0	0	1	1	1	3	6	9	11	15
	300	0	0	1	1	1	3	5	7	10	13
	350	0	0	1	1	1	2	4	6	9	11
	400	0	0	0	1	1	1	4	6	8	10
	500	0	0	0	1	1	1	3	5	6	8
	600	0	0	0	1	1	1	2	4	5	7
	700	0	0	0	1	1	1	2	3	4	6
	750	0	0	0	0	1	1	1	3	4	5
	800	0	0	0	0	1	1	1	3	4	5
	900	0	0	0	0	1	1	1	3	3	4
	1,000	0	0	0	0	1	1	1	2	3	4

Continued on next page >>

Table 8.1–Maximum Number of Conductors in Electrical Metallic Tubing (EMT) (Continued)

Type Letters	Conductor Size	Conduit or Tubing Trade Size (in.)									
	(AWG/kcmil)	½	¾	1	1¼	1½	2	2½	3	3½	4
TFN, TFFN	18	22	38	63	108	148	244	–	–	–	–
	16	17	29	48	83	113	186	–	–	–	–
XHHW, XHHW-2	14	8	15	25	43	58	96	168	254	332	424
	12	6	11	19	33	45	74	129	195	255	326
	10	5	8	14	24	33	55	96	145	190	243
8	2	5	8	13	18	30	53	81	105	135	
	6	1	3	6	10	14	22	39	60	78	100
	4	1	2	4	7	10	16	28	43	56	72
	3	1	1	3	6	8	14	24	36	48	61
	2	1	1	3	5	7	11	20	31	40	51
	1	1	1	1	4	5	8	15	23	30	38
	1/0	1	1	1	3	4	7	13	19	25	32
	2/0	0	1	1	2	3	6	10	16	21	27
	3/0	0	1	1	1	3	5	9	13	17	22
	4/0	0	1	1	1	2	4	7	11	14	18
	250	0	0	1	1	1	3	6	9	12	15
	300	0	0	1	1	1	3	5	8	10	13
	350	0	0	1	1	1	2	4	7	9	11
	400	0	0	0	1	1	1	4	6	8	10
	500	0	0	0	1	1	1	3	5	6	8
	600	0	0	0	1	1	1	2	4	5	6
	700	0	0	0	0	1	1	2	3	4	6
	750	0	0	0	0	1	1	1	3	4	5
	800	0	0	0	0	1	1	1	3	4	5
	900	0	0	0	0	1	1	1	3	3	4
	1,000	0	0	0	0	0	1	1	2	3	4

Source: 2008 NEC, Annex C, Table C.1

Table 8.2–Maximum Cable Diameters for Permissible Conduit Fill

No. of Wires or Cables	Conduit Trade Size (in.)									
	½	¾	1	1¼	1½	2	2½	3	3½	4
	Actual ID of Conduit (in.)									
	0.622	**0.824**	**1.05**	**1.38**	**1.61**	**2.07**	**2.47**	**3.07**	**3.55**	**4.03**
	Max. Diam. of Wires or Cables in Conduit (in.)									
1	0.462	0.610	0.778	1.02	1.19	1.53	1.83	2.27	2.63	3.00
2	0.249	0.380	0.420	0.552	0.644	0.828	0.988	1.23	1.42	1.61
3	0.231	0.305	0.390	0.511	0.596	0.767	0.914	1.14	1.31	1.49
4	0.201	0.266	0.339	0.445	0.519	0.668	0.796	0.999	1.14	1.30
5	0.178	0.235	0.300	0.395	0.460	0.591	0.706	0.876	1.01	1.15
6	0.164	0.216	0.276	0.364	0.424	0.545	0.650	0.807	0.935	1.06
7	0.152	0.201	0.256	0.337	0.392	0.505	0.603	0.749	0.868	0.984
8	0.141	0.187	0.239	0.314	0.366	0.470	0.562	0.698	0.806	0.916
9	0.133	0.175	0.224	0.294	0.342	0.441	0.525	0.653	0.755	0.858
10	0.125	0.165	0.210	0.276	0.322	0.414	0.494	0.614	0.710	0.807
11	0.119	0.157	0.200	0.263	0.307	0.394	0.471	0.584	0.676	0.768
12	0.114	0.151	0.193	0.254	0.296	0.380	0.453	0.563	0.651	0.740
13	0.109	0.144	0.184	0.242	0.283	0.363	0.435	0.538	0.623	0.707
14	0.106	0.139	0.178	0.234	0.273	0.351	0.418	0.520	0.601	0.683
15	0.102	0.135	0.172	0.226	0.264	0.339	0.405	0.503	0.583	0.661

Source: 2008 NEC, Chapter 9, Table 1

Table 8.3–Dimensions and Maximum Allowable Percent Fill of Electrical Metallic Tubing (EMT)

Trade Size	Internal Dia.	Total Area	Allowable Fill (sq. in.)		
			1 Cond.	2 Cond.	Over 2 Cond.
(in.)	(in.)	(sq. in.)	53 percent Fill	31 percent Fill	40 percent Fill
1/2	0.622	0.30	0.16	0.09	0.12
1 3/4	0.824	0.53	0.28	0.16	0.21
1	1.049	0.86	0.46	0.27	0.34
1 1/4	1.380	1.50	0.80	0.47	0.60
11/2	1.610	2.04	1.08	0.63	0.82
2	2.067	3.36	1.78	1.04	1.34
2 1/2	2.469	4.79	2.54	1.48	1.92
3	3.068	7.38	3.91	2.26	2.95
4	4.026	12.72	6.74	3.94	5.09

Source: 2008 National Electrical Code, Chapter 9, Table 4

For other conduit types, please refer to Table 4 in Chapter 9 of the NEC. The general equation for calculating wire or cable area is:

$$\text{Area in square inches} = \frac{\pi \times OD^2}{4} \times n$$

Where:

$\pi = 3.14$

OD — overall diameter of each single-conductor wire or multiconductor cable

n = number of wires or cables of that diameter

Example:

Pulling (3) 2/0 15 kV cables, each cable has an overall diameter of 1.20 inches.

Using the formula, solve as follows: $\frac{3.14 \times 1.2^2}{4} = 3.39$ sq. in.

Referring to the table, minimum conduit size would be 4 inches.

8.3 PULLING

8.3.1 Methods of Gripping Cables

In general, insulated cables may be gripped either directly by the conductors or by a basket-weave pulling grip applied over the cables. The appropriate method to use depends on the anticipated maximum pulling tension. When pulls are relatively light a basket-weave grip can often be used. Heavier pulls usually require connecting directly to the conductor either by means of pulling eyes or by forming a loop with the conductor itself. In some instances it is desirable to use a grip over the outer covering in addition to the conductor connection to prevent any slippage of one with respect to the other.

Nonmetallic Sheathed Cables

Smaller sizes of nonmetallic sheathed cables can usually be gripped directly by the conductors by forming them into a loop to which the pull wire or rope can be attached. The insulation on each conductor is removed before the loop is formed. Larger sizes are more easily handled by applying a pulling grip over the cable or cables provided the pull is not too severe. If more than one cable is involved, the ends should be bound together with electrical tape before applying the grip overall. Long, hard pulls will necessitate the use of pulling eyes.

Interlocked Armor Cables

When pulling interlocked armor cable it is usually necessary to grip both the armor and the conductors. This can be accomplished in a number of ways. One method requires that a portion of the armor be removed. Electrical tape is then applied over the armor and down over the conductors and a long basket-weave grip is applied such that it grips both the armor and the conductors. Another method requires that two holes be drilled through the cable (armor and conductors) at right angles to each other and a loop formed by passing steel wires through the holes and out over the end of the cable. A third approach is to use a pulling eye and a grip together, the grip being applied over the armor to prevent it from slipping back. This latter approach provides the greatest strength.

Preassembled Aerial Cable

This type of cable should always be gripped by the messenger that is usually attached to a pulling swivel. In addition, a basket grip should be applied over the conductors to prevent any slippage and to facilitate guiding the conductors through the pulleys.

8.3.2 Tension Limitations

When the pulling force is applied directly to the conductor (e.g., when pulling eyes are used or when the conductor is formed into a loop) it should be limited to 0.008 lb. per circular mil area of cross-section for copper and 0.006 lb. per circular mil for aluminum.

When a grip is applied over nonmetallic sheathed cables, the pulling force should be limited to 1,000 pounds provided this is not in excess of the force calculated above using the 0.008 or 0.006 factors.

To limit the sidewall pressure to a safe value at bends in duct and conduit runs, the pulling force in pounds should not exceed 300 to 500 times the radius of the bend in feet.

The above limits are maximum values that should not be exceeded. However, it is possible to damage cables while applying lower tensions if, for example, there are sharp projections in a poorly constructed duct bank, or if an interlocked armor cable is pulled around too small a sheave. Every installation detail cannot be covered here but staying within the above tension limits will help ensure a successful installation.

8.3.3 Helpful Hints

The following suggestions – though not all-inclusive – will give greater assurance of success.

(1) Be sure there is adequate clearance between conduit and cable. Clearance refers to the distance between the uppermost cable in the conduit and the inner top of the conduit. Clearance should be 1/4 inch at minimum and up to one inch for large cable installations or installations involving numerous bends. It is calculated as shown in Figure 8.1 where "D" is the inner diameter of the conduit and "d" is the outer diameter of the cable. When calculating clearance, ensure all cable diameters are equal. Use the triplexed configuration formula if you are in doubt. The cables may be of single or multiple conductor construction. Do not exceed recommended "conduit fill" requirements.

No. of Conductors/Cables	Configuration	Formula
1	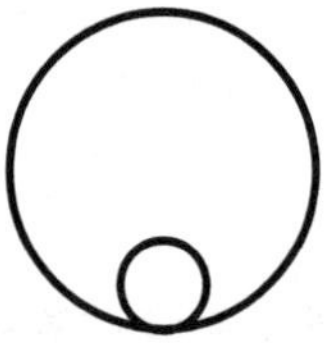	$\text{Clearance} = D - d$
3	 Cradled	$\text{Clearance} = \frac{D}{2} - 1.366d + \frac{D-d}{2}\sqrt{1-\left(\frac{d}{(D-d)}\right)^2}$
3	 Triplexed	$\text{Clearance} = \frac{D}{2} - \frac{d}{2} + \frac{D-d}{2}\sqrt{1-\left(\frac{d}{2(D-d)}\right)^2}$

Figure 8.1–How to Calculate Clearance

(2) Jamming is the wedging of three cables lying side by side in a conduit. This usually occurs when cables are being pulled around bends or when cables twist.

The jam ratio is calculated by slightly modifying the ratio D/d. A value of 1.05D is used for the inner diameter of the conduit because bending a cylinder creates an oval cross-section in the bend.

- If 1.05D/d is larger than 3.0, jamming is impossible.
- If 1.05D/d is between 2.8 and 3.0, serious jamming is probable.
- If 1.05D/d is less than 2.5, jamming is impossible but clearance should be checked.

Because there are manufacturing tolerances on cable, the actual overall diameter should be measured prior to computing the jam ratio.

(3) Use adequate lubrication of the proper type to reduce friction in conduit and duct pulls. Grease and oil type lubricants should not be used on nonmetallic sheathed cables. There are a number of commercially available wire pulling compounds (many of which are UL Listed) that are suitable for use with polymer jacketed cables. They usually consist of soap, talc, mica or the like, and are designed to have no deleterious effect on the cable. Graphite and other electrically conducting lubricants should not be used on nonshielded cables rated 2 kV and above. These materials can lead to tracking of the cable jacket.

(4) Avoid sharp bending of the cable at the first pulley in overhead installations by locating the payoff reel far enough away from the first pulley that the lead-in angle is kept relatively flat.

(5) After installation, check that end seals are still intact and have not been damaged to the point where water could enter. Apply plastic or rubber tape to help protect against invisible damage if the cable will be subjected to immersion or rain. This is particularly important if there will be a delay of some time between the pulling operation and splicing and terminating.

(6) When installing interlocked armor cables in cable tray, use a sufficient number of rollers to prevent the cable from dragging on the tray, which might result in excessive tension. Avoid sharp bends in the cable by using a conveyor sheave with multiple small rollers at all 45- and 90-degree bends.

(7) Keep adequate tension on the messenger in aerial cable installations to prevent sharp bends at pulleys. Do not release the tension on the messenger until it is secured to poles on both ends.

8.3.4 Pulling Tension Calculations

Pulling tension calculations are recommended in the design stage of all cable installations that are expected to fall in the moderate to difficult category. Software programs are commercially available that can perform sophisticated modeling of expected pulling tensions and sidewall pressures. These programs are recommended over manual methods. Below is an overview of the basic calculations. Additional information is available in IEEE 1185, IEEE 971, IEEE 576 and AEIC CG5.

(1) Maximum Pulling Tension

a. With pulling eye attached to copper conductors, the maximum pulling tension in pounds should not exceed 0.008 times the circular mil area.

b. With pulling eye attached to aluminum conductors, the maximum pulling tension in pounds should not exceed 0.006 times the circular mil area.

Example: For copper

$$T_M = 0.008 \times n \times CM$$

Where:

T_M = maximum tension, lb.

n = number of conductors

CM = circular mil area of each conductor

(2) Maximum Permissible Pulling Length:

$$L_M = \frac{T_M}{C \times W}$$

Where:

L_M = maximum pulling length, feet (valid only for straight sections)

T_M = maximum tension, lb.

W = weight of cable per foot, lb.

C = coefficient of friction (typically 0.5 but can vary from 0.2 to 1.0 depending on condition of the duct and the amount of lubricant used)

(3) Bend Multipliers
For a curved section, the multipliers given below are applied to the tension calculated for the straight section preceding the bend.

Table 8.4–Bend Multipliers for Pulling Tension Calculations

Bend Angle Degrees	Multiplier	Bend Angle Degrees	Multiplier
15	1.14	75	1.94
30	1.30	90	2.20
45	1.48	105	2.50
60	1.70	120	2.86

Note: These multipliers are based on a coefficient of friction of 0.5. If the coefficient of friction were 1.0 instead of 0.5, the multipliers would have to be squared. If the coefficient of friction were 0.75, the multipliers would be raised to the one and one-half power.

8.3.5 Pulling Lubricants

Many commercial lubricants are available and may be employed to reduce pulling tensions provided they do not affect electrical or mechanical characteristics of the cable. The primary function of a pulling lubricant is to reduce the tension on the cable as it is installed in a duct. This is accomplished by reducing the friction (technically the coefficient of friction) between the cable and the inside surface of the conduit, i.e., it makes the cable more slippery. Cable pulling lubricants should be formulated for the conditions of the pull, be safe for the environment, not degrade the cable jacket and be easy to work with. Some LSZH (Low Smoke Zero Halogen) cables require special pulling lubricants such as Polywater LZ to prevent chemical damage to the jacket.

The quantity of lubricant required depends on various factors: The pull length, the condition and size of the conduit and the difficulty of the pull. The recommended average quantity of lubricant per pull is equal to:

$$Q = 0.0015 \times L \times D$$

Where Q is the quantity of lubricant needed in gallons, L is the length of the pull in feet and D is the inner diameter of the conduit in inches.

The appropriate quantity to use can vary by ± 50 percent from the average depending on installation conditions. Follow the manufacturer's instructions for the conditions affecting each pull.

8.3.6 Sidewall Pressure (SWP)

To prevent damage to a cable from pressure that develops when a cable is pulled around a bend under tension, the pressure must be kept as low as possible and should not exceed specified values. Sidewall pressure = tension out of the bend divided by bend radius. Cable manufacturers generally recommend a maximum SWP of 500 lb./ft. for most 600 V and medium-voltage power cables.

8.3.7 Minimum Bending Radii

Power Cables without Metallic Shielding

The minimum bending radii for both single- and multiple-conductor cable without metallic shielding are as follows:

Table 8.5–Minimum Bending Radii for Cables without Metallic Shielding

Thickness of Conductor Insulation	Minimum Bending Radius as a Multiple of Cable Diameter		
	Overall Diameter of Cable in Inches		
(mils)	1.00 and less	1.01 to 2.00	2.01 and Greater
169 and less	4	5	6
170 and larger	5	6	7

Example:
If minimum bending radius is six times cable O.D. and cable O.D. is 2.0 inches, the minimum bending radius is 12 inches (minimum bending diameter is 24 inches).

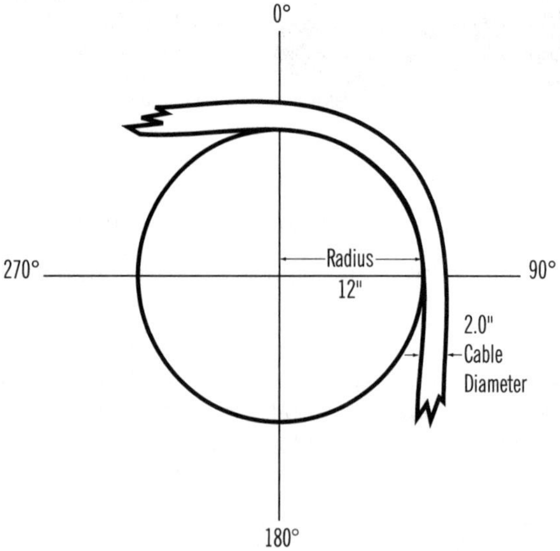

Figure 8.2–Calculating Minimum Bending Radius

Power Cables with Metallic Shielding
The minimum bending radius for all single conductor cables with metallic shielding is 12 times the overall diameter of the cable. For multiconductors, it is seven times the overall diameter or 12 times the individual conductor diameter, whichever is greater.

Portable Cables
The minimum bending radius for portable cables during installation and handling in service is six times the cable diameter for cables rated 5,000 volts and less. For cables rated over 5,000 volts use eight times the cable diameter. For flat twin cables, the minor diameter is used to determine the bending radius.

Fiber Optic Cables
Minimum bending radius for fiber optic cable is typically 10 times the cable diameter when under no tension and 15 times diameter at rated maximum tension. The manufacturer should be consulted for specific product limits.

Interlocked Armor or Corrugated Sheath (Type MC) Cables
The minimum bending radius for Type MC cable is seven times the external diameter of the metallic sheath.

Sources: NEC Articles 300.34, 330.24, 336.24 and 727.10
NEMA WC58 (ICEA S-75-381)
NEMA WC74 (ICEA S-93-639)
IEEE 1185
NEMA WC70 (ICEA S-95-658)

8.4 INSTALLATION METHODS

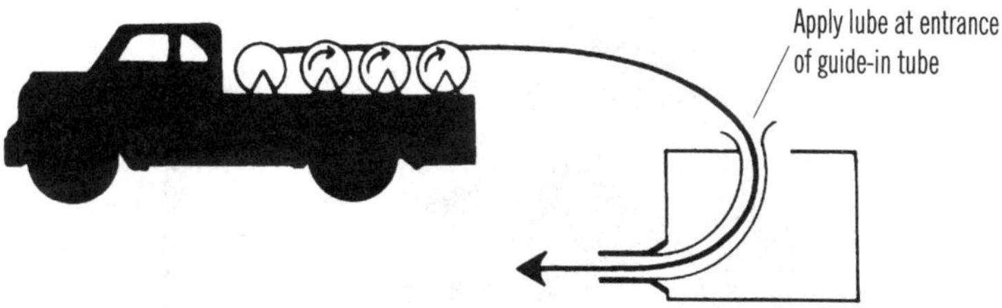

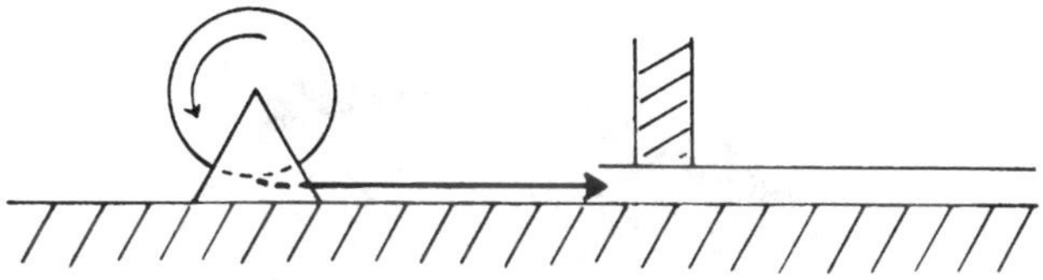

Setup for duct close to floor

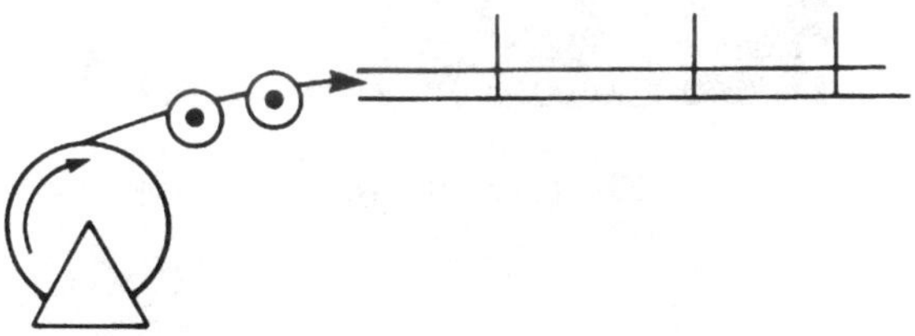

Setup for overhead into tray

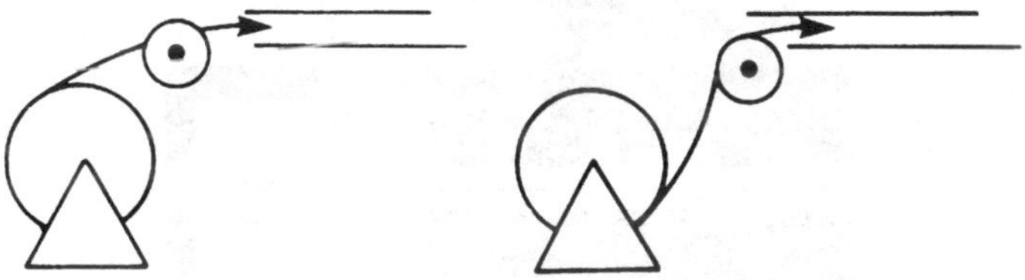

(a) Proper **(b)** Improper

The feed-in setup should unreel the cable with the natural curvature (a) as opposed to a reverse "S" curvature (b).

Figure 8.3–Cable Pulling Setups

Continued on next page >>

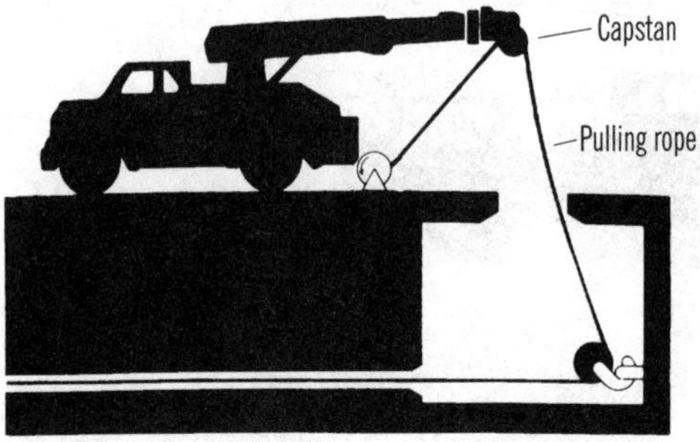

A setup with timbers because pulling eyes were not available

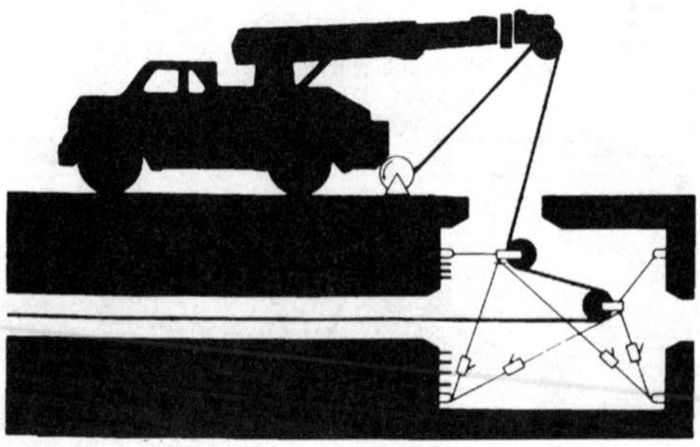

Figure 8.3–Cable Pulling Setups

Continued on next page >>

Single sheave

Single sheaves should be used only for guiding cables. Arrange multiple blocks if necessary to maintain minimum bending radii whenever cable is deflected.

Sheave assembly

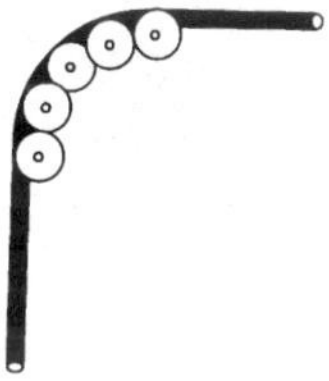

For pulling around bends, use multisheave assemblies (also called conveyor sheaves) of the appropriate radius.

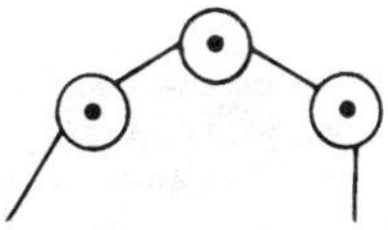

The pulleys must be positioned to ensure that the effective curvature is smooth and deflected evenly at each pulley. Never allow a polygon curvature to occur as shown.

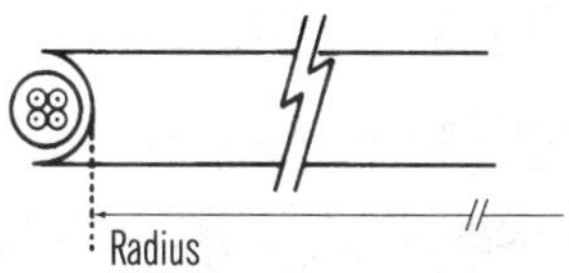

The fit of the pulley around the cable is also important when the pulling tension is high (for example, pulleys at the top of a vertical drop). Remember to use the radius of the surface over which the cable is bent, not the outside flange diameter of the pulley. A 10-inch cable sheave typically has an inside (bending) radius of 3 inches!

Figure 8.3–Cable Pulling Setups

8.5 OVERHEAD MESSENGERS

Table 8.6–Messenger Breaking Strength in lb.

Nominal Messenger Size	30 percent EHS* Copper-Clad Steel	Aluminum Clad Steel	EHS* Galvanized Steel	High-Strength Galvanized Steel	Type 316 Stainless Steel	Type 302 Stainless Steel
1/4 in. (7×12 AWG)	6,282	6,301	6,650	4,750	7,650	8,500
5/16 in. (7×10 AWG)	9,196	10,020	11,200	8,000	11,900	13,200
3/8 in. (7×8 AWG)	13,890	15,930	15,400	10,800	16,200	18,000
7/16 in. (7×7 AWG)	16,890	19,060	20,800	14,500	23,400	26,000
1/2 in. (7×6 AWG)	20,460	22,730	26,900	18,800	30,300	33,700

* Extra-High Strength

Table 8.7–Messenger Weight in lb./1,000 ft.

Nominal Messenger Size	30 Percent EHS Copper-Clad Steel	Aluminum Clad Steel	EHS Galvanized Steel	High-Strength Galvanized Steel	Type 316 Stainless Steel	Type 302 Stainless Steel
1/4 in.	139	104	121	121	136	132
5/16 in.	204	165	205	205	208	208
3/8 in.	324	262	273	273	278	278
7/16 in.	408	330	399	399	405	405
1/2 in.	515	416	517	517	525	525

Table 8.8–Maximum Core Weight in lb./ft. (Based on Final Sag of 30 in. at 60°F in a 150 ft. Span, 30 Percent of Breaking Strength)

Nominal Messenger Size	30 Percent EHS Copper-Clad Steel	Aluminum Clad Steel	EHS Galvanized Steel	High-Strength Galvanized Steel	Type 316 Stainless Steel	Type 302 Stainless Steel
1/4 in.	1.5	1.5	1.6	1.1	1.8	2.1
5/16 in.	2.2	2.4	2.7	1.9	2.9	3.3
3/8 in.	3.3	3.9	3.8	2.6	3.9	4.5
7/16 in.	4.0	4.7	5.1	3.4	5.8	6.5
1/2 in.	4.9	5.6	6.6	4.4	7.5	8.4

Table 8.9–Galvanized Steel Strand/Physical Specifications

Nominal Messenger Size (in.)	Grade	Weight lb./1,000 ft.	Minimum Strength lb.
3/16	Common	73	1,150
3/16	Utility 2.2 M	73	2,400
1/4	Common	121	1,900
1/4	Siemens Martin	121	3,150
1/4	High Strength	121	4,750
1/4	Extra High Strength	121	6,650
5/16	Common	205	3,200
5/16	Siemens Martin	205	5,350
5/16	Utilities Grade 6 M	225	6,000
5/16	High Strength	205	8,000
5/16	Extra High Strength	205	11,200
3/8	Common	273	4,250
3/8	Siemens Martin	273	6,950
3/8	Utility 10 M	273	11,500
3/8	High Strength	273	10,800
3/8	Extra High Strength	273	15,400
7/16	Siemens Martin	399	9,350
7/16	High Strength	399	14,500
7/16	Utility 16 M	399	18,000
1/2	Siemens Martin	517	12,100
1/2	High Strength	517	18,800
1/2	Utility 25 M	517	25,000

Class A: Minimum amount of zinc coating.
Class B: Twice the amount of zinc coating as Class A.
Class C: Three times the amount of zinc coating as Class A.

Additional information on overhead messengers is available in ICEA P-79-561 Guide for Selecting Aerial Cable Messengers and Lashing Wires.

8.6 VERTICAL SUSPENSION

8.6.1 Suspended By Clamping Around Cable

Table 8.10–Spacings for Conductor Supports

Maximum Support Spacing for Conductors in Vertical Raceways		
AWG or Circular Mil Size of Wire	Aluminum or Copper-Clad Aluminum	Copper
18 AWG through 8 AWG	100 ft.	100 ft.
6 AWG through 1/0 AWG	200 ft.	100 ft.
2/0 AWG through 4/0 AWG	180 ft.	80 ft.
over 4/0 AWG through 350 kcmil	135 ft.	60 ft.
over 350 kcmil through 500 kcmil	120 ft.	50 ft.
over 500 kcmil through 750 kcmil	95 ft.	40 ft.
over 750 kcmil	85 ft.	35 ft.

Source: NEC, Article 300.19

8.6.2 Suspended by Conductor

Additional information on vertically suspended cables is available in NEMA WC 70 (ICEA S-95-658) Section 4.4.1 and in NEMA WC 74 (ICEA S-93-639) Section 7.4.1.

$$F = \frac{A \times T}{W \times L}$$ Where

A = conductor area in sq. in.
T = conductor tensile strength in lb./sq. in.
W = cable weight in lb./ft.
L = length in feet
F = safety factor (must be at least 7.0 unless otherwise required by appropriate authority)

Example:

Suspend 470 ft. of cable having three 4/0 AWG (211,600 circular mils each) soft-drawn copper conductors, total cable weight is 3,240 lb./1,000 ft. or 1,080 lb./1,000 ft. per conductor, each conductor is supported at the top with a full tension terminal:

$$F = \frac{[(211{,}600)\ (\pi/4)\ /1{,}000{,}000]\ 36{,}000}{(1{,}080/1{,}000)\ 470} = 11.8 \text{ (OK)}$$

If the suspended cable is installed in a conduit elbow at the top, check sidewall loading.

8.7 HIPOT TESTING

Overview
This section provides an overview of high-potential DC testing of power cables. For more details, see IEEE Standards 400 and 400.1. All tests made after cable installation and during the guarantee period should be made in accordance with applicable specifications. All safety precautions must be observed during testing at high voltage. Read and understand and follow the operator's manual for the particular test set being used!

It should be also noted that other field tests are growing in popularity–including VLF (very low frequency) and PD (partial discharge) test methods. IEEE 400.2 and 400.3 contain additional details.

8.7.1 Test Equipment
Direct current test equipment is commercially available with a wide range of voltages. Accessory equipment is necessary to safely conduct high-voltage tests such as safety barriers, rubber gloves and nonconducting hard hats. Consult appropriate safety officer.

8.7.2 Test Procedure
Refer to IEEE Standard 400. Acceptable procedures, although varying slightly in technique, have more or less been standardized as either a "withstand test" or a "time-leakage current test."

Before performing any DC overpotential tests:

- All equipment must be disconnected from the cable circuit, e.g., disconnect transformers, switch taps, motors, circuit breakers, surge arrestors, etc. This will preclude damage to such equipment and will prevent test interruptions due to flashovers and/or trip-outs resulting from excessive leakage current.
- Establish adequate clearance between the circuit test ends and any grounded object, and to other equipment not under test (approximately 0.25 inches per kV).
- Ground all circuit conductors not under test and all cable shields as well as nearby equipment.
- Consult termination manufacturer for maximum test voltage recommendations and time limitations.

The direct current test may be applied either continuously or in predetermined steps to the maximum value in accordance with applicable specifications:

- Continuous Method – Apply test voltage at an approximate rise rate of 1 kV per second or 75 percent of the rated current input of the equipment, whichever is less. Some equipment will take longer to reach the maximum test voltage because of the amount of charging current.
- Step Method – Apply test voltage slowly in five to seven increments of equal value, to the maximum specified. Allow sufficient time at each step for the leakage current to stabilize. Normally this requires only a few seconds unless cable circuits of high capacitance are involved. Record leakage current at each step.
- Maintain the test voltage at the prescribed value for the time designated in applicable specifications.
- At the end of the test period, set the test set voltage control to zero. Allow the residual voltage on the circuit to decay then ground the conductor just tested.
- Caution – It should be recognized that DC charges on cable can build up to potentially dangerous levels if grounds are removed too quickly. Maintain solid grounds after the test on the cable for at least four times the duration of the test. It is a good safety practice to maintain these grounds longer and while reconnecting circuit components.

Acceptance Testing – After installation and before the cable is placed in regular service the specified test voltage is applied for 15 consecutive minutes.

Proof Testing – At any time during the period of guarantee, the cable circuit may be removed from service and tested at a reduced voltage (normally 65 percent of the original acceptance value) for five consecutive minutes.

Record the leakage current at one minute intervals for the duration of the test. A constant or decreasing leakage current with respect to time at maximum test voltage is the usual acceptance criterion for DC hipot testing.

Additional Considerations

High-potential testing of medium-voltage power cables is usually performed with negative polarity connected to the conductor.

High-potential testing is a tool for determining insulation resistance at high voltages. Effective insulation resistance of the cable system may be calculated by means of Ohm's Law: R = V/I. Restated another way the relation is:

$$\text{Megohms} = \frac{\text{Kilovolts}}{\text{Microamperes}} \times 1{,}000$$

Insulation resistance (IR) may also be measured with instruments that give a direct reading at 500 volts (or higher, depending on the model). IR in general has little or no direct relationship to breakdown strength.

The significance of conducting DC high-voltage tests on nonshielded, nonmetallic sheathed cable is dependent upon the environment in which it is installed because the characteristics of the return circuits are unknown. The environment must be carefully considered or test results may not be significant. In fact, these tests can result in damage to the cable insulation.

Humidity, condensation or actual precipitation on the surface of a cable termination can increase the leakage current by several orders of magnitude. Humidity also increases the termination leakage current, which is included in the total leakage current. Wind prevents the accumulation of space charges at all bare energized terminals. This results in an increase of corona. It is desirable to reduce or eliminate corona current at the bare metal extremities of cable or terminations. This may be accomplished by covering these areas with plastic envelopes, plastic or glass containers, plastic wrap (e.g., Saran® or Handiwrap®) or suitable electrical putty.

Routine periodic DC maintenance testing of cable for the evaluation of the insulation strength is not a common practice. Some power cable users have adopted a program of testing circuits during planned outages, preferring possible breakdowns during testing rather than experiencing a service outage. It is nearly impossible to recommend test voltage values for maintenance. An arbitrary test voltage level could break down a cable circuit that would otherwise render long trouble-free service at normal operating AC voltage.

One advantage of DC high-voltage testing is that it can detect conducting particles left on the creepage surface during splicing or termination.

Test equipment should be supplied from a stable, constant voltage source. Do not use the same source that is supplying arc welders or other equipment causing line voltage fluctuations. The output voltage of the test set must be filtered and regulated. Consider using a portable motor driven alternator to energize the test set.

Common Testing Problems

High-leakage current can be caused by:

- Failure to guard against corona
- Failure to clean insulation surface
- Failure to keep cable ends dry (high relative humidity, dampness, dew, fog, wind, snow)
- Failure to provide adequate clearance to ground
- Improper shield termination.

Erratic readings can be caused by:

- Fluctuating voltage to test set
- Improper test leads.

8.7.3 Test Voltage

DC hipot test voltages are specified by ICEA and NEMA for tests conducted during and after installation as follows:

- At any time during installation, a DC proof test may be made at a voltage not exceeding the test voltage specified below, applied for five consecutive minutes.
- After the cable has been completely installed and placed in service, a DC proof test may be made at any time within the first five years at the test voltage specified below, applied for five consecutive minutes. After that time, DC testing is not recommended.

Table 8.11–Maximum DC Test Voltages for Shielded Power Cables

Rated Voltage	Maximum DC Field Test Voltages in kV			
	During Installation		First 5 Years	
Phase-to-Phase (kV)	100 Percent (Grounded)	133 Percent (Ungrounded)	100 Percent (Grounded)	133 Percent (Ungrounded)
5	28	36	9	11
8	36	44	11	14
15	56	64	18	20
25	80	96	25	30
28	84	100	26	31
35	100	124	31	39
46	132	172	41	54

Sources: ICEA S-94-649 Appendix E, NEMA WC 74 (ICEA S-93-639) Appendix F and ICEA S-97-682 Appendix E

8.7.4 Evaluation of Results

The test current will momentarily increase for each voltage increment due to the charging of capacitance and dielectric absorption characteristics of the cable–ultimately leaving only the conduction current plus any external surface leakage or corona currents. The time required to reach steady-state current depends on insulation temperature and material. If, without any increase in applied voltage, the current starts to increase slowly at first but at an increasing rate, gradual insulation failure may be in progress. This process will probably continue until eventual failure of the cable unless the voltage is rapidly reduced. Rubber and nonpressurized impregnated paper insulations will usually exhibit this type of insulation failure; other insulations rarely exhibit this type of failure.

If at any time during the test, a violent increase in current occurs accompanied by tripping of the test set, failure or flashover has probably occurred in the cable, a splice or termination. A failure can be confirmed by the inability to sustain the second application of the test voltage.

8.8 FAULT LOCATING

One of the many types of fault locating equipment is the Time Domain Reflectometer (TDR). These units are portable, commercially available devices that can be used in the field to locate some types of conductor breaks or shorts. Connected to the end of a cable, the device functions much like radar, sending out low-voltage pulses that travel the length of the cable and echo back when an open, short or tap is encountered. The device can usually locate faults within ± 2 percent of the cable length. However, TDRs are only capable of locating breaks or shorts having an impedance different than that of the cable. For most cables, this includes shorts having a resistance of less than a few ohms and opens having a resistance greater than several hundred ohms. Splices, taps, etc., sometimes distort the echo and can mask the fault. Nevertheless, the method is nondestructive and is used successfully on faults having characteristics within the capabilities of the method.

8.9 MEGGER TESTING

If the DC voltage applied during an insulation resistance (IR) test on power cables is relatively low (0.6 to 2.5 kV), the test is often referred to as a "Megger" test. Low-voltage IR tests are particularly useful in detecting shorts due to installation or handling damage to 600-volt-rated cables.

An inherent limitation of low-voltage IR tests is their interpretation. The readings obtained from such testing on nonshielded, nonmetallic-sheathed cable is very dependent upon the environment because the environment determines the characteristics of the return circuit. Low resistance readings may be caused by contaminated or moist cable ends, high humidity, etc. Failure to clean water based cable pulling lubricants from the cable test ends has caused erroneous rejection of good cable. Refer to the figures below for suggested hookup.

Reminders:

- **Safety** – Follow the test equipment supplier's instructions. Stay clear of energized cable. Operators must know the equipment. Be sure shields are grounded! Remember that insulated conductors are capacitors.
- **Voltages** – Check cable and termination manufacturer's guidelines.
- **Records** – Keep detailed records and provide a copy to the owner.

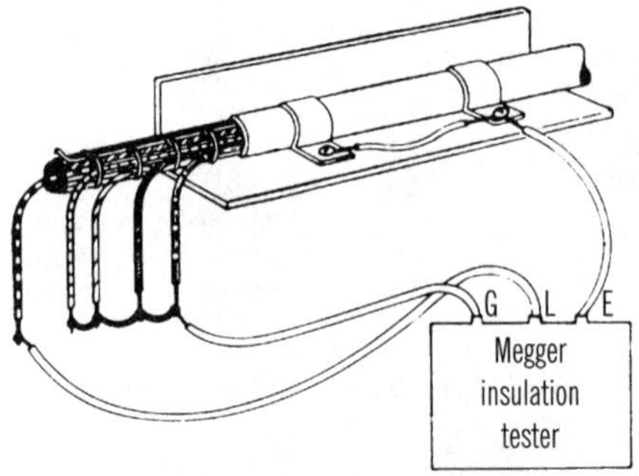

Figure 8.4–Connections for Testing Insulation Resistance Between One Wire and Ground, Without Being Affected by Leakage to Other Wires. Note Use of the Guard (G) Connection

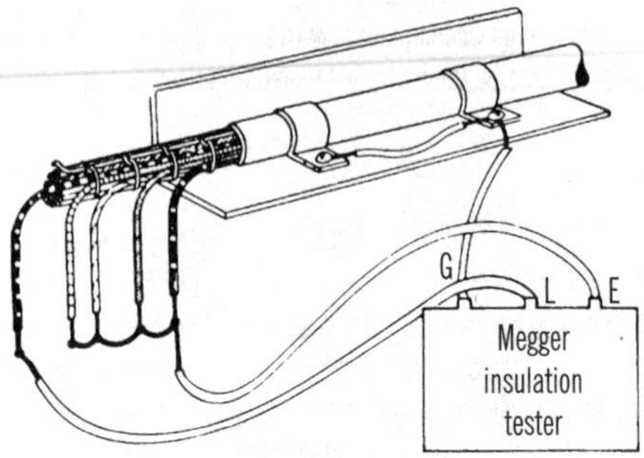

Figure 8.5–Connections for Testing Insulation Resistance Between One Wire and All Other Wires, Without Being Affected by Leakage to Ground

8.10 MOISTURE REMOVAL

8.10.1 Purging Water from Conductor Strand or Shield

Cables Not Yet Installed: Remove end seals. Position one cable end to its lowest possible elevation. At the cable end having the highest elevation, apply two layers of half-lapped HV insulating tape to act as a sealing cushion. Connect the cable ends to a dry nitrogen or dry air supply using hoses, valves, fittings and flow regulators as shown in Figure 8.6.

Attach a one-gallon plastic bag to the exhaust end of the cable. Secure the bag with tape or clamps. Make a small vent hole by clipping one bag corner.

As shown, several cables may be connected to the gas supply. Dry nitrogen is available from welding gas suppliers. Apply 15–25 psi (gauge). Maintain gas flow for at least eight hours after all indications of moisture have stopped.

Water vapor may be readily detected by sprinkling one tablespoon of anhydrous cupric sulfate in the plastic bag, which turns blue instead of "off" white when wet. The sulfate is available from scientific laboratory supply houses. A hardware store humidity gauge may also be used.

Installed Cables: The splices and terminations must be removed if they interfere with the flow of air or nitrogen. The cable can then be purged as described above.

All Cables: Purge the shield separately from the insulated strands; otherwise the nitrogen gas will only flow through the path offering the least resistance.

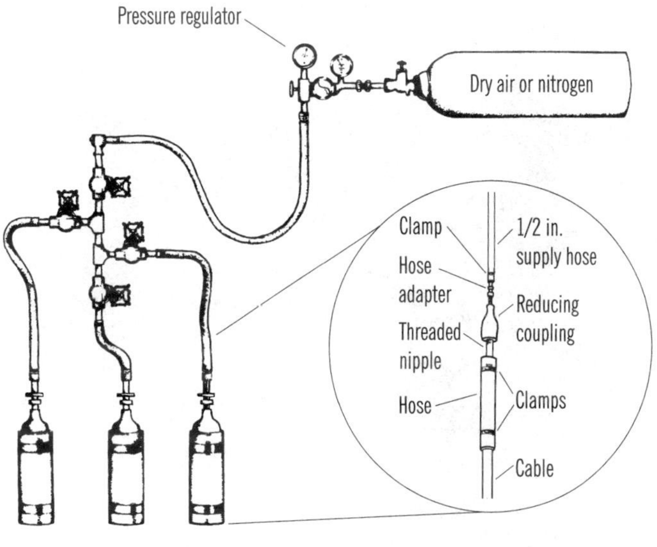

Figure 8.6–Moisture Removal Equipment

8.11 FIBER OPTIC TESTING

Testing a newly installed fiber optic system can increase the overall performance of a system, decrease the amount of downtime and reduce costs for the system owner. Attenuation is the parameter most frequently measured and includes the attenuation of the cable as well as that of attached connectors. Cable attenuation can be caused by microbending of the fiber, impurities in the fiber, excessive mechanical force on the cable or, of course, a broken fiber.

Handheld optical power meters and light sources are used to determine the total attenuation of the fiber including any splices or connectors. With this method the light source injects light with a known signal level (brightness) into one end of the fiber. The power meter is attached to the other end of the fiber and measures the light output at a specific wavelength. The difference is the attenuation and is usually reported in decibels.

Optical Time Domain Reflectometers (OTDRs) are used to locate faults and to measure attenuation of cables and connectors. A light pulse is sent down the fiber and as it encounters a fault, connector, splice, etc., a portion of the optical pulse is reflected back to the source. An OTDR is able to determine the distance to the reflection and the amount of signal loss at that point. OTDRs work on a radar-like principle.

Small optical microscopes are used to visually inspect the workmanship of installed fiber optic connectors.

8.12 LAN CABLE TESTING

With society's dependency on data networks around the world, the ability to maintain proper system operation is extremely important. There are several types of test equipment that are commonly used to evaluate local area network (LAN) unshielded twisted-pair (UTP) and shielded twisted-pair (STP) cabling.

Low-cost handheld LAN cable testers are available to certify the electrical performance, e.g., Category 3, 4, 5 or 6, of newly installed LAN cable. These devices typically characterize the installed system with regard to crosstalk, attenuation, impedance and other parameters.

Time Domain Reflectometers (TDRs) are devices used to locate faults, determine length and measure attenuation of the cable. The TDR sends a low-voltage pulse along the cable and then "looks" for reflections that result from impedance mismatches that are caused by shorts, opens or severely deformed cable. TDRs analyze the reflections and report the magnitude of the impedance mismatch and the location of faults.

9. CABLE ACCESSORIES

9.1 COAXIAL CONNECTORS

Coaxial connectors should appear electrically as extensions to the cable; in other words, they should connect to the cable with as little disruption of the electrical signal as possible. Thus, a connector is usually specified by its nominal impedance and its allowable voltage standing wave ratio (VSWR). The nominal impedance of the connector indicates its basic match to the nominal impedance of the cable. The VSWR indicates the quality of the match.

9.1.1 Selection

Just as MIL-DTL-17 covers the main types of coaxial cables, MIL-PRF-39012 covers many popular types of coaxial connectors. It includes mating and overall dimensions, materials, performance and testing procedures for each connector.

In selecting a connector, users generally consider cable size, frequency range and coupling method.

Cable size determines the connector series as subminiature, miniature, medium or large.

Frequency range determines the upper frequency limit of the application. Connectors can be used at frequencies below this range but are not recommended at frequencies above this range where performance (especially VSWR) becomes degraded. Both BNC and TNC series connectors, for instance, can be used with miniature cable. The TNC connector, however, is usable to 11 GHz, while the BNC is limited to 4 GHz. (This is due to the difference between bayonet and screw couplings.) If the highest frequency of the application is 2 GHz, either connector can be used. If the highest frequency is 8 GHz, the TNC is the obvious choice.

Coupling method determines the procedure for joining two mating connectors. The three common types are bayonet, screw and snap-on. Often the coupling method is the main difference between two series of connectors. For example, a BNC connector uses bayonet coupling; a TNC connector is essentially the same, but with a threaded coupling.

9.1.2 BNC

By far, BNC connectors are the most common for miniature cables because of the easy connection/disconnection offered by their bayonet coupling. In most versions, BNC connectors are 50-ohm connectors rated to 4 GHz. 75-ohm, 4 GHz connectors are now available to meet the demand and usage of 75-ohm coax cable.

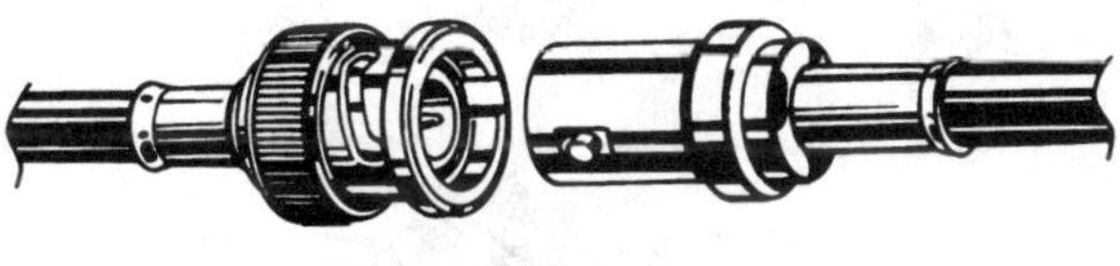

Figure 9.1–BNC Connectors

9.1.3 TNC

A TNC connector is virtually identical to a BNC connector, except it has a threaded rather than a bayonet coupling. The tight interface of the threads, especially when subjected to vibration, allows the connector to maintain a low VSWR up to 11 GHz with flexible cable and up to 15 GHz with semirigid cable.

9.1.4 SHV

For medium-size cables, SHV connectors are high-voltage connectors rated to 5,000 volts (rms). They have bayonet coupling but do not have a constant impedance. They were originally designed for high-energy physics applications.

9.1.5 SMA

Widely used in avionics, radar, military, and high-performance test equipment applications, SMA connectors are the most popular type for subminiature cable and offer the highest performance in their class. They meet MIL-PRF-39012 requirements up to 12.4 GHz when used on flexible cable and up to 18 GHz on semirigid cable.

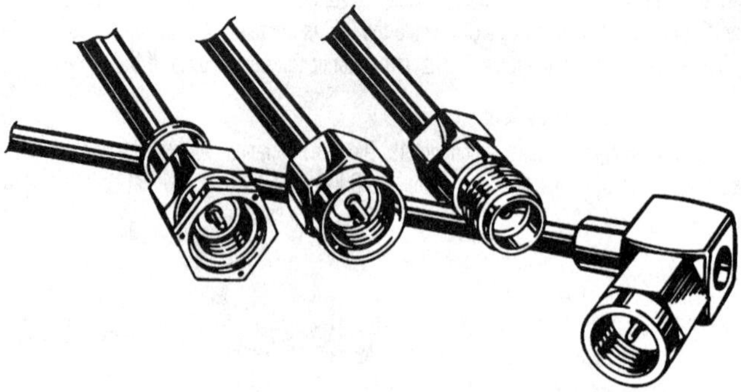

Figure 9.2–SMA Series Coax Connectors for Semirigid Cable

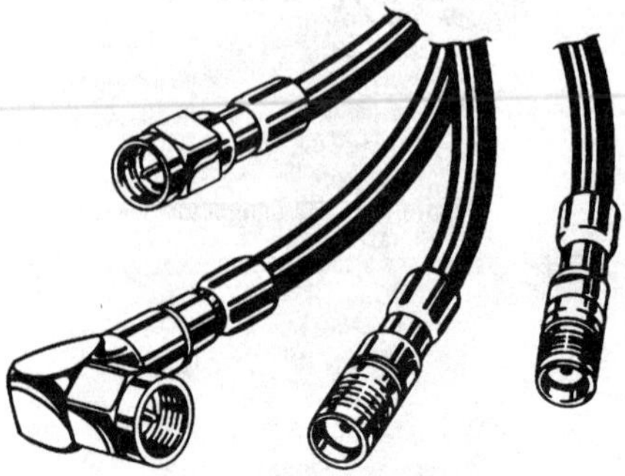

Figure 9.3–SMA Series Coax Connectors for Flexible Cable

9.1.6 UHF

The first coaxial connectors, designed in the 1930s, UHF connectors exhibit nonconstant impedance and a low upper-frequency limit of only 500 MHz, 2 GHz for the miniature version. Their main application is in cost sensitive consumer applications.

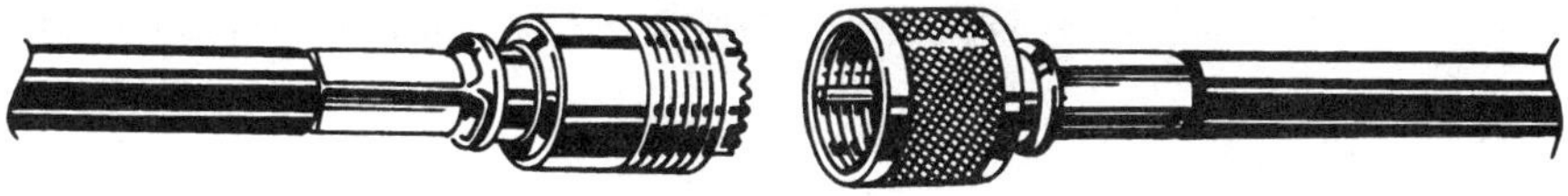

Figure 9.4–UHF Series Coax Connectors

9.1.7 N Series

These screw thread connectors were the first true RF connectors, developed during World War II to handle microwave frequencies up to 11 GHz. Despite the connector's age, it is still widely used, offering dual-crimp, low-cost commercial and 75-ohm versions in a variety of styles and materials. It is the standard coaxial connector for many coaxial cable based local area networks, including Ethernet and other IEEE 802 networks using medium-size coaxial cable.

Figure 9.5–N Series Coax Connectors

9.1.8 F Series

The F type connectors are 75-ohm, screw threaded couplers for RG-59, RG-6, and RG-11 type coaxial cables and are the standard for cable television systems. The F type connector is simple to install, economical and meets the specifications of CATV/MATV systems. Most connectors are terminated to the cable by a single crimp on the attached ferrule.

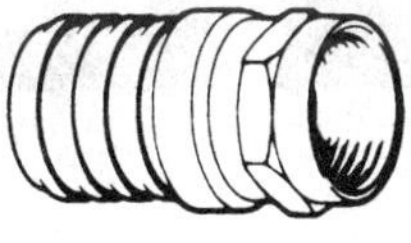

Figure 9.6–F Series Coax Connector

9.2 DATA CONNECTORS

9.2.1 RJ11 and RJ45 Modular

RJ11 and RJ45 modular plugs and jacks are widely used in communication applications. Some are designed for use with wires with solid conductors, others for stranded wire. The wiring configuration varies, depending on the wiring method selected for the system. The most used standards are ANSI/TIA/EIA 568-B (wiring methods "A" or "B"). With the locking tab down, the conductors are inserted into the rear of the plug in a specific pattern. The pins of the plug are numbered 1 through 8 as shown in Figure 9.7.

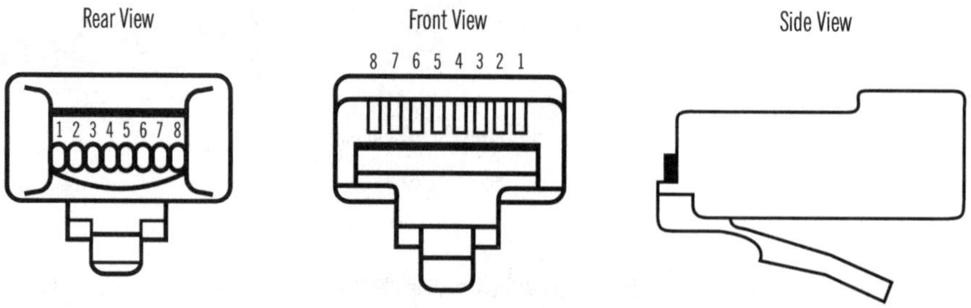

Figure 9.7–RJ45 (8 pin) Modular Plug

The individual pairs of 100-ohm UTP (unshielded twisted-pair) cabling are usually connected using one of the two pair assignments shown in Figure 9.8.

8 Position Modular Jack Pair Assignments for UTP

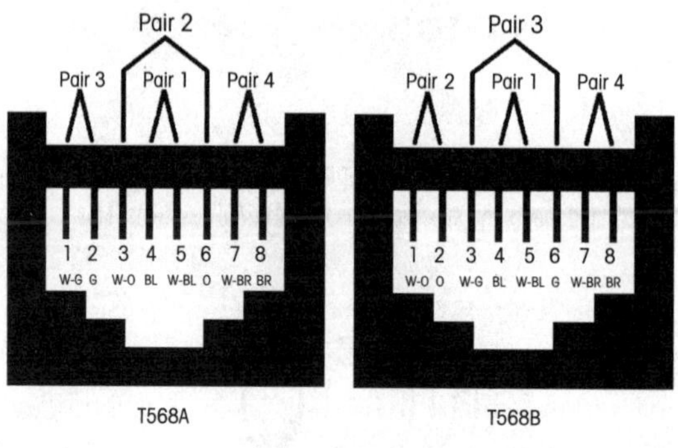

Figure 9.8–Pair Assignments A (left) and B (right) for an RJ45 Modular Jack

9.3 POWER CONNECTORS

9.3.1 Selection

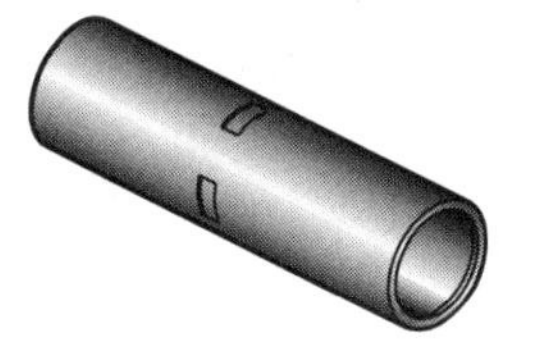

Figure 9.9–Typical Compression Connector

Compression Connectors
Compression connectors are designed for reliable and controllable electrical connections. Connectors must withstand a wide range of electrical and environmental conditions including current surges, temperatures, corrosion and vibration, for a wide variety of applications.

Copper compression connectors are normally manufactured from high-conductivity electrolytic copper. The connectors are normally tinplated, lead-plated, or plated with a proprietary finish to provide corrosion resistance.

Aluminum compression connectors are usually manufactured from high-conductivity, high-purity wrought aluminum. They are designed with sufficient mass and are electro-tin plated to minimize corrosion due to galvanic action between dissimilar metals. The connector barrels are typically prefilled with oxide inhibiting compound. Oxide inhibiting compound usually contains homogeneously suspended metallic particles that penetrate the wire's oxides to establish continuity between the individual strands and the connector barrel for a low-resistance connection. Connector designs are engineered to match the cable size to provide the necessary physical strength requirements for reliable electrical performance.

Selection and Use
Copper compression connectors are recommended for use on copper conductors. Aluminum compression connectors are recommended for use on aluminum conductors. Dual-rated aluminum compression connectors may be used on both copper and aluminum conductors.

Tooling
Tooling systems are essential for proper installation of a compression connector. Since connectors and dies are designed as a unit for specific wire sizes, only the recommended tools and dies should be used. Many aluminum and copper terminals and splices are marked with a die index number and are color-coded to identify the correct installation die. Dies marked with the matching die index number and color can be used to install the connector.

The tools include small plier types, full-cycle ratchet designs, hydraulically-powered heads and battery actuated tools. (See examples in Figure 9.10.) Some have permanent die grooves or adjustable dies, while others require a change of die sets or nest die for each connector size.

Manufacturers publish extensive tables of suitable connector, tool and die combinations to ensure a quality splice or termination. When properly installed, virtually all the air is removed leaving a tight homogeneous mass of connector and conductor.

Industry Standards
Compression terminals (See Figure 9.11), splices and tap connectors requiring third-party testing and approval are listed by Underwriters Laboratories, Inc. Many have also received CSA approval and are approved under SAE AS7928 (formerly MIL-T-7928) and other standards. All conform to applicable sections of the National Electrical Code.

Battery-operated crimping tool

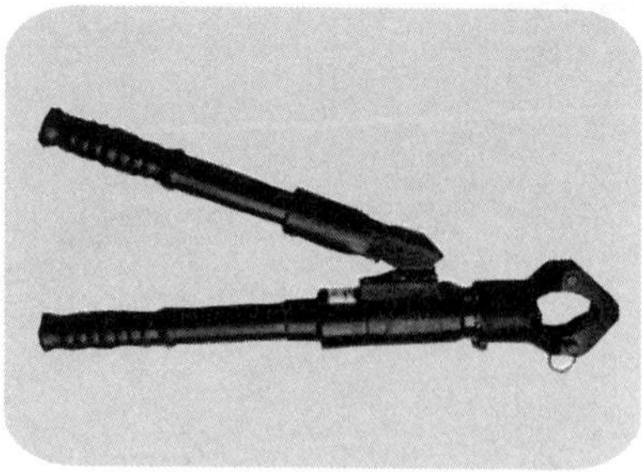

Hand-operated crimping tool

Figure 9.10 Typical Compression Tools

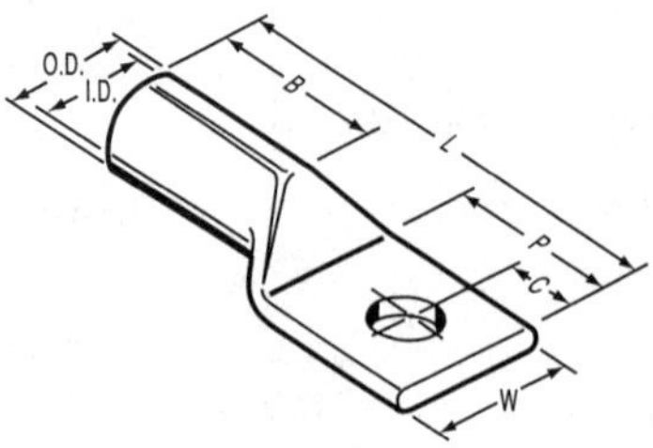

One-hole lug

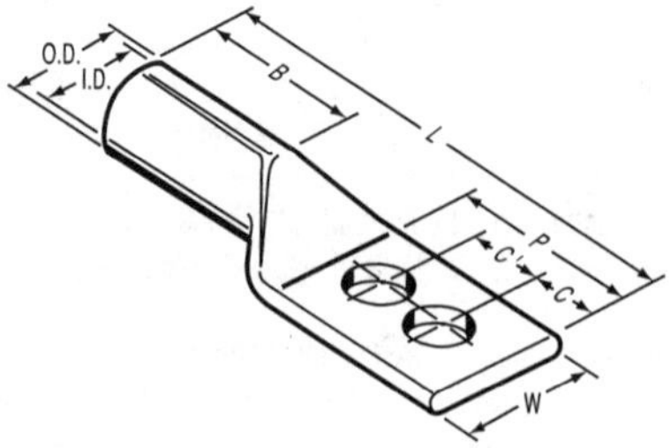

Two-hole lug

B = Length of barrel
C = Edge of tongue to center of stud hole
C^1= Stud hole spacing (two-hole lugs only)
I.D. = Inner diameter
L = Length of lug
O.D. = Outer diameter
P = Tongue length
W = Tongue width

Figure 9.11 Typical Compression Terminals (Lugs)

9.3.2 Stud Sizes

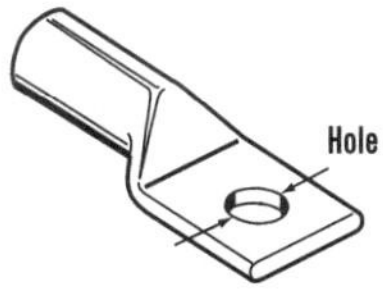

English			Metric		
Bolt Size	Bolt Diameter	Hole Diameter	Bolt Size	Bolt Diameter	Hole Diameter
	Inch	Inch		mm	mm
#2	.086	.095	M2	2.0	2.4
#4	.112	.120	M2.5	2.5	3.0
#5	.125	.148	M3	3.0	3.8
#6	.138	.148	M3.5	3.5	3.8
#8	.164	.174	M4	4.0	4.4
#10	.190	.200	M5	5.0	5.1
1/4"	.250	.265	M6	6.0	6.7
5/16"	.3125	.328	M8	8.0	8.3
3/8"	.375	.397	M10	10.0	10.1
7/16"	.4375	.450	M10	10.0	11.4
1/2"	.500	.515	M12	12.0	13.1
5/8"	.625	.656	M16	16.0	16.7
3/4"	.750	.781	M18	18.0	19.8

Source: ISO 263 for English stud sizes and ISO 262 for metric stud sizes. **Note: Bolt illustrations not drawn to scale.**

Figure 9.12–Terminal Stud Size Chart in English and Metric Units

9.4 FIBER OPTIC CONNECTORS

Fiber optic connectors are used at the ends of optical fiber. They allow fiber optic equipment and patching connections to be made easily and quickly. The connectors mechanically couple and align the cores of fibers so that light can pass through with a minimum amount of attenuation (loss). Many types of fiber optic connectors are available. The main differences among the types of connectors are their dimensions, methods of mechanical coupling, the number of fibers they contain and the particular termination methods (including tooling, consumable items and training) required to install them.

9.4.1 Types of Fiber Optic Connectors

The larger standard size fiber connectors, such as the ST and SC, have been around for many years and are still the most widely used. There are several varieties of newer small form factor (SFF) connectors that are much smaller than the ST and SC. Because of their small size, they allow a higher fiber port density on optical equipment and patch panels. Two common SFF connector types are the LC and the MT-RJ. For higher multi-fiber terminations (up to 12 fibers) there is the MTP/MPO connector. It is usually used with ribbon cable for high-density backbone, cross-connect and breakout applications.

9.4.1.1 ST Connector

The standard size ST connector was one of the first connector types widely implemented in fiber optic networking applications. Originally developed by AT&T, it stands for Straight Tip connector. ST connections use a 2.5 mm ferrule with a round plastic or metal body that holds a single fiber. The connector has a bayonet style "twist-on/twist-off" mechanism. The ST is still very popular for building applications but it is slowly being replaced by the SC and the smaller, denser, small form factor (SFF) connectors.

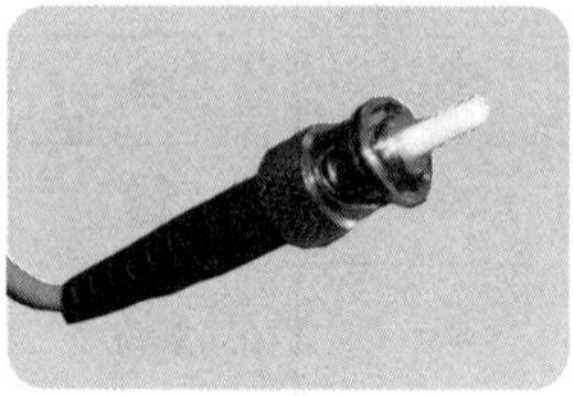

Figure 9.13–ST Connector

9.4.1.2 SC Connector

The standard size SC connector also has a round 2.5 mm ferrule to hold a single fiber. It is non-optically disconnecting and has a "keyed" insertion for performance reliability and to prevent tip rotation. The SC utilizes a push-on/pull-off mating mechanism, which is easier to use than the twist-style ST connector when in tight spaces. The connector body of an SC connector is square, and two SC connectors are often held together with a plastic clip making this a "duplex" connection. It was developed in Japan by NTT (the Japanese telecommunications company) and is believed to be an abbreviation for Subscriber Connector. This is the recommended standard size fiber connector for enterprise cabling installations.

Figure 9.14–SC Connector

9.4.1.3 LC Connector

The LC connector is the most commonly used small form factor (SFF) connector. It was developed by Lucent Technologies and it stands for Lucent Connector. It utilizes a push-pull mechanism, similar to the SC, and the connector body resembles the square shape of SC connectors but it is much smaller. Each connector holds one fiber. Two LC connectors are normally held together in a duplex configuration with a plastic clip. The joined duplex LC connector only takes as much space as one SC connector. The ferrule of an LC connector is 1.25 mm.

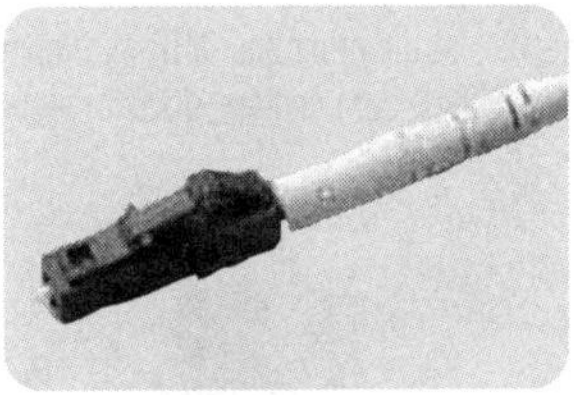

Figure 9.15–LC Connector

9.4.1.4 MT-RJ Connector

The MT-RJ is another small form factor (SFF) connector. It was developed by AMP/Tyco and Corning and MT-RJ stands for Mechanical Transfer-Registered Jack. The MT-RJ connector closely resembles an RJ-style modular plug and terminates two fibers in a single ferrule so it is always "duplex." This connector fits an adapter with the same footprint of a standard single-fiber connector. The connector locks into place with a tab (just like a modular RJ-style plug).

Figure 9.16–MT-RJ Connectors

9.4.1.5 MTP/MPO Connector

The MTP is a high-density multifiber optical connector. It is a trademark of US Conec and is an improvement on the original MPO (multi-fiber push-on) connector designed by NTT. The MTP connector contains up to twelve optical fibers within a single ferrule and is available in both multimode and single-mode versions. This connector is used with fiber ribbon cable to achieve very high density. The connection is held in place by a push-on/pull-off latch and has a pair of metal guide pins that protrude from the front of the connector for alignment. The MTP connector is often used on both ends of preterminated cable assemblies to facilitate the quick and easy interconnection of pre-connectorized patch panels. In addition, it is used at one end of cable assemblies that "break out" multifiber cables into separate single fiber connectors at the other end. MTP/MPO applications include horizontal zone cabling systems as well as high-density backbones and cross-connects used in large buildings, data centers, disaster recovery and industrial operations.

Figure 9.17–MTP/MPO Connector

9.4.1.6 FC Connector

The FC connector has a 2.5 mm ferrule tip with a threaded screw-on mechanism. It is keyed to prevent tip rotation and damage to mated fiber. These connectors are typically used for single-mode applications but multimode connectors are available. Not as popular as it once was, it is slowly being replaced by SC and LC connectors.

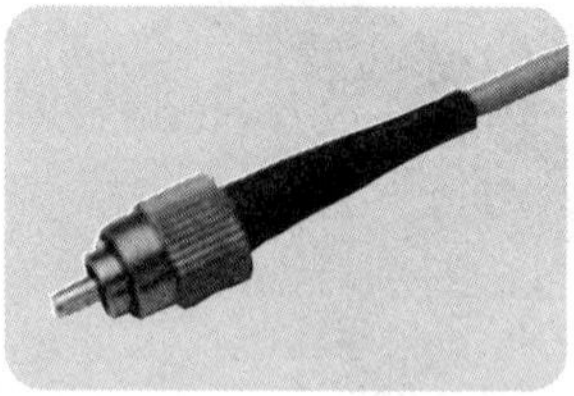

Figure 9.18–FC Connector

9.4.1.7 FDDI Connector

FDDI stands for fiber distributed data interface. It is actually a fiber-based network access method based on the token ring protocol and utilizes two fiber rings. One is the primary and one is for backup. The termination on the fiber optic cable itself is called an FDDI connector also known as an MIC (media interface connector) connector. It contains two ferrules in a large, bulky plastic housing that uses a squeeze-tab retention mechanism. The FDDI connector is designed to mate to its specific network. It is generally used to connect to equipment from a wall outlet while the rest of the network will have ST, SC or other SFF connectors. The FDDI connector has a keying system to prevent connections of incompatible network nodes. There are four receptacle keys: A, B, M, and S.

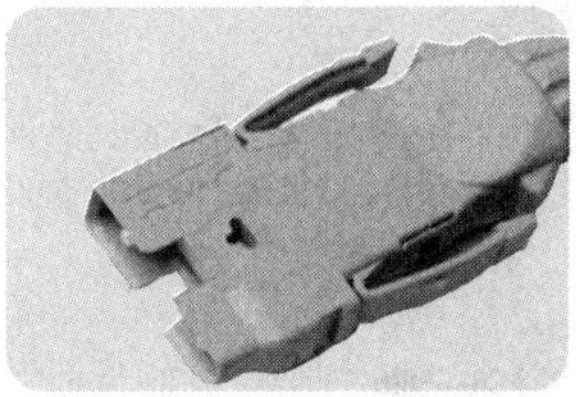

Figure 9.19–FDDI Connector

9.4.1.8 SMA Connector

This connector was designed by Amphenol and stands for "Subminiature A." It was originally designed for the military and was in use before the ST connector. It was the dominant connector in data and closed-circuit video applications. There are two types used: SMA 905, which has a ferrule that is the same diameter from the base to the tip, and SMA 906, which has an indented (smaller diameter) section on the tip side of the ferrule. Most SMA 906 type connectors come with a plastic ring on the tip so they can be used as a SMA 905 connector. While it is still found in predominantly military and industrial applications, the use of the SMA connector is diminishing. It is rarely used in building network applications.

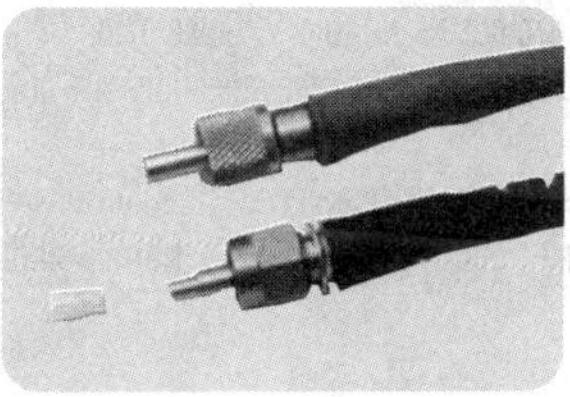

Figure 9.20–SMA Connector

9.4.2 Considerations for Selecting Fiber Optic Connectors

When selecting fiber optic connectors, the following must be taken into consideration:

- Type and construction of connector (SC, ST, LC, FC, MT-RJ, MTP etc.)
- Mode of fiber being connectorized (single-mode or multimode)
- Construction of fiber (tight buffer, loose tube, breakout, jumper cordage, etc.)
- Termination method (heat cure, UV cure, no epoxy/no polish, anaerobic, etc.)

9.4.2.1–Type and Construction of Connector

Choosing the right fiber connector means selecting the proper form factor, the specific style and in some cases the type of material used. The three most common form factors (sizes) are "standard" size, small form factor (SFF) and the high-density multi-fiber type connector.

Some considerations are:

- For new installations, SC is the recommended standard size connector followed by the ST in popularity. Other types are available
- LCs are the most common small form factor (SFF) connector. Other types, such as the MT-RJ connector, are also available
- The MPO/MTP type connector is used for high-density fiber applications usually with ribbon fiber
- When adding fiber and connectors to an existing facility, one should consider using the same type of connectors throughout the facility
- Many projects have a written specification that defines the particular type of fiber connector required
- If most opto-electronic equipment in a facility has predominantly one connector type, it might make sense to use the same type of connector on the fiber infrastructure and distribution cabinets
- Fiber patch cords with different connector types at each end can be used where the opto-electronic equipment connector does not match the connector on the fiber distribution equipment.

The type of material used in the tip (ferrule) of the connector is also important. These are commonly used materials:

- Ceramic – Connector tips made from ceramic are preferred because ceramic closely matches the thermal characteristics of glass. It is a hard, durable material that does not wear down even after a high number of reconnections; however, it is more expensive than other materials.
- Composite (polymer) – Connectors with composite tips are not as durable as those made from ceramic or stainless steel but offer a cost-effective solution and are suitable for many applications where a high number of reconnections are not anticipated. Polymer materials have the advantage of a lower cost without sacrificing performance.
- Stainless steel – Stainless steel tips offer durability. Connectors with stainless steel ferrules (when compared to ceramic or composite), however, have higher typical insertion loss, generate more debris during re-mating and do not perform well during thermal cycling or vibration testing.

9.4.2.2–Mode of the Fiber

The mode of the fiber is important when selecting a fiber connector. Fibers are either multimode or single-mode.

- Multimode – The most common size for multimode fiber is 62.5/125 μm. This indicates a core diameter of 62.5 μm surrounded by a cladding layer of glass making the overall diameter of the fiber 125 μm. Another type of multimode fiber gaining popularity is 50/125 μm. Note that the core size of the 50 μm fiber is smaller than the 62.5 μm fiber but provides a greater bandwidth. However, it has the same overall diameter of 125 μm as the 62.5 μm core because cladding layer of glass is thicker.
- Single-mode – The typical fiber size is 8/125 μm which indicates a core diameter of 8 μm surrounded by a cladding layer of glass making the overall diameter of the fiber 125 μm. Note that the overall fiber diameter, at 125 μm, is the same as for multimode.

Even though the overall fiber diameter size (core plus cladding) may be the same between single-mode and multimode fibers there is typically a different connector required for each within the same type (i.e., SC single-mode or SC multimode connector), Single-mode connectors must be manufactured to more precise tolerances; therefore, they are generally more expensive. This is because the proper alignment of the 8 μm single-mode core is more critical than on the larger 50 or 62.5 μm multimode cores.

9.4.2.3–Fiber Cable Construction

The construction of the fiber cable needs to be considered when selecting fiber optic connectors. It is important that the proper cable be selected based on the environment and application needed. Primary construction types are tight buffered, loose tube (loose buffer), ribbon cable, fanout/breakout cable and jumper cordage.

- Tight-buffered cable (Tight buffered multi-fiber distribution cable) – This type of fiber is used mainly indoors; however, there are now tight-buffered outdoor versions available. Tight-buffered cable has a "buffer" layer of plastic coating extruded onto the fiber. It is in direct contact with the fiber and surrounds it bringing the O.D. of the fiber up to 900 μm. Connectors can be directly installed on the ends of 900 μm tight-buffered fibers without the use of a fanout or breakout kit. Multi-fiber building cable has multiple tight-buffered fibers under a common jacket. It is available in both single-mode and multimode versions.
- Loose tube (loose buffer) cable – Loose tube cables are used for outdoor applications and contain multiple bare fibers that float freely within larger buffer tubes. They may contain multiple small tubes, called subunits, that contain several fibers each or there may be one "central" tube that contains all the fibers in the cable. These cables are water-blocked utilizing gel and/or dry water-swellable tapes or yarns within and around the buffer tube(s) inside the cable. They come in various constructions depending on whether they will be installed aerially, in duct or in direct buried applications. Loose tube cables require a buffer tube fanout or breakout kit in order to connectorize the fibers. Choose a connector that matches the breakout kit subunit type for a proper fit. Loose-tube cable is available in single-mode or multimode versions. Traditionally, most loose-tube cable was not UL Listed for indoor use so it had to be terminated or transitioned to an inside rated cable within 50 ft. of the building entrance point if run exposed. There are now versions available that are indoor/outdoor rated making it allowable to extend the cable further than 50 ft. from the building entrance point within buildings.
- Ribbon cable – A type of cable construction that provides the highest fiber density in the most compact cable size. Inside the cable the fibers are typically laid out in rows of 12 fibers each, one row on top of the other. It is ideal for mass-fusion splicing and for use with multifiber connectors such as the MTP/MPO. Ribbon fiber cable has become very popular as the cable of choice for deployment in campus, building, and data-center backbones where high-density fiber counts are needed. It is available in both single-mode and multimode versions and in indoor and outdoor constructions.
- Fanout/Breakout cable – Each fiber in a fanout/breakout cable is jacketed and protected with strength members. In effect, each fiber is a simplex jumper cord (patch cable). A connector can be installed directly on each jacketed fiber and the connectorized fibers can be patched directly into electronic equipment or patch panels etc. These cables can be single-mode or multimode are available in both indoor and outdoor versions. Connectors must be selected that fit the O.D. of the jacket on the cable.
- Jumper cordage (patch cable) – Jumper cordage is divided into four construction types: simplex, zipcord duplex, dual subunit duplex, and round duplex. Simplex, zipcord, and dual subunit cordages can be directly connectorized. Round duplex usually requires a breakout kit.

9.4.3 Connector Termination Methods

The final step in choosing a fiber connector, once the connector type, fiber mode and cable construction are known, is to select the desired termination method of the connector. Within the same connector types (ST, SC, LC, etc.), there are different termination methods to choose from based on the connector manufacturer's design and the particular methods used to to prepare, connect, hold and terminate the fibers within the connector itself. Each type of connector has its own procedures and requirements for the tools, accessories and in some cases consumable items, necessary for proper installation. These vary from manufacturer to manufacturer and even within connector types from the same manufacturer.

With technical specifications being equal, the termination method chosen usually comes down to a matter of cost based on the following considerations:

- Is a specific termination method specified?
- Is the installer already equipped with the tool kits and accessories needed for use with a particular type of connector termination method?
- How many connectors being installed?
- What is the training and experience level of the installer?
- What are the material costs? (connector itself, tooling etc.)
- What are the consumable costs? (adhesive/epoxy, polishing films, etc.)
- What is the cost of labor? (curing, polishing, setup and tear-down)

The following explains the most common termination methods:

9.4.3.1–No Epoxy/No Polish Connectors (Mechanical Splice Type)

This type of connector has become very popular. It is a connector with a polished fiber already factory-installed in the tip along with a mechanical splice type alignment system to facilitate attaching the connector to the end of a fiber. Single-mode, 50/125 μm multimode, laser-optimized 50/125 μm multimode and 62.5 /125 μm multimode versions are available for each connector.

When taking into account the material, labor and consumables costs of termination, these type connectors are often the most cost-effective solution.

- No epoxy or polishing
- No consumables and few tools needed
- No power source required
- Minimal setup required
- Connectors cost more
- Faster installation
- Reduced labor cost

9.4.3.2–Epoxy and Polish Connectors (Heat-Cured)

The fiber is secured in the connector using epoxy. This type of connector is a cost-effective way to make cable assemblies or to install in locations where a large number of fibers are terminated at one time.

- Heat-cured in an oven
- Batch termination
- Low connector cost
- Consumables required (epoxy and polishing paper, etc.)
- Requires polishing
- Longer installation time than no-epoxy/no polish

9.4.3.3–Epoxy and Polish Connectors (UV Cured)

This connector uses UV light to cure the epoxy. Usually a more cost-effective solution than heat-cured connectors.

- Epoxy cured by UV lamp – no heat generated
- Lower connector cost
- Lower consumables cost
- Requires polishing (easier polish than heat-cure)
- Faster cure and overall installation time than heat-cured connectors
- Higher yields/less scrap

9.4.3.4–Epoxy and Polish Connectors (Anaerobic)

No ovens or UV lamps are needed for curing. An adhesive is injected into the connector ferrule and it will not harden until mixed with a curing agent. The bare fiber is dipped into a primer and then pushed through the ferrule. This causes the primer and adhesive to mix and curing occurs.

- No oven or lamp needed
- No electrical source required
- Requires polishing
- Faster installation time than heat-cured

9.5 CABLE TRAY SYSTEMS

9.5.1 Support Span

The support span length is an important consideration as it affects the strength of the system and the length of the straight sections required. Tray types typically used for various span lengths include:

- Short span: 6- to 8-foot support spacing (use 12-foot sections)
- Intermediate span: 8- to 12-foot support spacing (use 12-foot sections)
- Long span: 16- to 20-foot support spacing (use 20-foot sections)
- Extra long span: over 20-foot to 30-foot support spacing (use 24- or 30-foot sections).

9.5.2 Working Load

The working load depends on tray size (width, loading depth and strength). Considerations include:

- Types and numbers of cables to support (total cable load in lb. per linear foot [lb./ft.])
- Power cables in a single layer – width is key issue (refer to applicable electrical code)
- Low-voltage cables in a stacked configuration – key issues are loading depth and width (refer to applicable electrical code).

9.5.3 Additional Load Considerations

- 200-lb. concentrated load – industrial installations
- Ice, wind, snow loads – outdoor installations

9.5.4. Installation Environment

Tray material and finish have a significant impact on tray performance in any given environment. Typical tray types used in various environments are shown below.

- Indoor dry (institutional, office, commercial, light industrial): aluminum, pregalvanized steel
- Indoor industrial (automotive, pulp and paper, power plants): aluminum, pregalvanized steel, possibly hot-dipped galvanized after fabrication (HDGAF)
- Outdoor industrial (petrochemical, automotive, power plants): aluminum, hot-dipped galvanized after fabrication (HDGAF)
- Outdoor marine (off shore platforms): aluminum, stainless steel, fiberglass
- Special (petrochemical, pulp and paper, environmental air): contact manufacturer

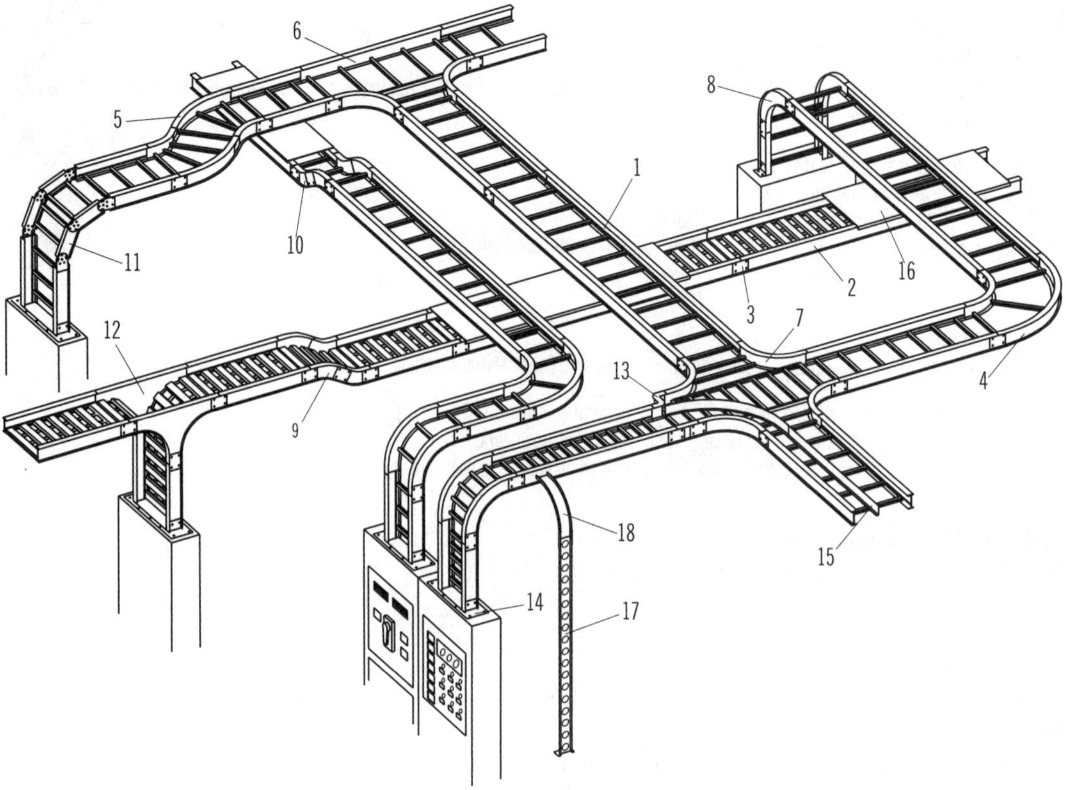

Figure 9.21–Cable Tray System

9.5.5 Nomenclature

The following items are keyed by number to the parts illustrated in Figure 9.21:

1. Ladder-type cable tray
2. Ventilated trough-type cable tray
3. Straight splice plate
4. 90 degrees horizontal bend, ladder-type cable tray
5. 45 degrees horizontal bend, ladder-type cable tray
6. Horizontal tee, ladder-type cable tray
7. Horizontal cross, ladder-type cable tray
8. 90 degrees vertical outside bend, ladder-type cable tray
9. 45 degrees vertical outside bend, ventilated-type cable tray
10. 30 degrees vertical inside bend, ladder-type cable tray
11. Vertical bend segment (VBS)
12. Vertical tee down, ventilated trough-type cable tray
13. Left-hand reducer, ladder type cable tray
14. Frame type box connector
15. Barrier strip straight section
16. Solid flanged tray cover
17. Ventilated channel straight section
18. Channel cable tray, 90 degrees vertical outside bend

9.5.6 Additional Information

Additional information on cable tray systems is contained in NEMA VE-1 "Metal Cable Tray Systems," NEMA VE-2 "Cable Tray Installation Guidelines," Article 392 of the National Electrical Code (NFPA 70) and on the Cable Tray Institute Web site at www.cabletrays.com.

9.6 NEMA PLUG AND RECEPTACLE CONFIGURATIONS

Table 19.1–NEMA Non-Locking Plug Configurations

15 AMP							
2-Pole 2-Wire	2-Pole, 3-Wire, Grounding			3-Pole 3-Wire	3-Pole, 4-Wire, Grounding		4-Wire
1-Phase 125 V	1-Phase 125 V	1-Phase 250 V	1-Phase 277 V	3-Phase 250 V	1-Phase 125/250 V	3-Phase 250 V	3-Phase 120/208 V
1-15	5-15	6-15	7-15	11-15	14-15	15-15	18-15

20 AMP								
2-Pole 2-Wire	2-Pole, 3-Wire, Grounding			3-Pole, 3-Wire		3-Pole, 4-Wire, Grounding		4-Wire
1-Phase 250 V	1-Phase 125 V	1-Phase 250 V	1-Phase 277 V	1-Phase 125/250 V	3-Phase 250 V	1-Phase 125/250 V	3-Phase 250 V	3-Phase 120/208 V
2-20	5-20	6-20	7-20	10-20	11-20	14-20	15-20	18-20

30 AMP								
2-Pole 2-Wire	2-Pole, 3-Wire, Grounding			3-Pole, 3-Wire		3-Pole, 4 Wire, Grounding		4-Wire
1-Phase 250 V	1-Phase 125 V	1-Phase 250 V	1-Phase 277 V	1-Phase 125/250 V	3-Phase 250 V	1-Phase 125/250 V	3-Phase 250 V	3-Phase 120/208 V
2-30	5-30	6-30	7-30	10-30	11-30	14-30	15-30	18-30

Continued on next page >>

Table 19.1–NEMA Non-Locking Plug Configurations (Continued)

50 AMP							
2-Pole, 3-Wire, Grounding			3-Pole, 3-Wire		3-Pole, 4-Wire, Grounding		4-Wire
1-Phase 125 V	1-Phase 250 V	1-Phase 277 V	1-Phase 125/250 V	3-Phase 250 V	1-Phase 125/250 V	3-Phase 250 V	3-Phase 120/208 V
5-50	6-50	7-50	10-50	11-50	14-50	15-50	18-50

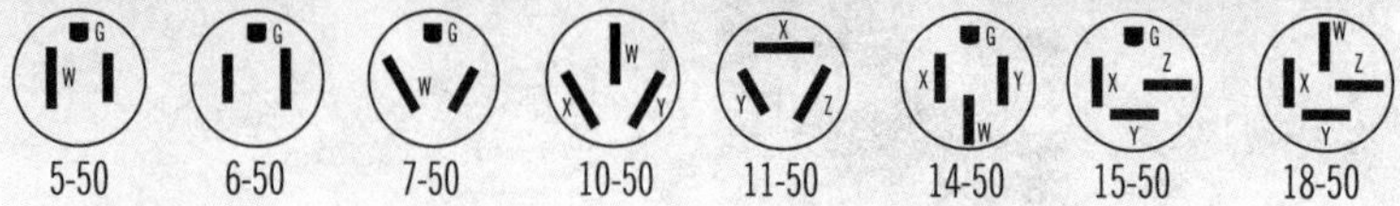

60 AMP		
3-Pole, 4-Wire, Grounding		4-Wire
1-Phase 125/250 V	3-Phase 250 V	3-Phase 120/208 V
14-60	15-60	18-60

Note: Receptacle configurations are a mirror image of the plug configurations shown.

Table 19.2–NEMA Locking Plug Configurations

15 AMP			
2-Pole, 2-Wire	2-Pole, 3-Wire, Grounding		
1-Phase 125 V	1-Phase 125 V	1-Phase 250 V	1-Phase 277 V
L1-15	L5-15	L6-15	L7-15

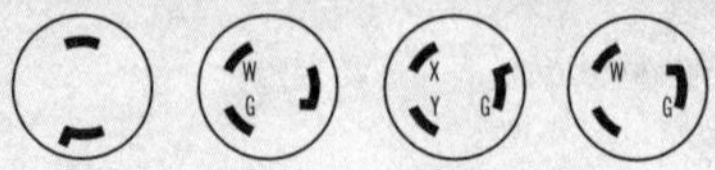

Continued on next page >>

Table 19.2–NEMA Locking Plug Configurations (Continued)

20 AMP								
2-Pole 2-Wire	2-Pole, 3-Wire, Grounding					3-Pole, 3-Wire		
1-Phase 250 V	1-Phase 125 V	1-Phase 250 V	1-Phase 277 V	1-Phase 480 V	1-Phase 600 V	1-Phase 125/250 V	3-Phase 250 V	3-Phase 480 V
L2-20	L5-20	L6-20	L7-20	L8-20	L9-20	L10-20	L11-20	L12-20

20 AMP								
3-Pole, 4-Wire, Grounding			4-Pole, 4-Wire			4-Pole, 5-Wire, Grounding		
1-Phase 125/250 V	3-Phase 250 V	3-Phase 480 V	3-Phase 120/208 V	3-Phase 277/480 V	3-Phase 347/600 V	3-Phase 120/208 V	3-Phase 277/480 V	3-Phase 347/600 V
L14-20	L15-20	L16-20	L18-20	L19-20	L20-20	L21-20	L22-20	L23-20

30 AMP								
2-Pole, 3-Wire, Grounding					3-Pole, 3-Wire			
1-Phase 125 V	1-Phase 250 V	1-Phase 277 V	1-Phase 480 V	1-Phase 600 V	1-Phase 125/250 V	3-Phase 250 V	3-Phase 480 V	3-Phase 600 V
L5-30	L6-30	L7-30	L8-30	L9-30	L10-30	L11-30	L12-30	L13-30

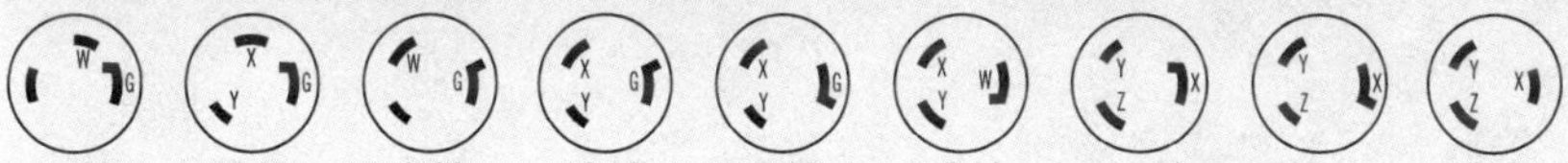

30 AMP									
3-Pole, 4-Wire, Grounding				4-Pole, 4-Wire			4-Pole, 5-Wire, Grounding		
1-Phase 125/250 V	3-Phase 250 V	3-Phase 480 V	3-Phase 600 V	3-Phase 120/208 V	3-Phase 120/208 V	3-Phase 277/480 V	3-Phase 120/208 V	3-Phase 277/480 V	3-Phase 347/600 V
L14-30	L15-30	L16-30	L17-30	L18-30	L19-30	L20-30	L21-30	L22-30	L23-30

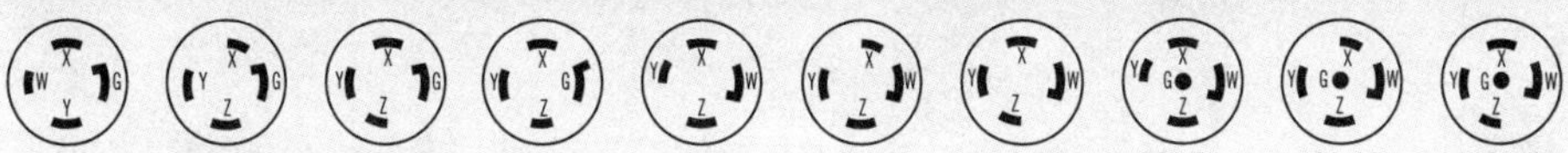

Note: Receptacle configurations are a mirror image of the plug configurations shown.

10. PACKAGING OF WIRE AND CABLE

10.1 REEL SIZE

Selection of proper reel (spool) size depends on the length and overall diameter (O.D.) of the cable or wire to be rewound. A reel not matched to the weight of the cable wound on it may be damaged during shipment.

All wire and cable has a minimum safe bending radius. If cable is subjected to bends sharper than the minimum radius, damage to the material is likely. The minimum drum (hub) diameters given in Section 10.1.2 should be observed.

10.1.1 Reel Terminology

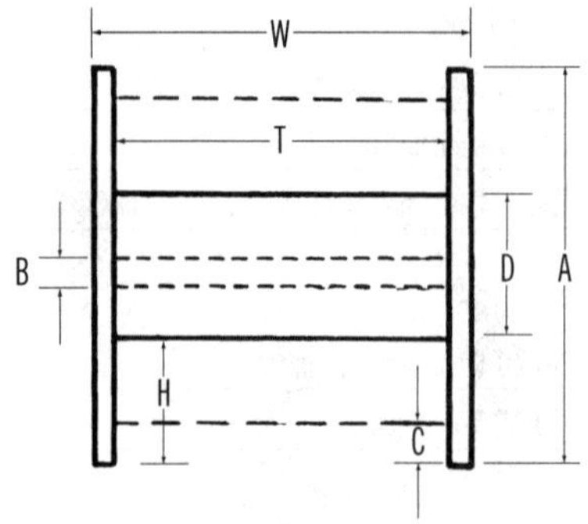

Figure 10.1–Reel Terminology

10.1.2 Minimum Drum Diameter

Table 10.1–Minimum Drum Diameter for Wire and Cable

Type of Cable	Minimum Drum Diameter as a Multiple of Outside Diameter of Cable
A. Single- and multiple-conductor nonmetallic-covered cable	
1. Nonshielded and wire shielded, including cable with concentric wires	
a. 0–2,000 volts	10
b. Over 2,000 volts	
(1) Nonjacketed with concentric wires	12
(2) All others	14
2. Tape shielded	
a. Helically applied	14
b. Longitudinally applied flat tape	20
c. Longitudinally applied corrugated tape	14

Continued on next page >>

Table 10.1 Minimum Drum Diameter for Wire and Cable (Continued)

Type of Cable	Minimum Drum Diameter as a Multiple of Outside Diameter of Cable
B. Single- and multiple-conductor metallic-covered cable	
1. Tubular metallic sheathed	
a. Lead	14
b. Aluminum	
(1) Outside diameter – 1.750" and less	25
(2) Outside diameter – 1.751" and larger	30
2. Wire armored	16
3. Flat tape armored	16
4. Corrugated metallic sheath	14
5. Interlocked armor	14
C. Multiple single conductors cabled together without common covering, including self-supporting cables – the circumscribing overall diameter shall be multiplied by the factor given in item A or B and then by the reduction factor of 0.75.	
D. Combinations – For combinations of the types described in items A, B and C, the highest factor for any component type shall be used.	
E. Single- and multiple-conductor cable in coilable nonmetallic duct	
Outside diameter of duct, inches – 0.0–0.50	26
0.51–1.00	24
1.01–1.25	22
1.26–1.50	21
Over 1.50	19
F. Fiber optic cables	20*
G. Bare conductor	20

Notes to Table:

1. When metallic-sheathed cables are covered only by a thermosetting or thermoplastic jacket, the outside diameter is the diameter over the metallic sheath itself. In all other cases, the outside diameter is the diameter outside of all the material on the cable in the state in which it is to be wound upon the reel.
2. For "flat-twin" cables (where the cable is placed upon the reel with its flat side against the drum), the minor outside diameter shall be multiplied by the appropriate factor to determine the minimum drum diameter.
3. The multiplying factors given for item E refer to the outside diameter of the duct.

* Some manufacturers recommend 30.

Source: NEMA WC 26 (EEMAC 201) Binational Wire and Cable Packaging Standard

10.1.3 Capacities and Dimensions of Shipping Reels

Table 10.2–Capacities of Typical Shipping Reels per NEMA WC 26

Flange Dia. (in.)	[illegible]	[illegible]	[illegible]	[illegible]	[illegible]	[illegible]	[illegible]	[illegible]	[illegible]	[illegible]
Traverse (in.)	10	12	12	12	12	16	18	18	24	24
Drum Dia. (in.)	12	6	8	10	12	10	12	12	16	16
Clearance (in.)	1	1	1	1	1	1	1	2	2	2
Cable O.D. (in.)										
0.10	3,406	12,576	20,436	30,182	26,724	40,243				
0.15	1,514	5,589	9,083	13,414	11,877	17,886	27,877	27,877	36,890	
0.20	852	3,144	5,109	7,546	6,681	10,061	15,681	15,681	20,750	30,182
0.25	545	2,012	3,270	4,829	4,276	6,439	10,036	10,036	13,280	19,317
0.30	378	1,397	2,271	3,354	2,969	4,471	6,969	6,969	9,222	13,414
0.35	278	1,027	1,668	2,464	2,182	3,285	5,120	5,120	6,776	9,855
0.40	213		1,277	1,886	1,670	2,515	3,920	3,920	5,188	7,546
0.45	168		1,009	1,490	1,320	1,987	3,097	3,097	4,099	5,962
0.50	136		817	1,207	1,069	1,610	2,509	2,509	3,320	4,829
0.55	113			998	883	1,330	2,073	2,073	2,744	3,991
0.60	95			838	742	1,118	1,742	1,742	2,306	3,354
0.65	81				633		1,485	1,485	1,965	2,858
0.70	70				545		1,280	1,280	1,694	2,464
0.75	61				475		1,115	1,115	1,476	2,146
0.80									1,297	1,886
0.85									1,149	1,671
0.90									1,025	1,490
0.95									920	1,338
1.00									830	1,207
1.05										
1.10										
1.15										
1.20										
1.25										
1.30										
1.35										
1.40										
1.45										
1.50										
1.55										
1.60										
1.65										
1.70										

The following formula from NEMA WC 26 can be used to calculate approximate cable capacity per reel:

$$\text{Footage} = \frac{0.262 \times T \times (H - C) \times (D + H - C)}{(\text{Wire OD})^2}$$

40	42	48	54	60	66	72	78	84	90	96
30	26	36	32	28	36	36	48	54	54	54
16	24	24	24	28	36	36	42	48	48	56
2	2	2	3	3	3	3	3	3	3	3
32,698	23,651									
22,707	16,424	35,632	40,243	43,445						
16,682	12,067	26,179	29,566	31,919	44,350					
12,773	9,239	20,043	22,637	24,438	33,955	45,097				
10,092	7,300	15,836	17,886	19,309	26,829	35,632				
8,174	5,913	12,828	14,488	15,640	21,731	28,862	43,010			
6,756	4,887	10,601	11,973	12,926	17,960	23,853	35,545	44,198		
5,677	4,106	8,908	10,061	10,861	15,091	20,043	29,868	37,139	46,688	48,771
4,837	3,499	7,590	8,573	9,255	12,859	17,078	25,450	31,645	39,782	41,557
4,171	3,017	6,545	7,392	7,980	11,087	14,725	21,944	27,285	34,302	35,832
3,633	2,628	5,701	6,439	6,951	9,658	12,828	19,116	23,769	29,881	31,214
3,193	2,310	5,011	5,659	6,110	8,489	11,274	16,801	20,890	26,262	27,434
2,829	2,046	4,439	5,013	5,412	7,519	9,987	14,882	18,505	23,263	24,301
2,523	1,825	3,959	4,471	4,827	6,707	8,908	13,275	16,506	20,750	21,676
2,264	1,638	3,553	4,013	4,333	6,020	7,995	11,914	14,814	18,624	19,454
2,044	1,478	3,207	3,622	3,910	5,433	7,215	10,752	13,370	16,808	17,558
	1,341	2,909	3,285	3,547	4,928	6,545	9,753	12,127	15,245	15,925
	1,222	2,650	2,993	3,231	4,490	5,963	8,886	11,049	13,891	14,510
	1,118	2,425	2,739	2,957	4,108	5,456	8,130	10,110	12,709	13,276
	1,027	2,227	2,515	2,715	3,773	5,011	7,467	9,285	11,672	12,193
	946	2,052	2,318	2,502	3,477	4,618	6,882	8,557	10,757	11,237
	875	1,898	2,143	2,314	3,215	4,270	6,362	7,911	9,945	10,389
	811	1,760	1,987	2,145	2,981	3,959	5,900	7,336	9,222	9,634
	754	1,636	1,848	1,995	2,772	3,681	5,486	6,821	8,575	8,958
	703	1,525	1,723	1,860	2,584	3,432	5,114	6,359	7,994	8,351
	657	1,425	1,610	1,738	2,415	3,207	4,779	5,942	7,470	7,803
				1,628	2,261	3,003	4,476	5,565	6,996	7,308
				1,527	2,122	2,819	4,200	5,223	6,566	6,858
				1,436	1,996	2,650	3,949	4,911	6,174	6,449
				1,353	1,880	2,497	3,721	4,626	5,816	6,075

Continued on next page >>

Table 10.2–Capacities of Typical Shipping Reels per NEMA WC 26 (Continued)

Flange Dia. (in.)	16	16	20	24	24	[illegible]	[illegible]	[illegible]		
Traverse (in.)	10	12	12	12	12	16	18	[illegible]	[illegible]	
Drum Dia. (in.)	12	[illegible]	[illegible]							
Clearance (in.)	1	1	1	1	1	1	1	2	2	2
Cable OD (in.)										
1.75										
1.80										
1.85										
1.90										
1.95										
2.00										
2.05										
2.10										
2.15										
2.20										
2.25										
2.30										
2.35										
2.40										
2.45										
2.50										
2.55										
2.60										
2.65										
2.70										
2.75										
2.80										
2.85										
2.90										
2.95										
3.00										
3.05										
3.10										
3.15										
3.20										
3.25										
3.30										
3.35										
3.40										
3.45										
3.50										

The following formula from NEMA WC 26 can be used to calculate approximate cable capacity per reel:

$$\text{Footage} = \frac{0.262 \times T \times (H - C) \times (D + H - C)}{(\text{Wire OD})^2}$$

40	42	48	54	60	66	72	78	84	90	96
30	26	36	32	28	36	36	48	54	54	54
16	24	24	24	28	36	36	42	48	48	56
2	2	2	3	3	3	3	3	3	3	3
				1,277	1,774	2,356	3,511	4,366	5,488	5,733
					1,677	2,227	3,319	4,127	5,188	5,419
					1,587	2,108	3,142	3,906	4,911	5,130
					1,505	1,999	2,979	3,704	4,656	4,864
					1,429	1,898	2,828	3,516	4,420	4,617
					1,358	1,804	2,688	3,342	4,202	4,389
					1,293	1,717	2,559	3,181	3,999	4,178
					1,232	1,636	2,438	3,032	3,811	3,981
					1,175	1,561	2,326	2,892	3,636	3,798
					1,122	1,491	2,222	2,762	3,473	3,628
							2,124	2,641	3,320	3,468
							2,033	2,527	3,177	3,319
							1,947	2,421	3,044	3,179
							1,867	2,321	2,918	3,048
							1,791	2,227	2,800	2,925
							1,720	2,139	2,689	2,809
							1,654	2,056	2,585	2,700
							1,591	1,978	2,486	2,597
								1,904	2,393	2,500
								1,834	2,306	2,408
								1,768	2,223	2,322
								1,705	2,144	2,239
								1,646	2,069	2,162
								1,590	1,999	2,088
								1,536	1,931	2,018
										1,951
										1,887
										1,827
										1,769
										1,715
										1,662
										1,612
										1,565
										1,519
										1,475
										1,433

Table 10.3–Typical Small Reel Dimensions

Reel	Reel dimensions (See Figure 10.2 below)					Approx. Reel Weight	Material
	A Flange Diameter (in.)	B Arbor Hole (in.)	D Drum Diameter (in.)	T Traverse (in.)	W Overall Width (in.)	(lb.)	
1	12	1½	6	6	6¾	2½	Plywood
2	13	1½	5	12¾	12¾	3¾	Plywood
3	16	1½	5	12	12¾	4½	Plywood
4	24	2	12	14	15	11	Plywood
5	30	2	12	14	15	21	Plywood
6	11⅝	2	8	3⅝	4	2	Plastic
7	16	2	12	3¼	3¾	4¼	Fiberboard
8	6½	¾	1$\frac{15}{16}$	1½	1¾	½	Metal
9	6½	¾	1$\frac{15}{16}$	2	2¼	½	Metal
10	6½	¾	1$\frac{15}{16}$	3¼	3½	½	Metal
11	6½	¾	1$\frac{15}{16}$	7½	7¾	¾	Metal
12	10½	1⅝	3½	2¾	3	1¼	Metal
13	10½	1⅝	3½	8	8¼	1¾	Metal

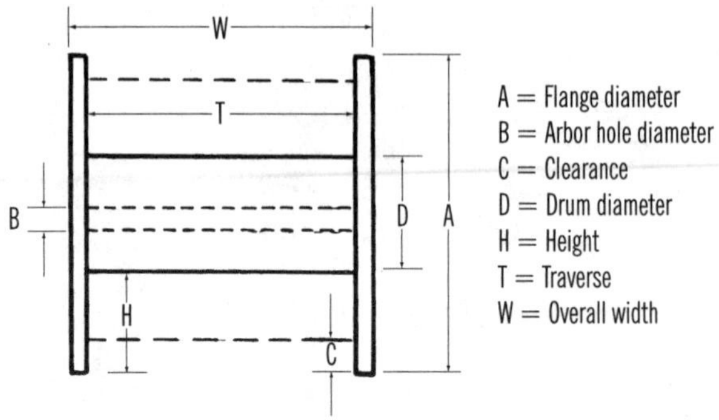

Figure 10.2–Reel Dimensions

10.2 REEL HANDLING

10.2.1 Storage and Shipment

Except for reels less than two feet in diameter and weighing less than 200 pounds, reels should be stored and shipped upright, i.e., resting on both flanges. Do not store or ship reels on their side. Storage or shipment of the reel while lying on its side greatly increases the likelihood of tangling and damage to the cable.

Both cable ends should be sealed against the entrance of moisture. Cables larger than 1/2 inch in diameter should be sealed with tight-fitting heat-shrinkable or hot-dipped (peel coat) end caps designed for the purpose. Smaller diameter cables should be sealed with PVC tape such as 3M Scotch 33 or with end caps (end caps preferred).

CAUTION: Make sure staples are shorter than flange thickness so that they cannot extend through the flange and damage the cable. Caution must also be used to prevent damage to the cable end as it is frequently utilized for hipot, continuity, or other tests. Be sure all staples and nails that might damage the cable are removed.

If reels of cable will be stored for longer than one month, they should be protected from rain and direct exposure to sunlight to maximize service life.

10.2.2 Moving and Lifting

Cradle both reel flanges between forks.

Lower reels from truck using hydraulic gate, hoist or fork lift. Lower carefully.

Upended heavy reels will often arrive damaged. Refuse or receive subject to inspection for hidden damage.

Never allow fork to touch cable surface or reel wrap.

Reels can be hoisted with a shaft extending through both flanges.

Always load with flanges on edge and chock and block securely.

Do not lift by top flange. Cables or reel will be damaged.

Never drop reels.

Figure 10.3–Proper Handling of Cable Reels

11. INDUSTRY STANDARDS

11.2 Fire Safety Tests

11.3 Regulatory and Approval Agencies

11.1 INDUSTRY STANDARDS LIST

11.1.1 AAR
Association of American Railroads
www.aar.org or www.arema.org

Document No.	Title
581.3	Specification for Single Conductor, Clean Stripping Rubber Insulated, 0–600 Volt, Neoprene Jacketed Cable for Locomotive and Car Equipment
S-501	Wiring and Cable Specification
S-502	Specification for Single Conductor, Chlorosulfonated Polyethylene Integral Insulated-Jacketed, 0–300 Volt, 0–600 Volt Cable for Locomotive and Car Equipment
S-503	Specification for Single Conductor, Silicone Rubber Insulation, 0–300 Volt, 0–600 Volt, Glass Polyester Braided, 125°C Cable for High Temperature Use on Locomotive and Car Equipment
S-506	Specification for Single Conductor, Clean Stripping Ethylene Rubber Insulated, 0–600 Volt, Chlorosulfonated Polyethylene Jacketed Cable for Locomotive and Car Equipment
S-4210	ECP Brake System Cable
–	Communications and Signals Manual, Section 10, Wire and Cable

11.1.2 AEIC
Association of Edison Illuminating Companies
www.aeic.org

Document No.	Title
CG1	Guide for Application of AEIC Maximum Insulation Temperatures at the Conductor for Impregnated Paper-Insulated Cables
CG3	Installation of Pipe-Type Cable Systems
CG4	Installation of Extruded Dielectric Insulated Power Cable Systems Rated 69 kV through 138 kV
CG5	Underground Extruded Cable Pulling Guide
CG7	Guide for Replacement and Life Extension of Extruded Dielectric 5 through 35 kV Underground Distribution Cables
CG11	Reduced Diameter Extruded Dielectric Shielded Power Cables Rated 5 through 46 kV
CS1	Impregnated Paper-Insulated, Metallic-Sheathed Cable Solid Type
CS2	Impregnated Paper-Insulated Cable, High-Pressure Pipe Type
CS3	Impregnated Paper-Insulated, Metallic-Sheathed, Low Pressure, Gas-Filled
CS4	Impregnated Paper-Insulated Low and Medium-Pressure Self-Contained, Liquid-Filled Cable
CS7	Cross-Linked Polyethylene Insulated Shielded Power Cables, 69 through 138 kV
CS8	Extruded Dielectric Shielded Power Cables Rated 5 through 46 kV
CS9	Extruded Insulation Power Cables and Accessories Rated Above 46 kV through 345 kV
CS31	Electrically Insulating Low Viscosity Pipe Filling Liquids for High-Pressure Pipe-Type Cables

11.1.3 ANSI
American National Standards Institute
www.ansi.org

Document No.	Title
0337-D	Local Distributed Data Interface (LDDI) Network Layer Protocol
0338-D	Data-Link Layer Protocol for Local Distributed Data Interfaces
0382-D	Fiber Distributed Data Interface (FDDI) Network Layer Protocol
0503-D	Fiber Distributed Data Interface (FDDI) Station Management Standard
0684-D	FDDI—Media Access Control
719	Nonmetallic-Sheathed Cables
X3.129	Intelligent Peripheral Interface (IPI) Enhanced Physical Interface (withdrawn)
X3.148	Fiber Distributed Data Interface (FDDI) Physical Layer (replaced by document #INCITS 148)
X3.184	Fiber Distributed Data Interface (FDDI) Single-Mode Fiber Physical Layer Medium Dependent (replaced by document #INCITS 184)

11.1.4 ASTM
American Society for Testing and Material
www.astm.org

Document No.	Title
B1	Hard-Drawn Copper Wire
B2	Medium-Hard-Drawn Copper Wire
B3	Soft or Annealed Copper Wire
B8	Concentric-Lay-Stranded Copper Conductors, Hard, Medium-Hard, or Soft
B33	Tinned Soft or Annealed Copper Wire
B47	Copper Trolley Wire
B49	Hot-Rolled Copper Rods
B105	Hard-Drawn Copper Alloy Wires for Electrical Conductors
B172	Rope-Lay-Stranded Copper Conductors (Bunch Stranded Members)
B173	Rope-Lay-Stranded Copper Conductors (Concentric Stranded Members)
B174	Bunch-Stranded Copper Conductors
B189	Lead-Alloy-Coated Soft Copper Wire
B193	Resistivity of Electrical Conductor Materials
B227	Hard-Drawn Copper Clad Steel Wire
B228	Concentric-Lay-Stranded Copper-Clad Steel Conductors
B229	Concentric-Lay-Stranded Copper and Copper-Clad Steel Composite Conductors
B230	Aluminum 1350-H19 Wire, for Electrical Purposes
B231	Concentric-Lay-Stranded Aluminum 1350 Conductors
B232	Concentric-Lay-Stranded Aluminum Conductors, Coated-Steel Reinforced (ACSR)
B233	Aluminum 1350 Drawing Stock for Electrical Purposes
B246	Tinned Hard-Drawn and Medium-Hard-Drawn Copper Wire
B258	Standard Nominal Diameters and Cross-Sectional Areas of AWG Sizes of Solid Round Wire Used as Electrical Conductors
B263	Determination of Cross-Sectional Area of Stranded Conductors
B298	Silver-Coated Soft or Annealed Copper Wire
B324	Aluminum Rectangular and Square Wire
B399	Concentric-Lay-Stranded Aluminum Alloy 6201-T81 Conductors
B400	Compact Round Concentric-Lay-Stranded Aluminum 1350 Conductors

Continued on next page >>

11.1.4 ASTM (Continued)

Document No.	Title
B401	Compact-Round Concentric-Lay-Stranded Aluminum Conductors, Steel Reinforced (ASCR/COMP)
B452	Copper-Clad Steel Wire for Electronic Application
B496	Compact Round Concentric-Lay-Stranded Copper Conductors
B500	Metallic Coated Stranded Steel Core for Aluminum Conductors, Steel Reinforced (ACSR)
B549	Concentric-Lay-Stranded Aluminum Conductors, Aluminum Clad Steel Reinforced (ACASR/AW)
B566	Copper-Clad Aluminum Wire
B609	Aluminum 1350 Round Wire, Annealed and Intermediate Tempers
B624	High Strength, High Conductivity, Copper Alloy Wire
B694	Copper, Copper Alloy, Copper-Clad Bronze, Copper-Clad Stainless Steel and Strip for Electrical Cable Shielding
B736	Aluminum; Aluminum Alloy, Aluminum Clad Steel Cable Shielding Stock
B800	8000 Series Aluminum Alloy Wire
B801	Concentric-Lay-Stranded Conductors of 8000 Series Aluminum Alloy
D470	Test Methods for Cross-Linked Insulations and Jackets for Wire and Cable
D866	Styrene-Butadiene (SBR) Synthetic Rubber Jacket for Wire and Cable
D1047	Polyvinyl Chloride Jacket for Wire and Cable
D1351	Polyethylene-Insulated Wire and Cable
D1523	Synthetic Rubber Insulation for Wire and Cable, 90°C Operation
D1929	Test for Ignition Temperature of Plastics
D2219	Polyvinyl Chloride Insulation for Wire and Cable, 60°C Operation
D2220	Polyvinyl Chloride Insulation for Wire and Cable, 75°C Operation
D2308	Polyethylene Jacket for Electrical Insulated Wire and Cable
D2655	Cross-linked Polyethylene Insulation for Wire and Cable Rated 0 to 2,000 V
D2656	Cross-linked Polyethylene Insulation for Wire and Cable Rated 2,001 V to 35 kV
D2671	Test Methods for Heat-Shrinkable Tubing
D2768	General-Purpose Ethylene-Propylene Rubber Jacket for Wire and Cable
D2770	Ozone-Resisting Ethylene Propylene Rubber Integral Insulation and Jacket for Wire
D2802	Ozone Resistant Ethylene-Alkene Polymer Insulation for Wire and Cable
D2863	Test Method for Measuring the Minimum Oxygen Concentration to Support Candle-Like Combustion of Plastics (Oxygen Index)

Continued on next page >>

11.1.4 ASTM (Continued)

Document No.	Title
D3554	Track-Resistant Black Thermoplastic High Density Polyethylene Insulation for Wire and Cable
D3555	Track-Resistant Black Cross-linked Thermosetting Polyethylene Insulation for Wire and Cable
D4244	General Purpose, Heavy-Duty and Extra-Heavy-Duty NBR/PVC Jackets for Wire and Cable
D4245	Ozone-Resistant Thermoplastic Elastomer Insulation for Wire and Cable, 90°C Dry–75°C Wet Operation
D4246	Ozone-Resistant Thermoplastic Elastomer Insulation for Wire and Cable, 90°C Operation
D4314	Specification for General Purpose, Heavy-Duty and Extra-Heavy-Duty Cross-linked Chlorosulfonated Polyethylene Jackets for Wire and Cable
D4565	Test Methods for Physical and Environmental Performance Properties of Insulations and Jackets for Telecommunications Wire and Cable
D4566	Test Methods for Electrical Performance Properties of Insulations and Jackets for Telecommunications Wire and Cable
D5537	Heat Release, Flame Spread and Mass Loss Testing of Insulating Materials Contained in Electrical or Optical Fiber Cables When Burning in a Vertical Cable Tray Configuration
E574	Duplex, Base Metal Thermocouple Wire with Glass Fiber or Silica Fiber Insulation
E662	Specific Optical Density of Smoke Generated by Solid Materials
E1354	Heat and Visible Smoke Release Rates for Materials and Products Using an Oxygen Consumption Calorimeter

11.1.5 Telcordia (formerly Bellcore) www.telcordia.com

Document No.	Title
GR-20	Generic Requirements for Optical Fiber and Optical Fiber Cable
GR-63	Network Equipment-Building System Requirements: Physical Protection
GR-78	Generic Physical Design Requirements for Telecommunication Products and Equipment
GR-110	Generic Requirements for Thermoplastic Insulated Steam Resistant Cable
GR-111	Generic Requirements for Thermoplastic Insulated Riser Cable
GR-115	Inner-City PIC Screened Cable (Filled, AASP Bonded, STALPETH and Bonded PASP)
GR-126	Generic Requirements for Network Outdoor Customer Premises and Universal Cross-Connecting Wire
GR-135	Generic Requirements for Miniature Ribbon Connector and Cable Assembly
GR-136	Generic Requirements for Distributing Frame Wire
GR-137	Generic Requirements for Central Office Cable
GR-139	Generic Requirements for Central Office Coaxial Cable
GR-326	Generic Requirements for Singlemode Optical Connectors and Jumpers
GR-347	Generic Requirements for Telecommunications Power Cable
GR-356	Generic Requirements for Optical Cable Innerduct
GR-409	Generic Requirements for Premises Optical Fiber Cable
GR-421	Generic Requirements for Metallic Telecommunications Cable
GR-492	Generic Requirements for Metallic Telecommunication Wire
GR-1398	Generic Requirements for Coaxial Drop Cable
GR-1399	Generic Requirements for Coaxial Distribution Cable
TR-NWT-000123	Single Pair Buried Distribution Wire
TR-NWT-000124	Multiple Pair Buried Wire
TR-NWT-000131	Generic Requirements for Network Plenum Cable/Wire
TR-NWT-000132	Generic Requirements for Network Shielded Station Wire
TR-NWT-000133	Generic Requirements for Inside Wiring Cable (3 to 125 Pair Sizes)
TR-NWT-000134	Generic Requirements for Two Pair Station Wire
TR-TSY-000103	Pulp Bonded STALPETH Cable
TR-TSY-000104	Pulp Bonded PASP Cable

Continued on next page >>

11.1.5 Telcordia (Continued)

Document No.	Title
TR-TSY-000105	Pulp Bonded Steam Resistance Cable
TA-TSY-000120	Customer Premises or Network Ground Wire
TA-TSY-000121	Generic Requirements for One-Pair Aerial Service Wire
TA-TSY-000122	Generic Requirements for Multiple-Pair Aerial Service Wire
TA-TSY-000125	Rural Aerial Distribution Wire
TA-TSY-000127	Network Aerial Block Wire
TA-TSY-000128	Bridle Wire
TA-TSY-000129	Tree Wire
TA-TSY-000140	Standard Shielded Polyethylene Insulated Twisted Pair Cable
TA-TSY-000141	Terminating Cable
TA-TSY-000142	Central Office Hook-up Wire

11.1.6 CANENA
Council for the Harmonization of Electrical Standards of the Americas
www.canena.org

THC (Technical Harmonization Committee) #20 is responsible for wire and cable products.

11.1.7 ECA
Electronic Components, Assemblies and Materials Association
www.ec-central.org

Document No.	Title
ECA-199-A	Solid and Semisolid Dielectric Transmission Lines
ECA-215	Broadcast Microphone Cables
ECA-230	Color Marking of Thermoplastic Covered Wire
ECA-280-C	Solderless Wrapped Electrical Connections
ECA-297-A	Cable Connectors for Audio Facilities for Radio Broadcasting
ECA-364	Electrical Connector Test Procedures
ECA-403-A	Precision Coaxial Connectors for CATV 75 Ohms
ECA-IS-43	Omnibus Specification–Local Area Network Twisted Pair Data Communication Cable
ECA-IS-43AA	Cable for LAN Twisted Pair Data Communications–Detail Specification for Type 1, Outdoor Cable
ECA-IS-43AB	Cable for LAN Twisted Pair Data Communications–Detail Specification for Type 1, Non-Plenum Cable
ECA-IS-43AC	Cable for LAN Twisted Pair Data Communications–Detail Specification for Type 1, Riser Cable
ECA-IS-43AD	Cable for LAN Twisted Pair Data Communications–Detail Specification for Type 1, Plenum Cable
ECA-IS-43AE	Cable for LAN Twisted Pair Data Communications–Detail Specification for Type 2, Non-Plenum Cable
ECA-IS-43AF	Cable for LAN Twisted Pair Data Communications–Detail Specification for Type 2, Plenum Cable
ECA-IS-43AG	Cable for LAN Twisted Pair Data Communications–Detail Specification for Type 6, Office Cable
ECA-IS-43AH	Cable for LAN Twisted Pair Data Communications–Detail Specification for Type 8, Undercarpet Cable
ECA-IS-43AJ	Cable for LAN Twisted Pair Data Communications–Detail Specification for Type 9, Plenum Cable

11.1.8 EIA
Electronic Industries Alliance
www.eia.org

EIA documents are available from Global Engineering Documents, Inc. www.global.ihs.com

Document No.	Title
EIA-492A000	Specification for Multimode Optical Wave Guide Fibers
EIA-359-A	Standard Colors for Color Identification and Coding (Munsell Color)

11.1.9 FAA
Federal Aviation Administration
www.faa.gov

Document No.	Title
AC 25.869-1	Fire and Smoke Protection
AC 150/534-7	Underground Electrical Cables for Airport Lighting Circuits (L-824-A, B, C)
AC-150/5345-53	Airport Lighting Certification Program
AC-150/5370-10	Airport Construction Standards
FAA-E-2042	Control Cable, Exterior
FAA-E-2793	Power Cable, Exterior, 5 to 25 kV
FAA-701	Rubber-Insulated Cable (0–8,000 Volts)
FAR 14, 25.869 (a)(4)	Fire Retardance of Wire and Cable

11.1.10 ICEA
Insulated Cable Engineers Association
www.icea.net

ICEA documents are available from Global Engineering Documents, Inc. www.global.ihs.com

Document No.	Title
S-56-434	Polyolefin Insulated Communication Cables for Outdoor Use
S-61-402	Thermoplastic-Insulated Wire and Cable for the Transmission and Distribution of Electrical Energy (NEMA WC 5)
S-81-570	600 V Direct Burial Cable Single Electrical Conductors and Assemblies with Ruggedized Extruded Insulation
P-32-382	Short Circuit Characteristics of Insulated Cable
P-43-457	Conductor Resistances and Ampacities at 400 and 800 Hz
P-45-482	Short Circuit Performance of Metallic Shields and Sheaths on Insulated Cable
P-46-426	Power Cable Ampacities (Replaced by IEEE 835)
P-51-432	Copper Conductors, Bare and Weather Resistant
P-53-426	Ampacities, Including Effect of Shield Losses for Single-Conductor Solid-Dielectric Power Cable 15 kV Through 69 kV (NEMA WC 50)
P-54-440	Ampacities of Cables in Open-Top Cable Trays (NEMA WC 51)
P-56-520	Cable Tray Flame Test
S-65-375	Varnished-Cloth-Insulated Wire and Cable for the Transmission and Distribution of Electrical Energy (NEMA WC 4)
S-70-547	Weather-Resistant Polyethylene Covered Conductors

Continued on next page >>

11.1.10 ICEA (Continued)

Document No.	Title
S-73-532	Standard for Control Cables (NEMA WC 57)
S-75-381	Portable and Power Feeder Cables for Use in Mines and Similar Applications (NEMA WC 58)
S-80-576	Category 1 and 2 Unshielded Twisted Pair Communications Cable for Wiring Premises
S-82-552	Instrumentation Cables and Thermocouple Wire (NEMA WC 55)
S-83-596	Optical Fiber Premises Distribution Cable
S-87-640	Fiber Optic Outside Plant Communications Cable
S-94-649	Concentric Neutral Cables Rated 5 through 46 kV
S-97-682	Utility Shielded Power Cables Rated 5 through 46 kV
S-104-696	Indoor-Outdoor Optical Fiber Cable
S-110-717	Optical Fiber Drop Cable
T-22-294	Test Procedures for Extended Time-Testing of Insulation for Service in Wet Locations
T-25-425	Guide for Establishing Stability of Volume Resistivity for Conducting Polymeric Components of Power Cables
T-26-465	Guide for Frequency of Sampling Extruded Dielectric Power, Control, Instrumentation and Portable Cables for Test (NEMA WC 54)
T-27-581	Standard Test Methods for Extruded Dielectric Power, Control, Instrumentation and Portable Cables (NEMA WC 53)
T-29-520	Conducting Vertical Cable Tray Flame Tests (210,000 BTU/Hour)
T-30-520	Conducting Vertical Cable Tray Flame Tests (70,000 BTU/Hour)
T-31-610	Water Penetration Resistance Test, Sealed Conductor
T-33-655	Low-Smoke, Halogen-Free (LSHF) Polymeric Cable Jackets

11.1.11 IEC
International Electrotechnical Commission
webstore.iec.ch

Document No.	Title
24764	Information Technology—Generic Cabling for Data Center Premises (not yet published)
60050-461	International Electrotechnical Vocabulary. Chapter 461. Electric cables.
60055-2	Paper-insulated metal-sheathed cables for rated voltages up to 18/30 kV (with copper or aluminum conductors and excluding gas pressure and oil-filled cables)
60079	Electrical apparatus for explosive atmospheres (hazardous locations)
60092	Electrical installations in ships
60096	Radio-frequency cables
60141	Tests on oil-filled and gas-pressure cables and their accessories
60169	Radio-frequency connectors
60173	Colors of the cores of flexible cables and cords
60183	Guide to selection of high-voltage cables
60189	Low-frequency cables and wires with PVC insulation and PVC sheath
60204	Safety of machinery—electrical equipment of machines (industrial)
60227	PVC insulated cables of rated voltages up to and including 450/750 V
60228	Conductors of insulated cables
60229	Tests on cable oversheaths which have a special protective function and are applied by extrusion
60230	Impulse tests on cables and their accessories
60245	Rubber insulated cables of rated voltages up to and including 450/750 V
60287	Calculations of the continuous current rating of cables (100% load factor)
60304	Standard colors for insulation for low-frequency cables and wires
60331	Tests for electric cables under fire conditions—circuit integrity
60332	Tests on electric and optical fiber cables under fire conditions
60339	General purpose rigid coaxial transmission lines and their associated flange connectors
TR60344	Calculation of DC resistance of plain and coated copper conductors of low-frequency cables and wires
60364-1	Low-voltage electrical installations—Part 1: Fundamental principles, assessment of global characteristics, definitions
60446	Identification of conductors by colors or alphanumeric
60457	Rigid precision coaxial lines and their associated precision connectors

Continued on next page >>

11.1.11 IEC (Continued)

Document No.	Title
60502	Extruded solid dielectric insulated power cables for rated voltages from 1 kV to 30 kV
60541	Comparative information on IEC and North American flexible cord types
TR60649	Calculation of maximum external diameter of cables for indoor installations
60695	Fire hazard testing
60702	Mineral insulated cables with a rated voltage not exceeding 750 V
60708	Low-frequency cables with polyolefin insulation and moisture barrier polyolefin sheath
60719	Calculation of the lower and upper limits for the average outer dimensions of cables with circular copper conductors and of rated voltages up to and including 450/750 V
60724	Short-circuit temperature limits of electric cables with rated voltages 1 kV and 3 kV
60728	Cable networks for sound and television signals
60754	Tests on gases evolved during combustion of electric cables
60757	Code for designation of colors
60800	Heating cables with a rated voltage of 300/500 V for comfort heating and prevention of ice formation
60811	Common tests methods for insulating and sheathing materials of electric and optical cables
60834	Performance and testing of teleprotection equipment of power systems
60840	Tests for power cables with extruded insulation for rated voltages above 30 kV up to 150 kV
60851	Winding wires–test methods
60885	Electrical test methods for electric cables (including partial discharge)
60949	Calculation of thermally permissible short-circuit currents
60966	Radio-frequency and coaxial cable assemblies
61034	Measurement of smoke density of cables burning under defined conditions (3 meter cube smoke apparatus)
61158	Fieldbus for use in industrial control systems
61883	Consumer audio/video equipment–Digital interface
61936-1	Power installations exceeding 1 kV AC Part 1: Common rules

11.1.12 IEEE
Institute of Electrical and Electronic Engineers, Inc.
www.ieee.org

Document No.	Title
45	Recommended Practice for Electrical Installations on Shipboard
48	Test Procedures and Requirements for AC Cable Terminations 2.5 kV through 765 kV
82	Test Procedure for Impulse Voltage Tests on Insulated Conductors
101	Guide for the Statistical Analysis of Thermal Life Test Data
120	Master Test Guide for Electrical Measurements in Power Circuits
323	Qualifying Class 1E Equipment for Nuclear Power Generating Stations
383	Qualifying Class 1E Electric Cables, Field Splices and Connections for Nuclear Power Generating Stations
400	Guide for Field Testing and Evaluation of the Insulation of Shielded Power Cable Systems
404	Standard for Extruded and Laminated Dielectric Shielded Cable Joints Rated 2.5 kV through 500 kV
422	Guide for Design and Installation of Cable Systems in Power Generating Stations (withdrawn)
510	Recommended Practices for Safety in High Voltage and High Power Testing
524	Guide to the Installation of Overhead Transmission Line Conductors
525	Guide for the Design and Installation of Cable Systems in Substations
532	Guide for Selecting and Testing Jackets for Underground Cables
539	Definitions of Terms Relating to Corona and Field Effects of Overhead Powerlines
575	Guide for the Application of Sheath-Bonding Methods for Single Conductor Cables and the Calculation of Induced Voltages and Currents in Cable Sheaths
576	Recommended Practice for Installation, Termination, and Testing of Insulated Power Cable as Used in Industrial and Commercial Applications
634	Standard Cable Penetration Fire Stop Qualification Test
635	Guide for Selection and Design of Aluminum Sheaths for Power Cables
690	Standard for the Design and Installation of Cable Systems for Class 1E Circuits in Nuclear Power Generating Stations
738	Standard for Calculating the Current-Temperature of Bare Overhead Conductors
802	Local and Metropolitan Area Networks: Overview and Architecture
816	Guide for Determining the Smoke Generation of Solid Materials Used for Insulations and Coverings of Electrical Wire and Cable (withdrawn)
835	Power Cable Ampacity Tables
848	Standard Procedure for the Determination of the Ampacity Derating of Fire-Protected Cables

Continued on next page >>

11.1.12 IEEE (Continued)

Document No.	Title
930	Statistical Analysis of Electrical Insulation Breakdown Data
1017	Field Testing Electric Submersible Pump Cable
1018	Specifying Electric Submersible Cable-Ethylene-Propylene Rubber Insulation
1019	Specifying Electric Submersible Pump Cable-Polypropylene Insulation
1143	Guide on Shielding Practice for Low Voltage Cables
1185	Guide for Installation Methods for Generating Station Cables
1202	Standard for Propagation Flame Testing of Wire and Cable
1394	Standard for a High Performance Serial Bus
1407	Guide for Accelerated Aging Test for MV Power Cables Using Water-Filled Tanks
C2	National Electrical Safety Code (NESC)
C62.41	Surge Voltages in Low-Voltage (1,000 V and less) AC Power Circuits
C62.92	Neutral Grounding in Electrical Utility Systems

11.1.13 ISA
Instrumentation, Systems and Automation Society
www.isa.org

Document No.	Title
RP 12.06.01	Wiring Methods for Hazardous (Classified) Locations Instrumentation Part 1: Intrinsic Safety
50.02	Fieldbus Standard for Use in Industrial Control Systems

11.1.14 ISO
International Organization for Standardization
www.iso.org

Document No.	Title
4589	Oxygen Index Test
5657	Radiant Cone Flame Test
TR9122	Toxicity Testing of Fire Effluents

11.1.15 ITU-T
International Telecommunication Union/Telecommunications Sector
www.itu.int

Document No.	Title
Blue Book, Facicle III.3	Transmission Media Characteristics

11.1.16 MSHA
Mine Safety and Health Administration
www.msha.gov

Document No.	Title
30 CFR 7.407	Flame Tests

11.1.17 NEMA
National Electrical Manufacturers Association
www.nema.org

Document No.	Title
HP 3	Electrical and Electronic PTFE Insulated High Temperature Hook-up Wire; Types ET (250 Volts), E (600 Volts) and EE (1,000 Volts)
HP 4	Electrical and Electronic FEP Insulated High Temperature Hook-up Wire: Types KT (250 Volts), K (600 Volts) and KK (1,000 Volts)
HP 100	High Temperature Instrumentation and Control Cable
HP 100.1	High Temperature Instrumentation and Control Cables Insulated and Jacketed with FEP Fluorocarbons
HP 100.2	High Temperature Instrumentation and Control Cables Insulated and Jacketed with ETFE Fluoropolymers
HP 100.3	High Temperature Instrumentation and Control Cables Insulated and Jacketed with Cross-Linked (Thermoset) Polyolefin (XLPO)
HP 100.4	High Temperature Instrumentation and Control Cables Insulated and Jacketed with ECTFE Fluoropolymers
WC 26	Wire and Cable Packaging
WC 50	Ampacities, Including Effect of Shield Losses for Single-Conductor Solid-Dielectric Power Cable 15 kV through 69 kV (ICEA P-53-426)
WC 51	Ampacities of Cables Installed in Cable Trays
WC 52	High-Temperature and Electronic Insulated Wire Impulse Dielectric Testing
WC 53	Standard Test Methods for Extruded Dielectric Power, Control, Instrumentation and Portable Cables (ICEA T-27-581)
WC 54	Guide for Frequency of Sampling Extruded Dielectric Power, Control, Instrumentation and Portable Cables for Test
WC 56	3.0 kHz Insulation Continuity Proof Testing of Hookup Wire
WC 57	Standard for Control Cables (ICEA S-73-532)
WC 58	Portable and Power Feeder Cables for Use in Mines and Similar Applications (ICEA S-75-381)
WC 61	Transfer Impedance Testing
WC 62	Repeated Spark/Impulse Dielectric Testing
WC 63.1	Twisted Pair Premise Voice and Data Communications Cables
WC 63.2	Coaxial Premise Data Communication Cable
WC 66	Category 6 and Category 7 100 Ohm Shielded and Unshielded Twisted Pair Cables
WC 67	Uninsulated Conductors
WC 70	Nonshielded Power Cables Rated 2,000 Volts or Less
WC 71	Nonshielded Cables Rated 2,001–5,000 Volts
WC 27500	Aerospace and Industrial Electrical Cable

11.1.18 NFPA National Fire Protection Association www.nfpa.org

Document No.	Title
70	NEC® (National Electrical Code)
75	Protection of Electronic Computer/Data Processing Equipment
79	Electrical Standard for Industrial Machinery
99	Health Care Facilities Handbook
258	Recommended Practice for Determining Smoke Generation of Solid Materials
262	Test for Flame Travel and Smoke of Wires and Cables for Use in Air-Handling Spaces

11.1.19 RUS Rural Utilities Service (formerly REA) www.usda.gov/rus/telecom

Document No.	Title
1753F-150	Construction of Direct Buried Plant
1753F-151	Construction of Underground Plant
1753F-152	Construction of Aerial Plant
1753F-204	Aerial Service Wires (PE-7)
1753F-205	Filled Telephone Cable (PE-39)
1753F-206	Filled Buried Wires (PE-86)
1753F-208	Filled Telephone Cable with Expanded Insulation (PE-89)
1753F-601	Filled Fiber Optic Cables (PE-90)

11.1.20 SAE International (formerly Society of Automotive Engineers)
www.sae.org

Document No.	Title
J156	Fusible Links
J378	Marine Engine Wiring
J1127	Low Voltage Battery Cable
J1128	Low Voltage Primary Cable
J1292	Automobile Truck, Truck-Tractor, Trailer, and Motor Coach Wiring
J1654	High Voltage Primary Cable
J1678	Low Voltage Ultra Thin Wall Primary Cable
J1939	Serial Control and Communications for Vehicle Network
J2394	Seven-Conductor Cable for ABS Power
AS22759	Fluoropolymer Insulated Electrical Wire
AS50861	PVC Insulated, Copper or Copper Alloy Wire
AS50881	Wiring Aerospace Vehicle
AS81044	Wire, Electric, Cross-linked Polyalkene, Cross-linked Alkane-Imide Polymer, or Polyarylene Insulated, Copper or Copper Alloy

11.1.21 TIA
Telecommunication Industries Association
www.tiaonline.org

Document No.	Title
TIA-225	Rigid Coaxial Transmission Lines 50 Ohms
TIA-232-F	Interface Between Data Terminal Equipment and Data Communication Equipment Employing Serial Binary Data Interchange
TIA-259	Rigid Coaxial Transmission Lines and Connectors, 75 Ohms
TIA-422-B	Electrical Characteristics of Balanced Voltage Digital Interface Circuits
TIA/EIA-423-B	Electrical Characteristics of Unbalanced Voltage Digital Interface Circuits
TIA-440-B	Fiber Optic Terminology
TIA-485-A	Generators and Receivers for Balanced Digital Multipoint Systems
TIA/EIA-455-B	Standard Test Procedures for Fiber Optic Fibers, Cables, Transducers, Connecting and Termination
TIA-492AAAA-A	Detail Specification for 62.5 μm Core Diameter/125 μm Cladding Diameter Class 1a Multimode, Graded-Index Optical Waveguide Fibers
TIA/EIA-568-B.1	Commercial Building Telecommunications Cabling Standard
TIA-569-B	Commercial Building Standard for Telecommunications Pathways and Spaces
TIA-570-B	Residential Telecommunications Infrastructure Standard
TIA/EIA-606-A	Administration Standard for the Telecommunications Infrastructure
TIA-942	Telecommunications Infrastructure Standard for Data Centers
TIA-472C000	Sectional Specification for Fiber Optic Premises Distribution Cables
TIA-472D000	Sectional Specification for Fiber Optic Cables for Outside Plant Use
TIA-475C000	Specification for Fiber Optic Type FSMA Connectors
TIA-515B000	Specification for Optical Fiber Cable Splice Closures

11.1.22 UL
Underwriters Laboratories, Inc.
www.ul.com

Document No.	Title
4	Armored Cable (Type AC)
13	Power-Limited Circuit Cable (Types CL3P, CL2P, CL3R, CL2R, CL3, CL3X, PLTC)
44	Thermoset-Insulated Wires & Cables (Types XHHW, XHHW-2, RHH, RHW, RHW-2, RH, SA, SIS)
62	Flexible Cord & Fixture Wire (Types SO, SOW, SOW-A, SJ, SJO, SPT-1, TFN, TFFN, etc.)
66	Fixture Wire
83	Thermoplastic-Insulated Wires and Cables (Types TW, THW, THW-2, THWN, THWN-2, THHN, THHW, TA, TBS, TFE, FEP, FEPB)
183	Manufactured Wiring Systems
444	Communication Cables (Types CMX, CM, CMR, CMP)
486A-486B	Wire Connectors
486C	Splicing Wire Connectors
486D	Sealed Wire Connector Systems
486E	Equipment Wiring Terminals for Use with Aluminum and/or Copper Conductors
493	Thermoplastic Insulated Underground Feeder & Branch Circuit Cables (Types UF, UF-B)
497	Protectors for Communication Circuits
719	Nonmetallic-Sheath Cables (Types NM-B, NMC-B)
723	Tests for Surface Burning Characteristics of Building Materials
758	Appliance Wiring Material (Type AWM)
814	Gas-Tube-Sign Cable
817	Cord Sets and Power-Supply Cords
854	Service Entrance Cables (Types USE, SE, SE-U, SE-R, USE-2)
1063	Machine Tool Wires and Cables (Type MTW)
1072	Medium Voltage Power Cable (Type MV)
1276	Welding Cables
1277	Power and Control Tray Cables (Type TC)
1309	Marine Shipboard Cable
1424	Cables for Power-Limited Fire-Alarm Circuits
1425	Cables for Non-Power-Limited Fire-Alarm Circuits

Continued on next page >>

11.1.22 UL (Continued)

Document No.	Title
1426	Electric Cables for Boats
1459	Telephone Equipment
1565	Positioning Devices
1581	Reference Standard for Electrical Wires, Cables, and Flexible Cords
1604	Electrical Equipment for Use in Class I and II, Division 2, and Class III Hazardous (Classified) Locations
1650	Portable Power Cable (Types G, G-GC, W, PPE)
1651	Optical Fiber Cable
1666	Standard Test for Flame Propagation Height of Electrical and Optical Fiber Cables Installed Vertically in Shafts
1680	Stage Lighting Cables
1685	Vertical-Tray Fire-Propagation and Smoke-Release Test for Electrical and Optical-Fiber Cables
1690	Data-Processing Cable (Type DP)
1740	Industrial Robots
1863	Communication Circuit Accessories
2196	Fire Resistive Cables ("CI" Rated)
2225	Cables for Use in Hazardous Locations (Type MC-HL)
2250	Instrumentation Tray Cable (Type ITC)
2556	Wire and Cable Test Methods (Trinational)
2261	Cables for Network-Powered Broadband Systems
2273	Festoon Cable
2276	Recreational Vehicle Cable
2424	Limited-Combustible Cable
60079	Electrical Apparatus for Explosive Gas Atmospheres

11.1.23 U.S. Government Specifications http://assist.daps.dla.mil/quicksearch

Document No.	Title
AA-59544	Wire & Cable Electrical Power
A-A-59551	Electrical Copper Wire, Uninsulated
J-C-145C	Weather-Resistant Power & Cable
J-C-580B	Flexible Cord and Fixture Wire (Replaced by UL 62)
MIL-DTL-17H	General Specifications for Cables, Radio Frequency, Flexible and Semirigid
MIL-DTL-915G	General Specification for Cable and Cord, Electrical, for Shipboard Use
MIL-DTL-3432H	Cable (Power & Special Purpose) and Wire, Electrical (300–600 Volts)
MIL-DTL-8777D	Wire, Electrical, Silicone-Insulated, Copper, 600 V, 200°C
MIL-DTL-13777H	General Specifications for Cable, Special Purpose, Electrical
MIL-DTL-16878G	General Specifications for Wire, Electrical, Insulated
MIL-DTL-23806B	General Specification for Cable, Radio Frequency, Coaxial, Semirigid, Foam Dielectric
MIL-DTL-24640G	General Specification for Cable, Electrical, Lightweight for Shipboard Use
MIL-DTL-24643B	General Specification for Cable and Cord, Electrical, Low Smoke, for Shipboard Use
MIL-DTL-25038H	Wire, Electrical, High Temperature, Fire Resistant and Flight Critical
MIL-DTL-27072F	Special Purpose, Electrical, Multiconductor and Single Shielded Power Cable
MIL-DTL-28830D	General Specification for Cable, Radio Frequency, Coaxial, Semirigid, Corrugated Outer Conductor
MIL-DTL-38359C	Cable, Power, Electrical, Airport Lighting, Cross-Linked, Polyethylene XLP
MIL-DTL-49055D	General Specifications for Cables, Power, Electrical (Flexible, Flat, Unshielded), Round Conductor
MIL-DTL-55021C	General Specification for Cables Twisted Pairs and Triples, Internal Hookup
MIL-DTL-81381C	Wire-Electric, Polymide-Insulated Copper or Copper Alloy
MIL-C-83522D	General Specification for Connectors, Fiber Optic, Single Terminus
MIL-DTL-83526C	General Specifications for Connectors, Fiber Optic, Circular, Environmental Resistant, Hermaphroditic
MIL-HDBK-299	Cable Comparison Handbook Data Pertaining to Electric Shipboard Cable
MIL-S-81824	Splices, Electric, Crimp Style, Copper, Insulated, Environment Resistant
MIL-W-76D	Wire and Cable, Hookup, Electrical, Insulated
MIL-W-5846C	Wire, Electrical, Chromel and Alumel Thermocouple
QPL-AS5756-I	Cable and Wire, Power, Electric, Portable

11.2 FIRE SAFETY TESTS

11.2.1 Fire Safety Test Methods

Table 11.1–Fire Safety Test Methods

Some common fire safety test methods used in the wire and cable industry are listed below:

Fire Hazard	North America	Worldwide
Ignitability	ASTM D2863	IEC 60332
Propagation	UL 1685 and IEEE 1202	IEC 60332
Smoke	UL 1685 and ASTM E662	IEC 61034-2
Toxicity	University of Pittsburgh	ISO TR 9122
Corrosivity	IEC 60754	IEC 60754-2
Halogen Content	MIL-DTL-24643	IEC 60754-1

11.2.2 NEC Fire Test Summary

Table 11.2–NEC Fire Test Summary

National Electrical Code Article	Plenum (NPFA 262)	Riser (UL 1666)	General Use (UL 1581 Vertical Tray)	Limited Use (UL 1581 Vertical Wire)
645 Under Raised Floor of IT Room	All types shown below	All types shown below	DP, MC, AC	–
725, Class 2 Power-Limited	CL2P	CL2R	CL2	CL2X
725, Class 3 Power-Limited	CL3P	CL3R	CL3	CL3X
725 Power-Limited Tray Cable	No listing	No listing	PLTC	No listing
727 Instrumentation Tray Cable	No listing	No listing	ITC	No listing
760 Fire Protective Power-Limited	FPLP	FPLR	FPL	No listing
760 Fire Protective Non-Power-Limited	NPLFP	NPLFR	NPLF	No listing
770 Optical Fiber Nonconductive	OFNP	OFNR	OFN or OFNG	No listing
770 Optical Fiber Conductive	OFCP	OFCR	OFC or OFCG	No listing
800 Communication	CMP	CMR	CM or CMG	CMX
800 Undercarpet Communication	No listing	No listing	No listing	CMUC
820 Cable TV	CATVP	CATVR	CATV	CATVX

Cable Application	Common Names	Flame Energy
Plenum Space	NFPA 262, Steiner Tunnel, CSA FT6	300,000 Btu/hr
Riser Shaft	UL 1666, Riser Test	527,000 Btu/hr
General Use	UL 1581 Vertical Tray, IEEE 1202, CSA FT4, UL 1685	70,000 Btu/hr
Limited Use	UL 1581 Vertical Wire, VW-1, CSA FT1	3,000 Btu/hr

11.2.3 Comparison of Vertical Cable Tray Tests

Table 11.3–Comparison of Vertical Cable Tray Tests

	IEEE 383 UL 1581	ICEA T-29-520	CSA FT4	IEEE 1202	UL 1685 /UL[a]	UL 1685 /IEEE[b]	IEC 60332-3
Burner power (kW)	21	62	20	20	21	21	20
Time of flame (min)	20	20	20	20	20	20	20, 40[g]
Alternate source	Yes, oily rag[d]	No	No	No	No	No	No
Burner placement[c]	600 mm[e] 75 mm in back	300 mm 200 mm in back	300 mm 75 mm in front	300 mm 75 mm in front	457 mm 75 mm in back	457 mm 75 mm in front	600 mm 75 mm in front
Angle of burner	Horiz.	Horiz.	20° up	20° up	Horiz.	20° up	Horiz.
Tray length (m)	2.4	2.4	3.0	2.4	2.4	2.4	3.5
Tray width (m)	0.3	0.3	0.3	0.3	0.3	0.3	0.5
Sample length (m)	2.4	2.4	2.3	2.3	2.4	2.4	3.5
Width of tray used for cables (m)	0.15 Front only	0.15	0.25 Front only	Full Front only	0.15 Front only	Full Front only	0.30 front or front + back
Thin-size cables to be bundled	No	No	if D< 13 mm	if D< 13 mm	No	if D< 13 mm	Mounted flush, with no spaces
Test enclosure specified	No	No	Yes	Yes	Yes	Yes	Yes
Required air flow rate	N/A	N/A	> 0.17 m^3/s	0.65 m^3/s	5 m^3/s	5m^3/s	[h]
Test runs needed	3	2	2	2 × 2[f]	1	1	1
Max. char length (m, from bottom)	2.4	2.4	1.786	1.786	2.4	1.786	3.1
Peak smoke release rate ($m^2 s^{-1}$)	N/A	N/A	N/A	N/A	0.25	0.40	N/A
Total smoke released (m^2)	N/A	N/A	N/A	N/A	95	150	N/A

[a]=Version with UL 1581 flame exposure; [b]=Version with CSA FT-4/IEEE 1202 flame exposure; [c]=Height above bottom of tray and distance from specimen surface, respectively; [d]=Not applicable in the UL 1581 version; [e]=This dimension is 457 mm in the UL 1581 version; [f]=Two each on two different sizes of specimens; [g]=Time is 20 minutes for Category C, 40 minutes for Categories A and B; [h]=Not yet specified; [i]=Depends on amount of cable loading

Source: NIST Technical Note 1291

11.2.4 NFPA 262 Steiner Tunnel Test for Plenum Rated Cable

The NFPA 262 "Steiner Tunnel" Flame Test (formerly UL 910) measures flame spread and smoke generation in a simulated air handling plenum. A 25-foot long Steiner Tunnel is used for the test, with intake and exhaust ducts and a means of regulating flow velocity of air through the tunnel. Windows at 1-foot intervals allow for flame spread measurements, and an optical device in the exhaust of the chamber measures smoke density. The cable samples are mounted in a cable tray in one layer in the tunnel and the tunnel is sealed. Two circular burners are mounted vertically at the intake end of the tunnel just in front of the cable tray. Methane is burned, along with a 240 ft./min forced draft through the tunnel for twenty minutes, and the flame is extinguished. Flame spread and smoke density are monitored throughout the test. A cable is listed for plenum use if flame spread is less than 5 feet from the end of the ignition flame, and optical density is less than 0.5 maximum peak, and 0.15 maximum average. The output of the burner is 300,000 Btu/hr and the energy consumed for the test is 100,000 Btus. The Canadian version of this test is known as the CSA FT6 fire test.

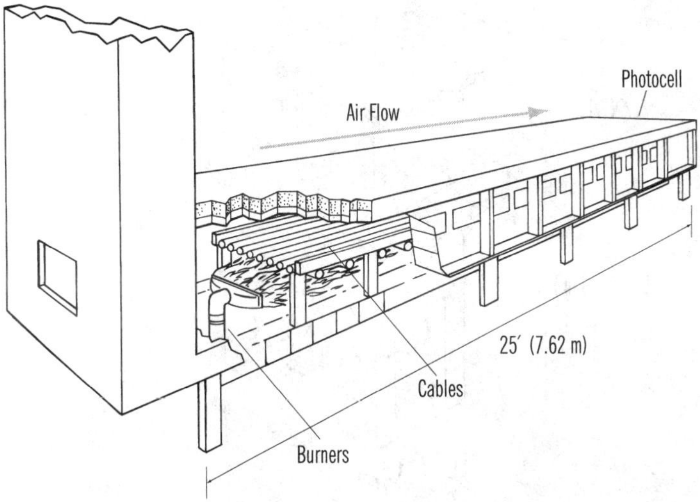

Figure 11.1–NFPA Steiner Tunnel Flame Test

11.2.5 UL 1666 Riser Flame Test

The Riser Flame Test, as described in Underwriters Laboratories Standard 1666, was developed to test cable flammability in riser applications. This test simulates a fire in a nonflame stopped riser within a high-rise building. The chamber for the test is a three story block construction design. Steel fire doors provide access to the second and third levels for installing cables, and 1-foot × 2-foot rectangular holes in both the second and third level floors allow cable to be installed in racks extending between the first and third levels. A burner is made up of a 1/4 in. gas pipe with 90 degree elbow mounted below a 1-foot square drilled steel plate. The burner is mounted on the edge of the riser hole on the floor of the second level. A mixture of air and propane is burned for thirty minutes and then shut off, extinguishing the burner flame. A cable may be listed as riser cable if the flame does not propagate up to the floor of the third level. The energy output of the burner is 527,500 Btu/hr, or a consumed test energy of 263,750 Btus.

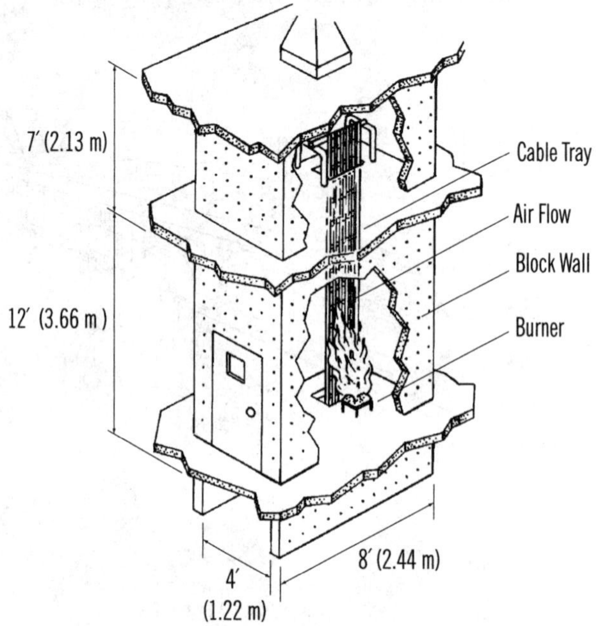

Figure 11.2–UL 1666 Riser Flame Test

11.2.6 UL 1581 Vertical Tray Flame Test

The Vertical Tray Flame Test is used as a good approximation of flame spread in cables run in groups. A steel ladder type tray 12 inches wide × 3 inches deep and 8 feet long with 1-inch × 1/2-inch rungs spaced 9 inches apart is mounted vertically on the floor of the test chamber. The center 6 inches of the tray is filled with cable samples in one layer spaced 1/2 cable diameter apart. A 6 to 1 mixture of air to propane is burned using a 10-inch wide ribbon burner. The burner is placed horizontally 3 inches from the back of the tray, 2 feet from the floor and midway between two rungs. The flame is applied for twenty minutes and then removed. A cable passes the vertical tray test if it does not propagate flame to the top of the tray (6 ft.). A cable may continue to burn after the burner is shut off; however, the test is not complete until the cable stops burning. The energy output of the burner is 70,000 Btu/hr and the cable is subjected to 23,333 Btus for the test. This test method is also contained in IEEE 383 and in UL 1685. See IEEE 1202 or CSA FT4 for a similar but slightly harsher test method.

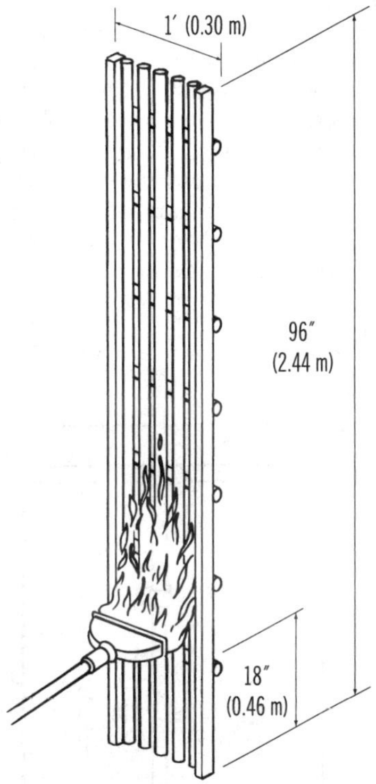

Figure 11.3–UL 1581 Vertical Tray Flame Test

11.2.7 ICEA T-29-520

A variation on the UL 1581 (UL 1685) Vertical Tray Test is the 210,000 Btu flame test specified in ICEA Standard T-29-520. In the 210,000 Btu test, the setup is essentially the same as with the 70,000 Btu test except the gas flow is increased to generate 210,000 Btu/hr instead of 70,000 Btu/hr of flame energy and the burner-to-cable spacing is increased to 200 millimeters. See Section 11.2.3 for more details. This test method appears to be losing favor in the industry. IEEE 1202 or CSA FT4 are often used instead.

11.2.8 CSA FT4

A flame test very similar to the IEEE 1202 test is CSA's FT4 test. This test is more stringent than the IEEE 383 test. This test method is also one of two methods contained in UL 1685.

11.2.9 IEEE 1202

The IEEE 1202 flame test is virtually identical in test severity to the Canadian FT4 test. This test method is also one of two contained in UL 1685.

11.2.10 UL 1685

UL Standard 1685 contains both the UL 1581 (IEEE 383) and FT4 (IEEE 1202) fire test methods. An optional smoke emission test method is also described. A cable passing this smoke test can be given an "LS" (Limited Smoke) rating.

11.2.11 UL 1581 VW-1 (Vertical-Specimen) Flame Test

The VW-1 Flame Test was the first flame test developed for studying flame spread on wire and cable. The test measures relative flame propagation of a single wire or cable. The test procedure is detailed in Underwriters Laboratories Standard 1581, but a general overview of the test is as follows. The fixture used is a bench-mountable 12-inch wide, 14-inch deep and 24-inch high steel box open at the front and top. Clamps hold a single specimen vertically in the center of the box. A Tirrill burner (similar to a Bunsen burner) is mounted on a 20 degree angle block and has a flame 4- to 5-inches high with a 1/2-inch inner blue cone. The burner is placed so the inner cone meets the test sample surface. Ten inches above this point a kraft paper "flag" is placed on the sample facing away from the burner, and cotton batting covers the floor of the chamber to a height 9 inches below the point. The flame is applied to the sample for 15 seconds five times (total 75 seconds) with a minimum 15 seconds between flame applications or until burning ceases, whichever is longer. A sample "Passes VW-1" if less than 25 percent of the flag is burned away, the cable doesn't burn longer than 60 seconds after any flame application, and the cotton batting is not ignited by dripping particles. The energy output of the burner is less than 3,000 Btu/hr and the test energy is less than 65 Btus. The VW-1 test is very similar to CSA's FT1 flame test.

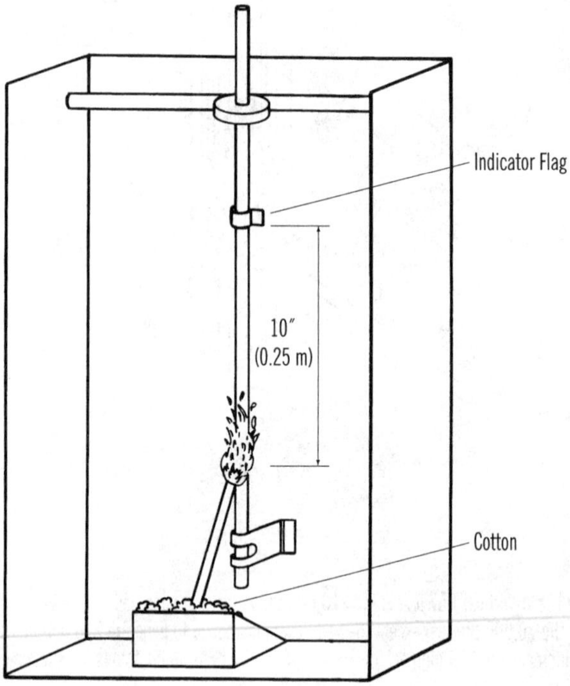

Figure 11.4–UL 1581 VW-1 Flame Test

11.3 REGULATORY AND APPROVAL AGENCIES

11.3.1 Underwriters Laboratories

Table 11.4–Summary of Wire and Cable Types Covered by UL Standards

UL Standard		UL Listing(s) Covered in the Standard
4	Armored Cable	AC
13	Power-Limited Circuit Cable	CL3P, CL2P, CL3R, CL2R, CL3, CL2, PLTC
44	Rubber Insulated Wires & Cables	XHHW, RHH, RHW, RH, SIS, RHW-2, XHHW-2
62	Flexible Cord & Fixture Wire	TFN, TFFN, TPT, TST, TS, S, SA, SE, SO, SEO, SOO, ST, STO, STOO, STOW, STOOW
83	Thermoplastic Insulated Wires	THW, THHN, THNN, FEP, FEPB, TFE, THW-2, THWN-2, Z, ZW
444	Communication Cables	CMP, CMR, CM, CMX
493	Thermoplastic Insulated Underground Feeder & Branch Circuit Cables	UF
719	Nonmetallic-Sheath Cables	NM, NMC
758	Appliance Wiring Material	AWM and all UL "Styles"
814	Gas-Tube-Sign Cable	GTO-5, GTO-10, GTO-15
854	Service-Entrance Cables	USE, SE, USE-2
1063	Machine-Tool Wires & Cables	MTW
1072	Medium Voltage Power Cable	MV
1276	Welding Cable	WELDING CABLE
1277	Electrical Power & Control Tray Cables with Optional Optical-Fiber Members	TC
1426	Electrical Cables for Boats	Boat Cable
1569	Metal Clad Cables	MC
1581	Reference Standard for Electrical Wires, Cables, and Flexible Cords	–
1650	Portable Power Cables	W, G, G-GC, PPE

Typical examples of UL's mark appear below:

Figure 11.5–Typical UL Marks

11.3.2 National Electrical Code (NEC)

History and Articles

The first NEC document was written in 1897 at the insistence of various insurance, electrical, architectural and other interested parties. Up to and including 2008, there have been a total of 51 editions. It is revised on a regular three year schedule. The National Electrical Code is divided into approximately 120 articles.

The Code is published by the National Fire Protection Association (NFPA) as a "recommended standard" and does not become law until it is officially adopted by state or local governments. Enforcement and interpretation of the Code is ultimately the responsibility of "the authority having jurisdiction," i.e., the local inspector.

The intent of the Code is to ensure the electrical and fire safety of electrical equipment. It does not attempt to ensure the reliability, performance, proper operation or long life of equipment – these considerations are beyond its scope.

National Electrical Code Articles related to the wire and cable industry include:

Article 100	Definitions
Article 110	Requirements for Electrical Installations
Article 200	Use and Identification of Grounded Conductors
Article 210	Branch Circuits
Article 215	Feeders
Article 220	Branch-Circuit, Feeder and Service Calculations
Article 225	Outside Branch Circuits and Feeders
Article 230	Services
Article 250	Grounding and Bonding
Article 300	Wiring Methods
Article 310	Conductors for General Wiring
Article 310.15	Ampacities for Conductors Rated 0–2000 Volts
Article 310.60 (B)	Ampacities of Conductors Rated 2001 to 35,000 Volts
Article 320	Armored Cable: Type AC
Article 324	Flat Conductor Cable: Type FCC
Article 328	Medium Voltage Cable: Type MV
Article 330	Metal-Clad Cable: Type MC
Article 332	Mineral-Insulated, Metal-Sheathed Cable: Type MI
Article 334	Nonmetallic Sheathed Cable Types NM, NMC and NMS
Article 336	Power and Control Tray Cable: Type TC
Article 338	Service-Entrance Cable: Types SE and USE
Article 340	Underground Feeder and Branch-Circuit Cable: Type UF
Article 344	Rigid Metal Conduit: Type RMC
Article 350	Liquid-Tight, Flexible Metal Conduit: Type LFMC
Article 356	Liquid-Tight, Flexible Nonmetallic Conduit Type: LFMC
Article 358	Electrical Metallic Tubing: Type EMT
Article 362	Electrical Nonmetallic Tubing: Type ENT
Article 392	Cable Trays
Article 396	Messenger Supported Wiring
Article 400	Flexible Cords and Cables

Continued on next page >>

(Continued)

Article 402	Fixture Wires
Article 409	Industrial Control Panels
Article 500	Hazardous (Classified) Locations, Classes I, II, and III, Divisions 1 and 2
Article 501	Class I Locations
Article 502	Class II Locations
Article 503	Class III Locations
Article 504	Intrinsically Safe Systems
Article 505	Class I, Zone 0, 1, and 2 Locations
Article 590	Temporary Installations
Article 604	Manufactured Wiring Systems
Article 610	Cranes and Hoists
Article 645	Information Technology Equipment
Article 725	Class 1, Class 2, Class 3, Remote-Control, Signaling and Power-Limited Circuits
Article 727	Instrumentation Tray Cable: Type ITC
Article 760	Fire Alarm Systems
Article 770	Optical Fiber Cables and Raceways
Article 800	Communication Circuits
Article 820	Community Antenna Television (CATV) and Radio Distribution Systems
Article 830	Network-Powered Broadband Communication Systems
Chapter 9	Tables (Conduit fill, conductor properties, etc.)

Table 11.5– NEC Article 725 – Summary of Remote Control, Signaling and Power-Limited Circuit Types

Circuit Type	Circuit Voltage	Maximum Current
Class 1 Remote Control and Signaling (Not Power-Limited)	0 through 600	Unlimited
Class 1 Power-Limited	0 through 30	33 Amps
Class 2 Power-Limited (Fire and Shock Safe)	0 through 30 30 through 150	8 Amps 0.005 Amps
Class 3 Power-Limited (Fire Safe Only)	30 through 150	10 Amps

Note: The above is a highly simplified overview only. See Article 725 of the NEC for complete requirements. Class 2 cables must be rated at least 150 volts and Class 3 cables must be rated at least 300 volts, but may not be so marked.

11.3.3 International

Table 11.6–Symbols of International Organizations

Agency	Country(ies) Represented	Symbol
CSA (Canadian Standards Association)	Canada	
CEBEC (Comite Electrotechnique Belge Service de la Marque)	Belgium	
DEMKO (Danmarks Electriske Materailkontrol)	Denmark	
SETI (Electrical Inspectorate Sakiniementie)	Finland	
CENELEC (European Committee for Electrotechnical Standards)	See Note 1.	
IEC (International Electrotechnical Commission)	More than 60 around the world	
ISO (International Standards Organization)	More than 60 around the world	
IMQ (Istituto Italiano del Marchio di Qualità)	Italy	
KEMA, KEUR (NV tot Keuring van Elektrotechnische Materialen)	Netherlands	
NEMKO (Norges Electriske Materallknotroll)	Norway	

Continued on next page >>

Table 11.6–Symbols of International Organizations (Continued)

Agency	Country(ies) Represented	Symbol
ÖVE (Österreichischer Verband für Elektrotechnik)	Austria	
SEMKO (Svenska Electriska Materielkontrollanstalten)	Sweden	
SEV (Schweizerischen Electrotechnischen Verein)	Switzerland	
UL (Underwriters Laboratories)	USA	
UTE (Union Technique de L'Électricité)	France	
VDE (Verband Deutscher Elektrotechnischer)	Germany	

Note 1:

Austria
Belgium
Bulgaria
Cyprus
Czech Republic
Denmark
Estonia
Finland
France
Germany
Greece
Hungary
Iceland
Ireland
Italy
Latvia
Lithuania
Luxembourg
Malta
Netherlands
Norway
Poland
Portugal
Romania
Slovakia
Slovenia
Spain
Sweden
Switzerland
United Kingdom

12. CONTINENTAL EUROPE

12.1 EUROPEAN UNION (EU) STANDARDS

12.1.1 CENELEC
European Committee for Electrotechnical Standardization
www.cenelec.org

Document No.	Title
EN 50143	Cables for Signs and Luminous–Discharge–Tube Installations Rated 1 through 10 kV
EN 50173	Information Technology–Generic Cabling Systems
EN 50214	Flat PVC Sheathed Flexible Cables
EN 50395	Electrical Test Methods for Low Voltage Cables
EN 50396	Non-electrical Test methods for Low Voltage Cables
EN 60079-11	Intrinsically Safe Electrical Apparatus for Explosive Atmospheres
EN 60079-14	Electrical Apparatus for Explosive Gas Atmospheres
EN 60228	Conductors of Insulated Cables
EN 61138	Cables for Portable Earthing and Short-Circuitting Equipment
HD 21	PVC Insulated Cables of Rated Voltages up to and Including 450/750 V
HD 22	Rubber Insulated Cables of Rated Voltages Up To and Including 450/750 V
HD 308	Identification and Use of Cores of Flexible Cables
HD 359	Flat PVC Sheathed Cables for Lifts and Similar Applications
HD 505	Test Methods for Insulating and Sheathing Materials of Electric Cables
HD 586	Mineral Insulated Cables with Rated Voltage Not Exceeding 750 V
HD 603	Distribution Cables of Rated Voltage 0.6/1 kV
HD 604	0.6/1 kV Power Cables with Special Fire Performance for Use in Power Stations

CENELEC has adopted common standards for the European community. CENELEC adopts existing IEC standards whenever possible. As a result, HD 21 and HD 22 CENELEC documents are based on relevant IEC specifications. The member countries adopt these standards without any fundamental changes. Each country's testing authority can do its own testing for a manufacturer to obtain HAR (Harmonized) approval. Additional information is available at www.eepca.org.

The countries in Table 12.1 are CENELEC members that recognize the HAR mark. When wire and cable is manufactured in a CENELEC country and is marked with the HAR approval on the jacket, it may be used interchangeably in the member countries.

The HAR identification mark is applied along with the marks of origin and testing authority, to at least one conductor or the outer jacket. For example, Siemens products are marked: SIEMENS ◁VDE▷ ◁HAR▷. In addition, there are country identification threads that are colored black, red and yellow. The different lengths of the colors indicate the nationality of the testing authority. For example, black 3 cm (1.2 inches) long, red 1 cm (0.4 inches) long, and yellow 1 cm (0.4 inches) long, indicates Germany.

Table 12.1–CENELEC Harmonized Approvals in the European Union

Member Country	Licensing Body/ Certification Agency	Imprint or Embossing on Jacket or Insulation	Black		Red		Yellow	
			in.	cm	in.	cm	in.	cm
Austria	Österreichischer Verband für Elektrotechnik (ÖVE)	◁ÖVE▷ ◁HAR▷	1.181	3	0.394	1	1.969	5
Belgium	Comité Electrotechnique Belge (CEBEC)	CEBEC ◁HAR▷	0.394	1	1.181	3	0.394	1
Denmark	Danmarks Elektriske Materielkontroll	◁DEMKO▷ ◁HAR▷	1.181	3	0.394	1	1.181	3
France	Union Technique de l'Electricité (UTE)	UTE ◁HAR▷	1.181	3	1.181	3	0.394	1
Germany	Verband Deutscher Elektrotechniker (VDE)	◁VDE▷ ◁HAR▷	1.181	3	0.394	1	0.394	1
Ireland	National Standards Authority of Ireland (NSAI)	◁IIRS▷ ◁HAR▷	1.181	3	1.181	3	1.969	5
Italy	Istituto Italiano del Marchio de Qualita (IMQ)	IMQ ◁HAR▷	0.394	1	1.181	3	1.969	5
Netherlands	N.V. tot Keuring van Elektrotechnische Materialen (KEMA)	KEMA ◁HAR▷	0.394	1	1.181	3	1.181	3
Norway	Norges Elektriske Materiellkontroll (NEMKO)	◁NEMKO▷ ◁HAR▷	0.394	1	0.394	1	2.756	7
Spain	Asociación Electrotécnica y Electrónica Española (AEE)	◇UNE◇ ◁HAR▷	1.181	3	0.394	1	2.756	7
Sweden	Svenska Elektriska Materielkontrollanstalter (SEMKO)	◁SEMKO▷ ◁HAR▷	0.394	1	0.394	1	1.963	5
United Kingdom	British Approvals Service for Cables	BASEC ◁HAR▷	0.394	1	0.394	1	1.181	3

12.1.2 CENELEC Cable Identification

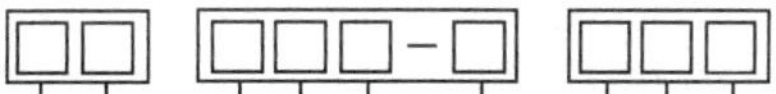

Type of Cable

- H Harmonized standards
- A Recognized national standards
- N Non-authorised national standards

Rated Voltage U_0/U

- 01 100 V
- 03 300/300 V
- 05 300/500 V
- 07 450/750 V

Insulating and sheathing material

- B EPR (ethylene propylene rubber)
- E4 PTFE (polytetrafluoroethylene)
- E6 ETFE (ethylene tetrafluoroethylene)
- E7 PP (polypropylene)
- Q4 PA (polyamide)

Special constructions

- H Flat construction - divisible
- H2 Flat construction - nondivisible

Conductor form

- U Rigid, round conductor, solid (Class 1)
- R Rigid, round conductor, stranded (Class 2)
- K Flexible conductor for fixed installations (Class 5)
- F Flexible conductor of a flexible cable (Class 5)
- H Highly flexible conductor of a flexible cable (Class 6)
- Y Tinsel conductor

Number of conductors

Green/yellow conductor for earthing

- X Without earthing conductor
- G With earthing conductor

Nominal cross section of the conductors

Figure 12.1–CENELEC Cable Identification Code

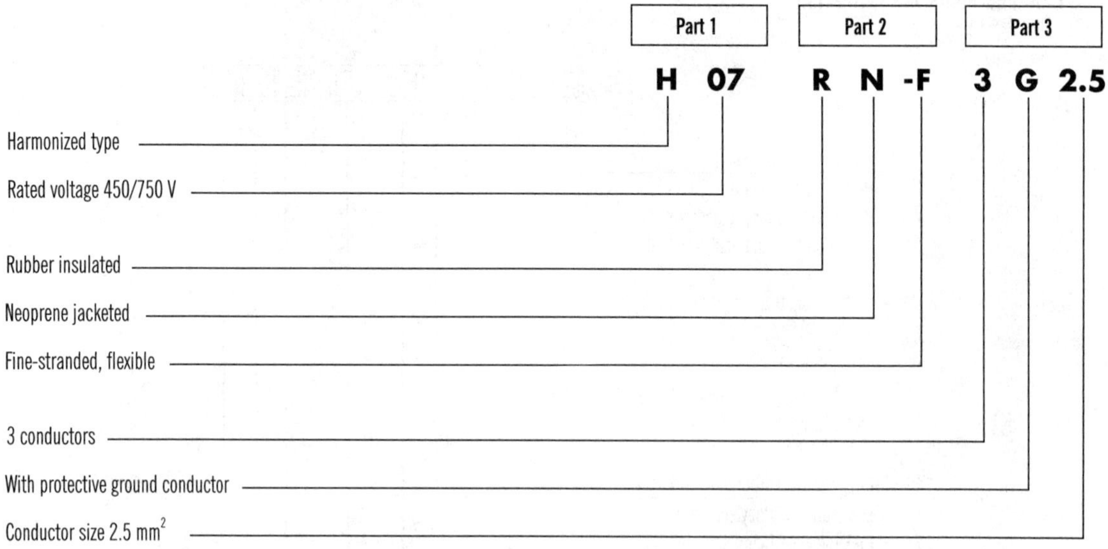

Figure 12.2–Example of a CENELEC Cable Identification Code

Below are CENELEC identification codes for some common cable types:

Cenelec Harmonized ID Code

H03VH-H	PVC Flexible Figure "8" Cable
H03VV-F2	PVC/PVC Flexible Mains Cable, 300 V, 2 Core
H03VV-F3	PVC/PVC Flexible Mains Cable, 300 V, 3 Core
H03VVH2-F2	PVC/PVC Flexible Mains Cable, 300 V, 2 Core Oval
H05RR-F2	Rubber/Rubber Flexible Mains Cable, 300/500 V, 2 Core
H05RR-F3	Rubber/Rubber Flexible Mains Cable, 300/500 V, 3 Core
H05RR-F4	Rubber/Rubber Flexible Mains Cable, 300/500 V, 4 Core
H05RR-F5	Rubber/Rubber Flexible Mains Cable, 300/500 V, 5 Core
H05V-U	PVC Single Conductor Building Wire, 450/750 V
H05VV-F2	PVC/PVC Flexible Mains Cable, 300/500 V, 2 Core
H05VV-F3	PVC/PVC Flexible Mains Cable, 300/500 V, 3 Core
H05VV-F4	PVC/PVC Flexible Mains Cable, 300/500 V, 4 Core
H05VV-F5	PVC/PVC Flexible Mains Cable, 300/500 V, 5 Core
H07V-K	PVC Appliance Wire, 450/750 V, Single Conductor
H07V-R	PVC Single Conductor Building Wire, 450/750 V
H07V-U	PVC Single Conductor Building Wire, 450/750 V, Solid Copper
H05VVH2-F	Flat Flexible Elevator Cable

12.1.3 CENELEC Color Codes

Through CENELEC, the EU countries have established the following color code for flexible cables as of 2006. "Green/yellow" indicates green insulation with a yellow stripe.

- **For Cables WITHOUT a Green/Yellow Earth Conductor:**
  - 1 Conductor — All colors except yellow, green, white or grey
  - 2 Conductor — Brown and blue
  - 3 Conductor — Brown, black and grey
  - 4 Conductor — Blue, brown, black and grey
  - 5 Conductor — Blue, brown, black, grey and black
  - >5 Conductors — All conductors black with white numbers
- **For Cables WITH a Green/Yellow Earth Conductor:**
  - 1 Conductor — All colors except yellow, green, white or grey
  - 2 Conductor — Not specified
  - 3 Conductor — Green/yellow, brown and blue
  - 4 Conductor — Green/yellow, brown, black and grey
  - 5 Conductor — Green/yellow, blue, brown, black and grey
  - >5 Conductors — One conductor green/yellow, others black with white numbers

Reference: CENELEC HD 308

12.1.4 CENELEC Copper Conductors

In the EU, the number of wires in a conductor (the fineness of the strand) is indicated by a numeral instead of by a letter as with the U.S. system. For example, a Class 6 conductor has more wires than a Class 5. Tables 12.2 through 12.5 give the DC resistance for some common conductor sizes and stranding types. CENELEC standard EN 60228 contains additional information.

Table 12.2–DC Resistance of Class 1 (Solid) Copper Conductors

Nominal Conductor Size	Minimum Number of Wires	Approximate Diameter of Wire	Maximum DC Resistance at 20°C
(mm^2)		(mm)	(ohms/km)
1.5	1	1.38	12.10
2.5	1	1.78	7.41
4.0	1	2.25	4.61
6.0	1	2.76	3.08
10.0	1	3.57	1.83
16.0	1	4.50	1.15
25.0	1	5.65	0.727
35.0	1	6.60	0.524

Table 12.3–DC Resistance and Stranding of Class 2 Copper Conductors

Nominal Conductor Size	Minimum Number of Wires		Maximum DC Resistance at 20°C
	Circular	Compact Circular or Sector Shaped	
(mm^2)			(ohms/km)
1.5	7	–	12.10
2.5	7	–	7.41
4.0	7	–	4.61
6.0	7	–	3.08
10.0	7	–	1.83
16.0	7	6	1.15
25.0	7	6	0.727
35.0	7	6	0.524
50.0	19	6	0.387
70.0	19	12	0.268
95.0	19	15	0.193
120.0	37	18	0.153
150.0	37	18	0.124
185.0	37	30	0.0991
240.0	61	34	0.0754
300.0	61	34	0.0601
400.0	61	53	0.0470
500.0	61	53	0.0366
630.0	91	53	0.0283
800.0	91	53	0.0221
1,000.0	91	53	0.0176

Table 12.4–DC Resistance and Stranding of Class 5 (Flexible) Copper Conductors

Nominal Conductor Size	Maximum Diameter of Wires	Maximum DC Resistance at 20°C	
		Plain Copper	Tinned Copper
(mm^2)	(mm)	(ohms/km)	(ohms/km)
0.5	0.21	39.0	40.1
0.75	0.21	26.0	26.7
1.0	0.21	19.5	20.0
1.5	0.26	13.3	13.7
2.5	0.26	7.98	8.21
4.0	0.31	4.95	5.09
6.0	0.31	3.30	3.39
10.0	0.41	1.91	1.95
16.0	0.41	1.21	1.24
25.0	0.41	0.780	0.795
35.0	0.41	0.554	0.565
50.0	0.41	0.386	0.393
70.0	0.51	0.272	0.277
95.0	0.51	0.206	0.210
120.0	0.51	0.161	0.164
150.0	0.51	0.129	0.132
185.0	0.51	0.106	0.108
240.0	0.51	0.0801	0.0817
300.0	0.51	0.0641	0.0654
400.0	0.51	0.0486	0.0495
500.0	0.61	0.0384	0.0391

Table 12.5–DC Resistance and Stranding of Class 6 (Highly Flexible) Copper Conductors

Nominal Conductor Size	Maximum Diameter of Wires	Maximum DC Resistance at 20°C	
		Plain Copper	Tinned Copper
(mm^2)	(mm)	(ohms/km)	(ohms/km)
0.5	0.16	39	40.1
0.75	0.16	26	26.7
1.0	0.16	19.5	20
1.5	0.16	13.3	13.7
2.5	0.16	7.98	8.21
4.0	0.16	4.95	5.09
6.0	0.21	3.30	3.39
10.0	0.21	1.91	1.95
16.0	0.21	1.21	1.24
25.0	0.21	0.780	0.795
35.0	0.21	0.554	0.565
50.0	0.31	0.386	0.393
70.0	0.31	0.272	0.277
95.0	0.31	0.206	0.210
120.0	0.31	0.161	0.164
150.0	0.31	0.129	0.132
185.0	0.31	0.106	0.108
240.0	0.31	0.0801	0.0817
300.0	0.31	0.0641	0.0654

12.1.5 CEN
European Committee for Standardization
www.cen.eu

12.1.6 Supply Voltages and Plug Configurations

Table 12.6–EU Supply Voltages

Frequency (Hz)	Voltage
50	220
50	250

Table 12.7–EU Power Plug Configurations

Jack	Plug	Description
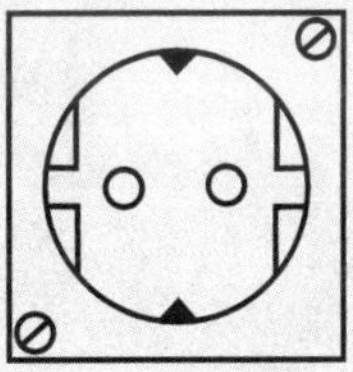		**"Schuko" European** CEE 7
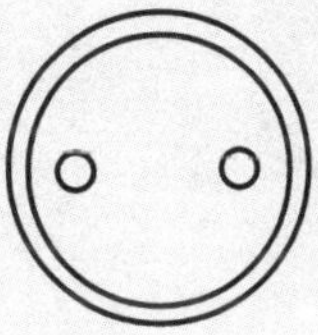		**Ungrounded Eurocord** CEE 7/16

12.2 AUSTRIAN STANDARDS

12.2.1 ÖVE
Österreichischer Verband für Elektrotechnik
www.ove.at

12.2.2 Supply Voltage and Plug Configuration

Table 12.8–Austrian Supply Voltage

Frequency (Hz)	Voltage
50	220/380

The neutral wire of the secondary distribution system is grounded. A grounding conductor is required in the electrical cord attached to appliances that are not double insulated.

Table 12.9–Austrian Plug Configuration

Jack	Plug	Description
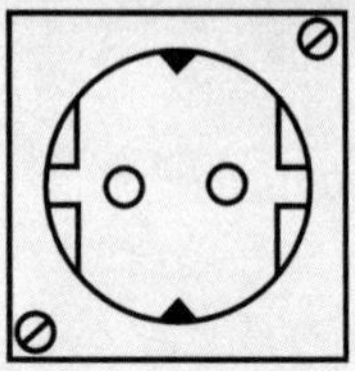	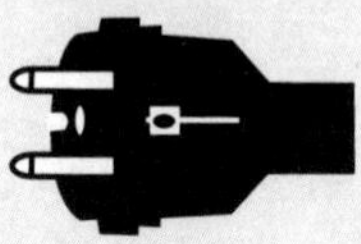	**"Schuko" European** CEE 7

12.3 BELGIAN STANDARDS

12.3.1 CEBEC
Comité Electrotechnique Belge
www.cebec.sgs.com

12.3.2IBN/NBT
Institut Belge de Normalisation
www.ibn.be

Document No.	Title
NBN C 33-111	Armored Cables Insulated with Impregnated Paper for the Transmission and Distribution of Electrical Energy
NBN C 33-121	Armored, PVC Insulated Cables for the Transmission and Distribution of Electrical Energy
NBN C 33-211	Armored, Aluminum Conductor Cables Insulated with Impregnated Paper for 1, 6, 12, and 15 kV
NBN C 33-221	Power Cables
NBN C 33-321	Power Cables–Insulated Overhead Transmission Lines Rated 0.6/1 kV
NBN C 33-322	Underground Power Cables Rated 1 kV
NBN 759	Power Cables

12.3.3 Supply Voltages and Plug Configurations

Table 12.10–Belgian Supply Voltages

Frequency (Hz)	Voltage
50	127/220
50	220/380

The neutral wire of the secondary distribution system is grounded. A grounding conductor is required in the electrical cord attached to appliances.

Table 12.11–Belgian Plug Configurations

Jack	Plug	Description
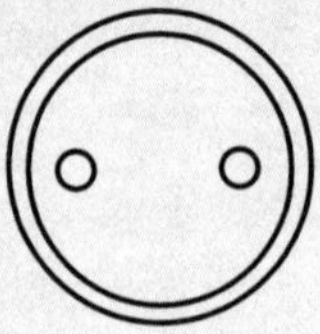		**Ungrounded Eurocord** CEE 7/16
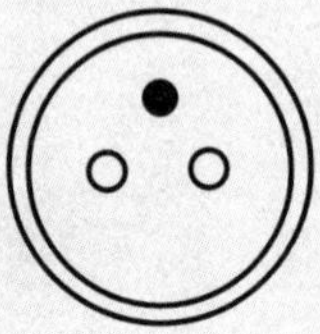		**Belgium Socket** CEE 7/7 plug

12.4 DANISH STANDARDS

12.4.1 DEMKO
Danmarks Elektriske Materielkontrol
www.demko.dk

12.4.2 Supply Voltage and Plug Configuration

Table 12.12–Danish Supply Voltage

Frequency (Hz)	Voltage
50	220/380

Table 12.13–Danish Plug Configuration

Jack	Plug	Description
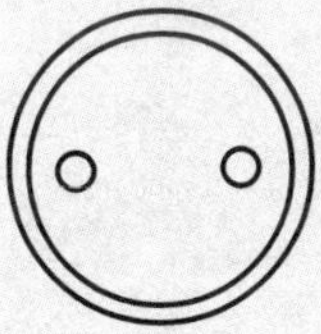		**Ungrounded Eurocord** CEE 7/16

12.9 DUTCH STANDARDS

12.9.1 KEMA
N.V. tot Keuring van Elektrotechnische Materialen
www.kema.com

Document No.	Title
KEMA 157	Cables for temporary installations rated 450/750 kV
KEMA 149	Cross-linked polyethylene insulated, screened and PVC sheathed EMC cables rated 0.6/1 kV
KEMA 250	Screened flame retardant EMC cables with XLP insulation and halogen-free thermoplastic sheaths rated 0.6/1 kV

12.9.2 NEC
Netherlands Electro-Technical Committee
www.nen.nl

12.9.3 Supply Voltage and Plug Configurations

Table 12.14–Dutch Supply Voltage

Frequency (Hz)	Voltage
50	220/380

The neutral wire of the secondary distribution system is grounded.

Table 12.15–Dutch Plug Configurations

Jack	Plug	Description
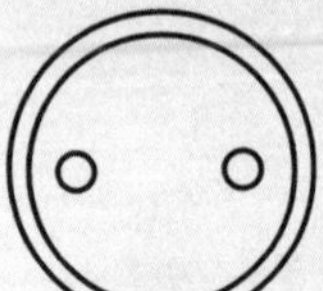		**Ungrounded Eurocord** CEE 7/16
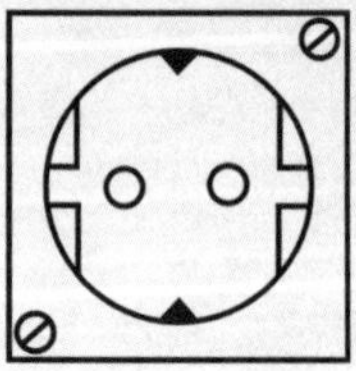	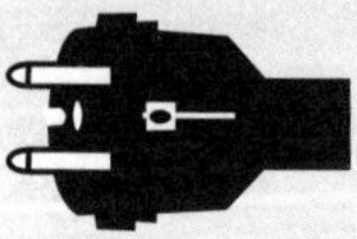	**"Schuko" European** CEE 7

12.5 FRENCH STANDARDS

12.5.1 UTE
Union Technique de l'Electricité
www.ute-fr.com

Document No.	Title
NF C32-207	Insulated Cables Covered with a Light PVC Sheath and Rated 300/500 V
NF C33-209	Bundle Assembled Cores for Overhead Systems Rated 0.6/1 kV

12.5.2 AFNOR
Association Française de Normalisation
www.afnor.org

12.5.3 CIGRE
Conference International des Grands Reseaux Electriques a Haute Tension
www.cigre.org

12.5.4 Supply Voltages and Plug Configurations

Table 12.16–French Supply Voltages

Frequency (Hz)	Voltage
50	220/230
50	127/220

Table 12.17–French Plug Configurations

Jack	Plug	Description
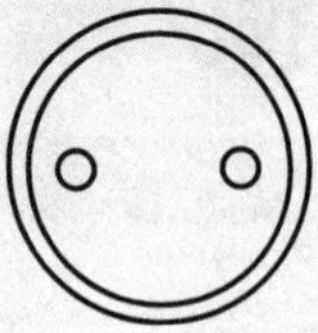		**Ungrounded Eurocord** CEE 7/16
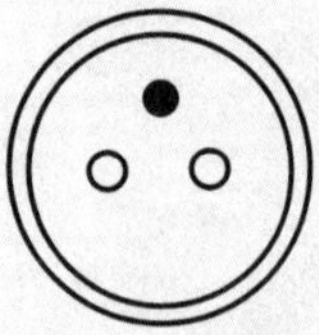		**French Socket** CEE 7/7 plug
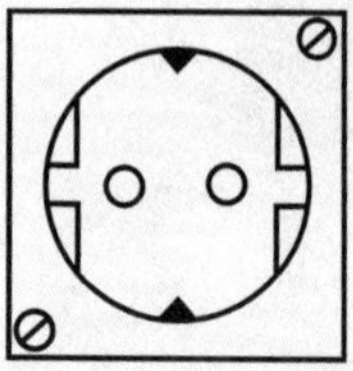		**"Schuko" European** CEE 7
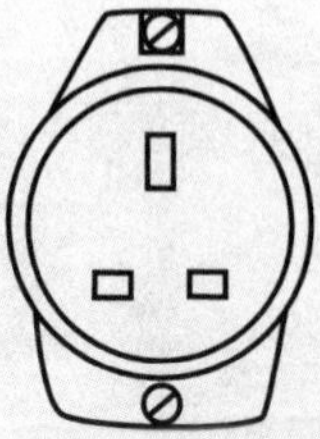	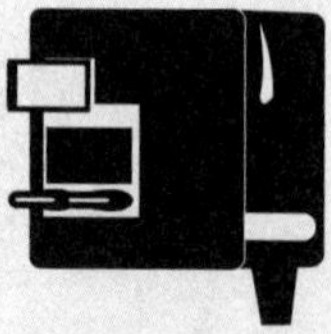	**British Standard** BS 1363

12.6 GERMAN STANDARDS

12.6.1 DIN
Deutsches Institut für Normung
www.din.de

Document No.	Title
VDE 0888 (EN 60793)	Optical Fibres–Measurement Methods and Test Procedures

Table 12.18–DIN 47100 Color Code for Single Conductors (WITH Color Repetition Above 44)

Conductor No.		Color	Conductor No.	Color
1	45	white	23	white/red
2	46	brown	24	brown/red
3	47	green	25	white/black
4	48	yellow	26	brown/black
5	49	grey	27	grey/green
6	50	pink	28	yellow/green
7	51	blue	29	pink/green
8	52	red	30	yellow/pink
9	53	black	31	green/blue
10	54	violet	32	yellow/blue
11	55	grey/pink	33	green/red
12	56	red/blue	34	yellow/red
13	57	white/green	35	green/black
14	58	brown/green	36	yellow/black
15	59	white/yellow	37	grey/blue
16	60	yellow/brown	38	pink/blue
17	61	white/grey	39	grey/red
18		grey/brown	40	pink/red
19		white/pink	41	grey/black
20		pink/brown	42	pink/black
21		white/blue	43	blue/black
22		brown/blue	44	red/black

Table 12.19–DIN 47100 Color Code for Paired Conductors (WITH Color Repetition Above 22)

Pair No.			Wire A	Wire B
1	23	45	white	brown
2	24	46	green	yellow
3	25	47	grey	pink
4	26	48	blue	red
5	27	49	black	violet
6	28	50	grey/pink	red/blue
7	29	51	white/green	brown/green
8	30	52	white/yellow	yellow/brown
9	31	53	white/grey	grey/brown
10	32	54	white/pink	pink/brown
11	33	55	white/blue	brown/blue
12	34		white/red	brown/red
13	35		white/black	brown/black
14	36		grey/green	yellow/grey
15	37		pink/green	yellow/pink
16	38		green/blue	yellow/blue
17	39		green/red	yellow/red
18	40		green/black	yellow/black
19	41		grey/blue	pink/blue
20	42		grey/red	pink/red
21	43		grey/black	pink/black
22	44		blue/black	red/black

12.6.2 VDE
Verband Deutscher Elektrotechniker
(German Electrotechnical Society)
www.vde-verlag.de

Document No.	Title
0250	Insulated Power Cables
0265	Polymer Insulated Lead Sheathed Power Cables
0266	Power Cables with Improved Behavior in Case of Fire Rated 0.6/1 kV
0271	High voltage cable for special applications rated 0.6/1 kV
0281	Wire and cable for power circuits with thermoplastic insulation rated to 450/750 V
0282	Wire and cable for power circuits with rubber insulation rated to 450/750 V
0293	Identification of conductors in cables and flexible cords used in power installations rated up to 1,000 V
0295	Conductors of cables, wires and flexible cords
0472-815	Tests for halogen content of wires and cables

12.6.3 DKE
Deutsche Kommission Elektrotechnik Informationstechnik
www.dke.de

12.6.4 Supply Voltage and Plug Configurations

Table 12.20–German Supply Voltage

Frequency (Hz)	Voltage
50	220/380

Table 12.21–German Plug Configurations

Jack	Plug	Description
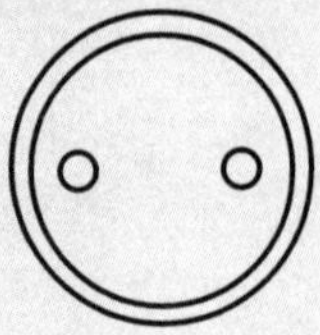		**Ungrounded Eurocord** CEE 7/16
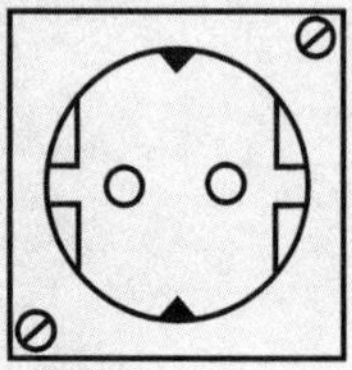	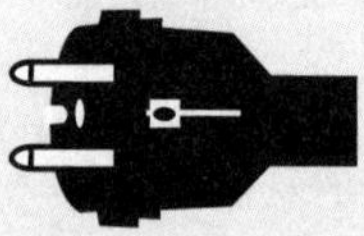	**"Schuko" European** CEE 7

12.7 IRISH STANDARDS

12.7.1 NSAI

National Standards Authority of Ireland
www.nsai.ie

12.7.2 Supply Voltage and Plug Configurations

Table 12.22–Irish Supply Voltage

Frequency (Hz)	Voltage
50	220/380

The neutral wire of the secondary distribution system is grounded. A grounding conductor is required in the electrical cord attached to appliances.

Table 12.23–Irish Plug Configurations

Jack	Plug	Description
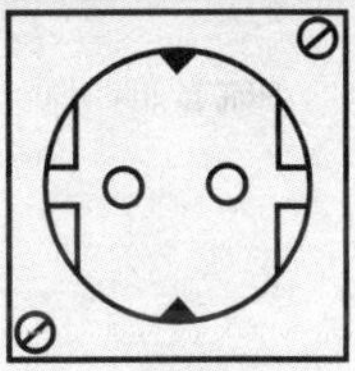		**"Schuko" European** CEE 7
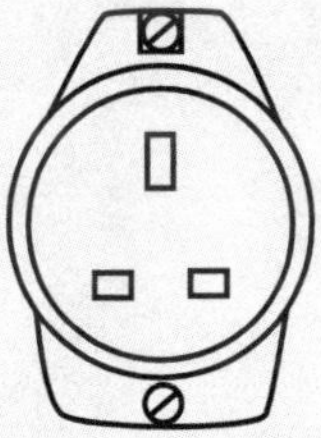		**British Standard** BS 1363

12.8 ITALIAN STANDARDS

12.8.1 IMQ
Istituto Italiano del Marchio de Qualita
www.imq.it

12.8.2 CEI
Comitato Elettrotecnico Italiano
www.ceiuni.it

12.8.3 CESI
Centro Elettrotecnico Sperimentale Italiano
www.cesi.it

12.8.4 Supply Voltage and Plug Configurations

Table 12.24–Italian Supply Voltage

Frequency (Hz)	Voltage
50	220/380

A grounding conductor is required in the electrical cord attached to appliances. The neutral wire of the secondary distribution system is grounded.

Frequency tolerance: ±2 percent; voltage tolerance: ±10 percent.

Table 12.25–Italian Plug Configurations

Jack	Plug	Description
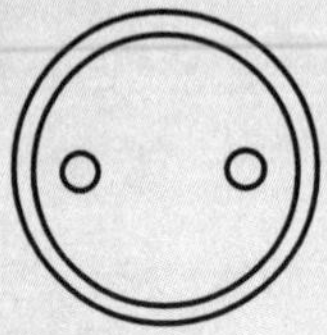		**Ungrounded Eurocord** CEE 7/16
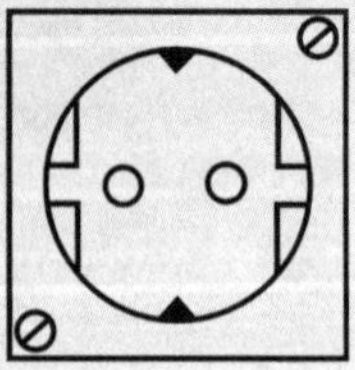		**"Schuko" European** CEE 7

12.10 NORWEGIAN STANDARDS

12.10.1 NEMKO
Norges Elektriske Materiellkontroll
www.nemko.com

12.10.2 NPT
Norwegian Post and Telecommunications Authority
www.npt.no

12.10.3 Supply Voltage and Plug Configurations

Table 12.26–Norwegian Supply Voltage

Frequency (Hz)	Voltage
50	230

Table 12.27–Norwegian Plug Configurations

Jack	Plug	Description
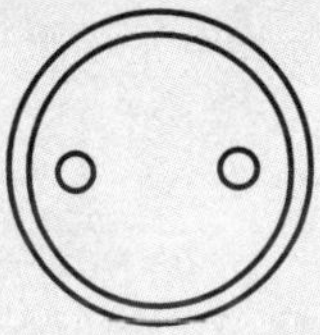		**Ungrounded Eurocord** CEE 7/16
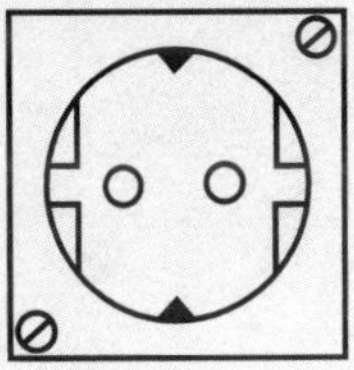	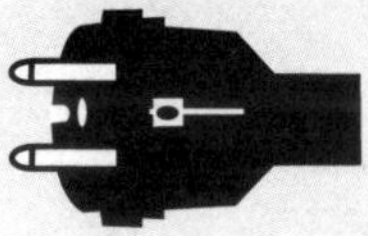	**"Schuko" European** CEE 7

12.11 PORTUGUESE STANDARDS

12.11.1 IPQ

Instituto Portugues da Qualidade
www.ipq.pt

12.11.2 Supply Voltage and Plug Configurations

Table 12.28–Portuguese Supply Voltage

Frequency (Hz)	Voltage
50	220/380

Table 12.29–Portuguese Plug Configurations

Jack	Plug	Description
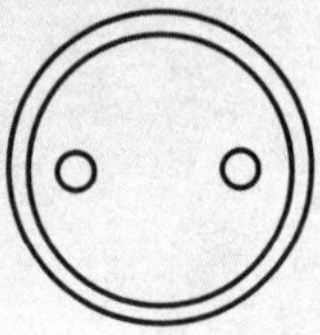		**Ungrounded Eurocord** CEE 7/16
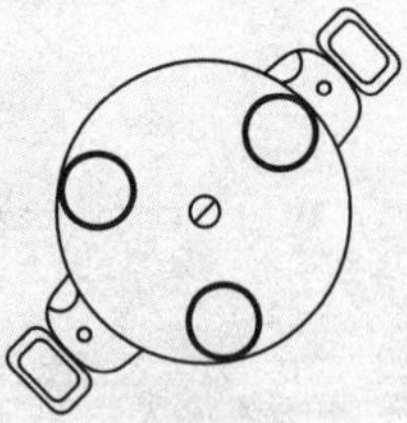	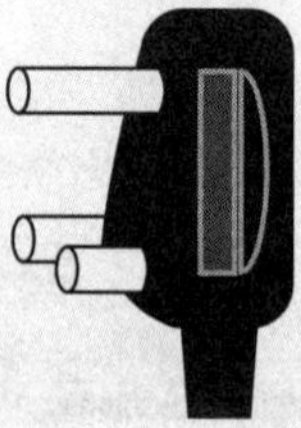	**Old British Standard** BS 546

12.12 SPANISH STANDARDS

12.12.1 AENOR
Asociación Española de Normalización y Certificación
www.aenor.es

Document No.	Title
UNE 20432	Tests on Electrical Cable Under Fire Conditions
UNE 21172	Measurement of Smoke Density of Electric Cable Burning in Defined Conditions

12.12.2 Supply Voltages and Plug Configurations

Table 12.30–Spanish Supply Voltage

Frequency (Hz)	Voltage
50	127/220
50	220/380

A grounding conductor is required for the 220/380 voltage.

Table 12.31–Spanish Plug Configurations

Jack	Plug	Description
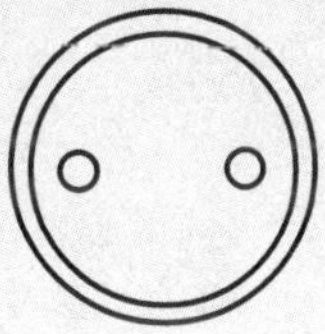		**Ungrounded Eurocord** CEE 7/16
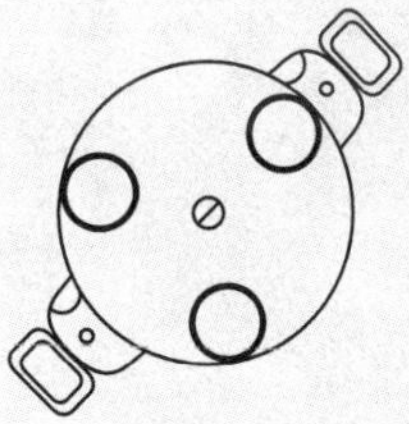	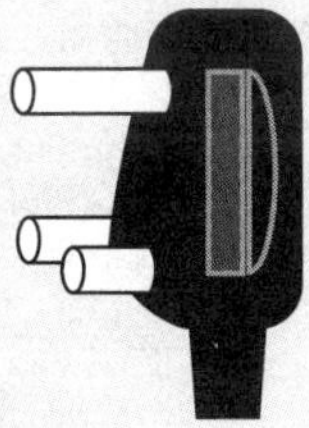	**Old British Standard** BS 546

12.13 SWEDISH STANDARDS

12.13.1 SEMKO
Svenska Electriska Materielkontrollanstalten
www.semko.se

12.13.2 ITS
Informationstekniska Standardiseringen
www.its.se

12.13.3 Supply Voltage and Plug Configurations

Table 12.32–Swedish Supply Voltage

Frequency (Hz)	Voltage
50	220/380

The neutral wire of the secondary distribution system is grounded. A grounding conductor is required in the electrical cord attached to appliances.

Table 12.33–Swedish Plug Configurations

Jack	Plug	Description
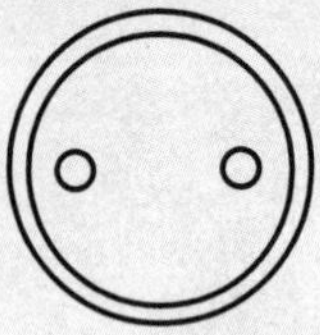		**Ungrounded Eurocord** CEE 7/16
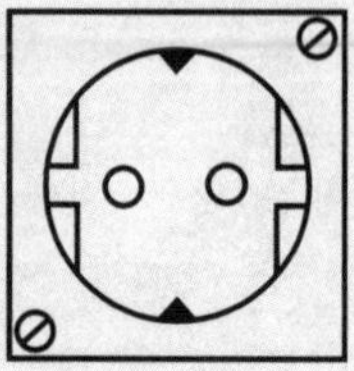		**"Schuko" European** CEE 7

12.14 SWISS STANDARDS

12.14.1 SNV
Schweizerische Normen-Vereinigung
www.snv.ch

12.14.2 ESTI (formerly SEV)
Eidgensisches Starkstrominspektorat
www.esti.ch

12.14.5 Supply Voltage and Plug Configurations

Table 12.34–Swiss Supply Voltage

Frequency (Hz)	Voltage
50	230/380

The neutral wire of the secondary distribution system is grounded. A grounding conductor is required in the electrical cord attached to appliances.

Table 12.35–Swiss Plug Configurations

Jack	Plug	Description
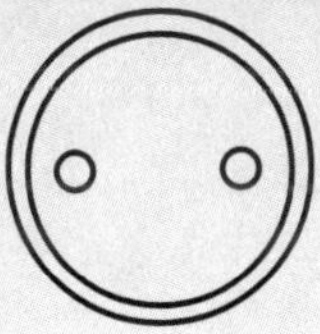		**Ungrounded Eurocord** CEE 7/16
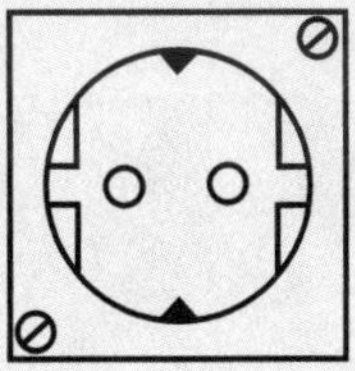		**"Schuko" European** CEE 7

13. UNITED KINGDOM

13.1 STANDARDS

13.1.1 BSI
British Standards
www.bsi-global.com

Document No.	Title
BS 3G 210	PTFE insulated equipment wires with silver-plated conductors
BS 115	Metallic resistance materials for electric purposes
BS 183	General purpose galvanized steel wire strand
BS 215	Aluminum conductors and aluminum conductors, steel-reinforced, for overhead power transmission
BS 638-1	Welding cables
BS 1117	Bare fine resistance wire for precision electrical equipment
BS 2316	Radio frequency cables
BS 2627	Wrought aluminum wire for electrical purposes
BS 3573	Polyethylene-insulated copper conductor telecommunication distribution cables
BS 4393	Tin or tin-lead coated copper wire
BS 4553	PVC-insulated split concentric cables with copper conductors for electric supply
BS 4808	L.F. cables and wires with PVC insulation and PVC sheath for telecommunication
BS 5099	Spark testing of electric cables
BS 5308	Instrumentation cables
BS 5467	Specification for armored cables with thermosetting insulation for electricity supply
BS 6004	PVC-insulated cables for electric power and lighting
BS 6007	Single core heat resisting cables
BS 6141	Replaced by specifications BS 6500, BS 6004, BS 6007 and BS 7919
BS 6195	Insulated flexible cables and cords for coil leads
BS 6207	Mineral-insulated cables
BS 6231	PVC-insulated flexible cables for switchgear and controlgear wiring rated 600/1,000 V
BS 6500	Insulated flexible cords
BS 6708	Specification for trailing cables for mining purposes
BS 6724	Thermoset insulated, armored cables having low emission of smoke and corrosive gases
BS 6746C	Color chart for insulation and sheath of electric cables

Continued on next page >>

13.1.1 BSI (Continued)

Document No.	Title
BS 6883	Elastomer insulated cable for fixed wiring in ships
BS 6899	Replaced by specification BS 7655
BS 6977	Replaced by specification BS EN 50214
BS 7655	Insulating and sheathing materials for cables
BS 7671	IEE Wiring Regulations
BS 7884	Copper and copper-cadmium conductors for overhead power transmission
BS 7919	Electric Cables-Flexible cables rated up to 450 V/750 V for use with appliances and equipment intended for industrial and similar environments
BS EN 10244-2	Zinc or zinc alloy coatings on steel wire
BS EN 10257	Galvanized steel wire for armoring land and submarine cables
BS EN 12166	Copper and copper alloys; wire for general purposes
BS EN 13601	Copper and copper alloys for electrical purposes; rod, bar and wire
BS EN 13602	Copper and copper alloys; drawn, round copper wire for electrical conductors
BS EN 50117	Specification for coaxial cables for wired distribution systems
BS EN 50143	Cables for signs and luminous-discharge-tube installations
BS EN 50149	Copper and copper alloy grooved contact wire for electric traction
BS EN 50183	Conductors for overhead lines: aluminum-magnesium-silicon alloy wires
BS EN 50189	Conductors for overhead lines: zinc coated steel wires
BS EN 50266	Test methods for cables under fire conditions; vertical flame spread
BS EN 50214	Flat Polyvinyl Chloride sheathed flexible cables
BS EN 50267	Test methods for cables under fire conditions; evolved gases
BS EN 60228	Conductors of insulated cables
BS EN 60332	Tests on electric and optical cables under fire conditions
BS EN 60811	Test methods for insulating and sheathing materials of electric cables
BS EN 61238	Mechanical and compression connectors for power cable
PD 6455	Cancelled

13.1.2 BASEC
British Approvals Service for Cables
www.basec.org.uk

BASEC is a product marking and certification agency similar to UL and CSA in North America

13.1.3 BBC
British Broadcasting Company
www.bbc.co.uk

Document No.	Title
PSF 1/2M	Video cable (similar to Belden 8281)
PSF 1/3M	Video cable
PSF 1/9M	Flexible camera cable
PSF 4/1M	Microphone cable

13.1.4 BNFL
British Nuclear Fuels
www.bnfl.com

Document No.	Title
PM73479	Cables for electricity supply and control having low emission of smoke and corrosive gases when affected by fire (see Section 13.1.1 "BS 6724" for additional information)

13.1.5 BRB
British Railways Board
www.brb.gov.uk

Document No.	Title
TDE/76/P/16	Single core cables for traction and rolling stock (cross-linked polyolefin types)

13.1.6 British Telecom (BT Group)
www.btplc.com

Document No.	Title
CW1109	Single, twin and triple jumper wire for electronic equipment
CW1128	Polyethylene insulated and sheathed, jelly-filled, twisted-pair telephone cable for outdoor use (up to and including 100 pairs)
CW1198	External telephone cable
CW1229	Coaxial cable
CW1236	Polyethylene insulated and sheathed, jelly-filled, twisted-pair telephone cable for outdoor use (above 100 pairs)
CW1252	Self-supporting aerial telephone cable
CW1257	Jumper wire
CW1293	Internal telephone cable
CW1308	PVC-insulated and sheathed telephone cable for indoor use
CW1311	Telephone cordage
CW1316	Undercarpet telephone cable
CW1378	Drop wire no. 10
CW1600	"LFH" insulated and sheathed telephone cable for indoor use

13.1.7 Department of the Environment
www.doeni.gov.uk

Document No.	Title
M&E 42	Aviation ground lighting. Single core 6 mm^2 (2,000 V) PVC-sheathed cable

13.1.8 Department for Transport
www.dft.gov.uk

Document No.	Title
TR 1173	Multipair communications cable, polyethylene-insulated, polyethylene-sheathed, armored
TR 1238	Power cable for motorway communication systems (split concentric, armored)
TR 2029	Inductive loop cable for vehicle detection systems

13.1.9 ERA Technology Ltd.
www.era.co.uk

Document No.	Title
ERA 69-30 Part I	Sustained current ratings for paper-insulated, lead-sheathed cables
ERA 69-30 Part III	Sustained current ratings for PVC insulated cables
ERA 69-30 Part V	Sustained current ratings for cables with thermosetting insulation
ERA 93-0233R	User's guide to power cable fault location
ERA 98-0668	Cabling installations: user friendly guide (EMI separation recommendations)

13.1.10 EA (Formerly ESI)
www.energynetworks.org

Document No.	Title
09-6	Auxiliary multicore and multipair cables
09-7	PVC-insulated concentric service cables with stranded copper conductors and copper concentric conductors
09-8	Impregnated paper insulated 600/1,000 V cable with three solid aluminum phase conductors and aluminum sheath/neutral conductor (CONSAC)
09-12	Impregnated paper insulated corrugated aluminum sheathed 6,350/11,000 V cable (PICAS)
43-13	Aerial bundled conductors (ABC) insulated with cross-linked polyethylene for low voltage overhead distribution

13.1.11 IET (formerly IEE)
www.theiet.org

Document No.	Title
BS7671	IEE Wiring Regulations

13.1.12 London Underground Limited
www.tfl.gov.uk/tube

Document No.	Title
G4727	2, 3 and 4 core SWA cable low voltage, limited fire hazard (low smoke nonhalogenated), 2.5 mm^2 to 300 mm^2
G4726	Single core cable insulated nonsheathed, limited fire hazard (low smoke nonhalogenated), 1.5 mm^2 to 300 mm^2
RSE/STD/024 part 6	Cables for use on rolling stock ("LFII" types)

13.1.13 Ministry of Defense (MODUK) British Defense Standards www.dstan.mod.uk

Document No.	Title
DEF STAN 02-525	Requirements for electric cables, thin-wall insulated, limited fire hazard
DEF STAN 02-526	Requirements for cables, electric, rubber insulated, limited fire hazard, sheathed for general services
DEF STAN 02-527	Requirements for cables, electric, fire survival, high temperature zones and limited fire hazard, sheathed
DEF STAN 02-711	Determination of the smoke index of the products of combustion
DEF STAN 02-713	Determination of the toxicity index of the products of combustion
DEF STAN 61-12 part 1	Cables, electrical (insulated flexible cords and flexible cables)
DEF STAN 61-12 part 2	Cables, electrical (for power and lighting)
DEF STAN 61-12 part 4	Cables, special purpose, electrical (subminiature electric cables)
DEF STAN 61-12 part 5	Cables, special purpose, electrical, and cables, power, electrical (small multicore cables)
DEF.STAN 61-12 part 6	Polyvinyl chloride (PVC), polyethylene, or silicone rubber insulated equipment wires
DEF STAN 61-12 part 8	PTFE insulated equipment wire
DEF STAN 61-12 part 9	Cables, radio frequency (coaxial)
DEF STAN 61-12 part 18	Equipment wires, low toxicity
DEF STAN 61-12 part 25	Cables, electrical, limited fire hazard, up to conductor size 2.5 mm^2

13.1.14 Network Rail
www.networkrail.co.uk

Document No.	Title
NR/PS/SIG/00005	Flame retardant compound insulated cables for railway signalling

13.1.15 RIA
Railway Industry Association
www.riagb.org.uk

Document No.	Title
NBRB/RIA 10	Twin and multicore jumper cables for diesel and electric vehicles
BRB/LUL/RIA 21	Single-core cables for installation on traction and rolling stock (rubber types)

13.2 SUPPLY VOLTAGE AND PLUG CONFIGURATIONS

Table 13.1 – United Kingdom Supply Voltage

Frequency (Hz)	Voltage
50	230/400

Table 13.2–United Kingdom Plug Configurations

Jack	Plug	Description
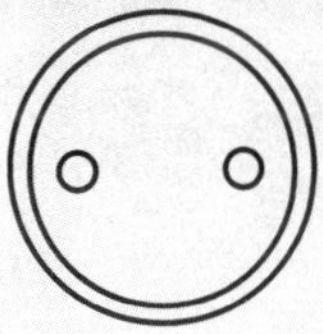		**Ungrounded Eurocord** CEE 7/16
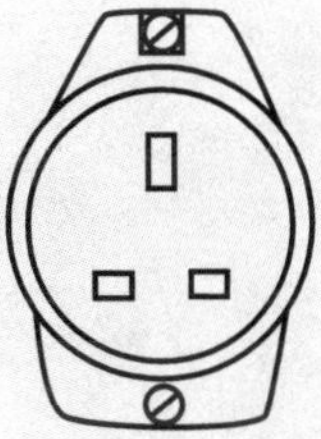		**British Standard** BS 1363

14. LATIN AND SOUTH AMERICA

14.1 MEXICAN STANDARDS

14.1.1 ANCE
Asociación de Normalización y Certificación
www.ance.org.mx

Document No.	Title
NOM-001-SEDE	Mexican Code Standard for Electrical Installations
NMX-E-034-SCFI	Plastics Industry–Carbon Black Contents on Polyethylene Materials–Test Method
NMX-J-008-ANCE	Tinned Soft or Annealed Copper Wire for Electrical Purposes
NMX-J-010-ANCE	Conductors with Thermoplastic Insulation Rated 600 V
NMX-J-012-ANCE	Concentric Lay Stranded Copper Conductors for Electrical Purposes
NMX-J-013-ANCE	Rope Lay Stranded Copper Conductors Having Concentric Stranded Members for Electrical Purposes
NMX-J-014-ANCE	Rope Lay Stranded Copper Conductors Having Bunch Stranded Members for Electrical Purposes
NMX-J-032-ANCE	Concentric Lay Stranded Aluminum Cable for Electrical Purposes
NMX-J-036-ANCE	Soft or Annealed Copper Wire for Electrical Purposes
NMX-J-040-ANCE	Determination of Moisture Absorption in Insulations and Jackets of Electrical Conductors–Test Method
NMX-J-066-ANCE	Determination of Diameters of Electrical Conductors–Test Method
NMX-J-177-ANCE	Determination of Thickness in Semiconductive Shielding, Insulations, and Jackets of Electrical Conductors–Test Method
NMX-J-178-ANCE	Ultimate Strength and Elongation of Insulation, Semiconductive Shielding, and Jackets of Electrical Conductors–Test Method
NMX-J-186-ANCE	Accelerated Aging in Forced Air Convection Oven of Semiconducting Shields, Insulations and Jackets of Electrical Conductors–Test Method
NMX-J-190-ANCE	Thermal Shock Resistance of PVT Insulations and PVT Protective Coverings of Electrical Conductors–Test Method
NMX-J-191-ANCE	Heat Distortion of Semiconductive Shielding, Insulations and Protective Coverings of Electrical Conductors–Test Method
NMX-J-192-ANCE	Flame Test on Electrical Wires–Test Method
NMX-J-193-ANCE	Cold Bend of Thermoplastic Insulation and Protective Jackets, Used on Insulated Wire and Cable–Test Method
NMX-J-212-ANCE	Electrical Resistance, Resistivity and Conductivity–Test Method
NMX-J-294-ANCE	Insulation Resistance–Test Method
NMX-J-312-ANCE	Tensile Strength and Elongation by Strain of Wires for Electrical Conductors–Test Method
NMX-J-417-ANCE	Convection Laboratory Ovens for Evaluation of Electrical Insulation–Specifications and Test Methods
NMX-J-437-ANCE	Determination of Light Absorption Coefficient of Polyethylene Pigmented with Carbon Black–Test Method
NMX-J-451-ANCE	Power Cables with XLP or EP Insulation Rated 600 V
NMX-J-498-ANCE	Vertical Tray–Flame Test–Test Method
NMX-J-556-ANCE	Wire and Cable Test Methods

14.1.2 Supply Voltage and Plug Configuration

Table 14.1–Mexican Supply Voltages

Frequency (Hz)	Voltage
50	127/220

Table 14.2–Mexican Plug Configuration

Jack	Plug	Description
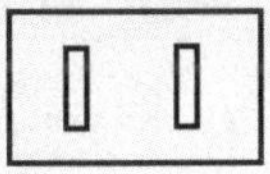	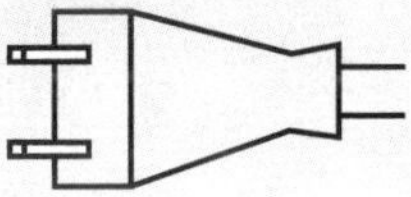	**North American Ungrounded**

14.2 VENEZUELAN STANDARDS

14.2.1 FONDONORMA
Fondo para la Normalización y Certificación de Calidad
www.fondonorma.org.ve or www.codelectra.org

Document No.	Title
0200	Venezuelan National Electrical Standard (Código Eléctrico Nacional) (Based on the U.S. National Electrical Code [NFPA 70])
0397	Wires and single-conductor cables with thermoplastic insulation
0457	Calculation of the permissible current in cables
0468	Sampling and methods of test to determine the cross-sectional area of conductors
0529	Bare copper cables. Specifications and methods test.
0530	Aluminum 1350/1370 for manufacture of electrical conductors
0533	Aluminum conductors 1350 with concentric stranding
0534	Conductors with twisted concentric aluminum layers, reinforced with steel
0541	Insulated wires and cables for distribution of electrical energy up to 2,000 V and control cables
0555	Compacted bare aluminum conductors for electrical use
0556	Round aluminum alloy 6201-T 81 wires for electrical use
0557	Conductors with aluminum alloy 6201 T-81 wires
0558	Flexible cords and wires
0777	Bare soft copper wires, rectangular
1110	Aluminum conductors with aluminum alloy reinforcement (ACAR)
2570	Wires and Cables. Determination of the level of extinction of partial discharges
2644	Compacted concentric round aluminum cables
2645	Round aluminum wire 1350, annealed and of intermediate tempers
3198	Multi-pair telephone cables for external plant
3484	Enameled coil wire

14.3 BRAZILIAN STANDARDS

14.3.1 ABNT
Associação Brazileira De Normas Técnicas
www.abnt.org.br

Document No.	Title
NBR 5410	Low Voltage Electrical Installations
NBR 11301	Calculation of the Continuous Current Ratings of Cables at 100 percent load factor
NBR 14039	Installation of Electrical Systems Rated 1.0 kV to 36.2 kV

14.4 COLOMBIAN STANDARDS

14.4.1 ICONTEC
Instituto Colombiano de Normas Técnicas
www.icontec.org.co

Document No.	Title
NTC 2050	Colombian National Electrical Standard (Based on the U.S. National Electrical Code [NFPA 70])

14.5 ARGENTINE STANDARDS

14.5.1 IRAM
Instituto Argentino de Normalización y Certificación
www.iram.com.ar

Document No.	Title
NM 247	PVC Insulated Cables Rated up to 450/750 V
NM 274	Silicone Insulated Cables Rated up to 450/750 V
NM 280	Conductors of Insulated Cables
2214	Cathodic Protection Cables
62266	Power and Control Cables with Low Smoke Emission and No Halogen (LSZH) Rated 1 kV

15. CANADA

15.1 STANDARDS

15.1.1 CSA
Canadian Standards Association
www.csa.ca and www.csa-international.org

Document No.	Title
C22.1	Canadian Electrical Code, Part I
C22.2 No. 0	General Requirements—Canadian Electrical Code, Part II
C22.2 No. 0.3	Test Methods for Electrical Wire and Cables
C22.2 No. 0.8	Safety Functions Incorporating Electronic Technology
C22.2 No. 0.12	Wiring Space and Wire Bending Space in Enclosures for Equipment Rated 750 V or Less
C22.2 No. 18	Outlet Boxes, Conduit Boxes and Fittings
C22.2 No. 21	Cord Sets and Power Supply Cords
C22.2 No. 26	Wireways, Auxiliary Gutters and Associated Fittings
C22.2 No. 35	Extra-Low-Voltage Control Circuit Cables, Low-Energy Control Cable and Extra-Low-Voltage Control Cable
C22.2 No. 38	Thermoset-Insulated Wires and Cables
C22.2 No. 41	Grounding and Bonding Equipment
C22.2 No. 42	General Use Receptacles, Attachment Plugs and Similar Wiring Devices
C22.2 No. 48	Nonmetallic Sheathed Cable
C22.2 No. 49	Flexible Cords and Cables
C22.2 No. 51	Armored Cables
C22.2 No. 52	Service-Entrance Cables
C22.2 No. 56	Flexible Metal Conduit and Liquid-Tight Flexible Metal Conduit
C22.2 No. 62	Surface Raceway Systems
C22.2 No. 65	Wire Connectors
C22.2 No. 75	Thermoplastic-Insulated Wires and Cables
C22.2 No. 96	Portable Power Cables
C22.2 No. 123	Aluminum-Sheathed Cables
C22.2 No. 124	Mineral-Insulated Cable
C22.2 No. 126	Cable Tray Systems
C22.2 No. 127	Equipment and Lead Wires
C22.2 No. 129	Neutral Supported Cable
C22.2 No. 130	Heat Cable Systems
C22.2 No. 131	Type TECK 90 Cable
C22.2 No. 138	Heat Tracing Cable and Cable Sets for Use in Hazardous Locations
C22.2 No. 174	Cables and Cable Glands for Use in Hazardous Locations
C22.2 No. 179	Airport Series Lighting Cables
C22.2 No. 188	Splicing Wire Connectors
C22.2 No. 197	PVC Insulating Tape

Continued on next page >>

Document No.	Title
C22.2 No. 198	Sealed Wire Connector Systems
C22.2 No. 208	Fire Alarm and Signal Cable
C22.2 No. 210	Appliance Wiring Material Products
C22.2 No. 211	Non-Metallic Conduit
C22.2 No. 214	Communications Cables
C22.2 No. 222	Type FCC Under-Carpet Wiring Systems
C22.2 No. 230	Tray Cables
C22.2 No. 232	Optical Fiber Cables
C22.2 No. 239	Control and Instrumentation Cables
C22.2 No. 245	Marine Shipboard Cable
C22.2 No. 262	Optical Fiber Cable and Communications Cable
C22.2 No. 2556	Wire and Cable Test Methods
C49.2	Compact Aluminum Conductors Steel Reinforced (ACSR)
C49.3	Aluminum Alloy 1350 Round Wire, All Tempers for Electrical Purposes
C49.5	Compact Round Concentric-Lay Aluminum-Stranded Conductors (Compact ASC)
C57	Electrical Power Connectors for use in Overhead Line Conductors
C68.1	Paper-Insulated Power Cable
C68.3	Concentric Neutral Power Cable Rated 5-45 kV
C83	Communication and Power Line Hardware
Z243	Information Processing Systems

15.1.2 SCC
Standards Council of Canada
www.scc.ca

Standardization is the development and application of standards—publications that establish accepted practices, technical requirements and terminologies for products, services and systems.

Standards help ensure better, safer and more efficient methods and products, and are an essential element of technology, innovation and trade.

The Standards Council carries out a variety of functions intended to ensure the effective and coordinated operation of standardization in Canada. It also represents Canada's interests on standards-related matters in foreign and international forums.

15.2 CABLE TYPES

Table 15.1—CSA Cable Types

Conditions of Use	Trade Designation	CSA Type Designation	Maximum Allowable Conductor Temperature °C
For exposed wiring in dry locations only	Armored cable	TECK90 AC90	90 90
For exposed wiring in dry locations where exposed to corrosive action, if suitable for corrosive conditions encountered	Armored cable	TECK90	90
For exposed wiring in dry locations where not exposed to mechanical injury	Non-metallic-sheathed cable	NMD90	90
For exposed wiring in dry locations and in Category 1 and 2 locations, where not exposed to mechanical injury	Non-metallic-sheathed cable	NMW, NMWU	60
For exposed wiring in dry or damp locations	Rubber- (thermoset-) insulated cable	R90	90
	Thermoplastic-insulated cable	TW	60
	Nylon jacketed thermoplastic-insulated cable	T90 NYLON	90
	Non-metallic-sheathed cable	NMD90	90

Continued on next page >>

Table 15.1–CSA Cable Types (Continued)

Conditions of Use	Trade Designation	CSA Type Designation	Maximum Allowable Conductor Temperature °C
For exposed wiring in wet locations	Armored cable	TECK90 ACWU90	90 90
	Rubber- (thermoset-) insulated cable	RW75 RL90, RW90	75 90
	Aluminum-sheathed cable	RA75 RA90	75 90
	Mineral-insulated cable	MI, LWMI	90
	Thermoplastic-insulated cable	TW TW75, TWN75	60 75
	Nonmetallic-sheathed cable	NMWU	60
For exposed wiring where subjected to the weather	Armored cable	TECK90	90
	Rubber- (thermoset-) insulated cable	RW75 R90, RW90	75 90
	Thermoplastic-insulated cable	TW, TWU TWU75	60 75
	Neutral-supported cable	NS-75 NS-90	75 90
	Nonmetallic-sheathed cable	NMWU	60
For concealed wiring in dry locations only	Armored cable	TECK90 AC90	90 90
For concealed wiring in dry and damp locations	Nonmetallic-sheathed cable	NMD90	90
For concealed wiring in dry locations and in Category 1 and 2 locations where not exposed to mechanical injury	Nonmetallic-sheathed cable	NMW, NMWU	60

Continued on next page >>

Table 15.1–CSA Cable Types (Continued)

Conditions of Use	Trade Designation	CSA Type Designation	Maximum Allowable Conductor Temperature °C
For concealed wiring in wet locations	Armored cable	TECK90 ACWU90	90 90
	Nonmetallic-sheathed cable	NMWU	60
	Aluminum-sheathed cable	RA75 RA90	75 90
	Mineral-insulated cable	MI, LWMI	90
For use in raceways, except cable trays, in dry or damp locations	Rubber- (thermoset-) insulated cable	R90	90
	Thermoplastic-insulated cable	TW	60
	Nylon jacketed thermoplastic-insulated cable	T90 NYLON	90
For use in raceways, except cable trays, in wet locations	Rubber- (thermoset-) Insulated cable	RW75, RWU75 RW90, RWU90	75 90
	Thermoplastic-insulated cable	TW, TWU TW75, TWN75, TWU75	60 75
For use in ventilated, nonventilated and ladder type cable trays in dry locations only	Armored cable	AC90 TECK90	90 90
For use in ventilated, nonventilated and ladder type cable trays in wet locations	Armored cable	TECK90 ACWU90	90 90
	Aluminum-sheathed cable	RA75 RA90	75 90
	Mineral-insulated cable	MI, LWMI	90
	Rubber- (thermoset-) insulated lead-sheathed cable	RL90	90

Continued on next page >>

Table 15.1–CSA Cable Types (Continued)

Conditions of Use	Trade Designation	CSA Type Designation	Maximum Allowable Conductor Temperature °C
For use in ventilated and nonventilated cable trays in vaults and switch rooms	Rubber- (thermoset-) insulated cable	RW75 RW90	75 90
For direct earth burial (with protection as required by inspection authority)	Armored cable	ACWU90 TECK90	90 90
	Nonmetallic-sheathed cable	NMWU	60
	Rubber- (thermoset-) insulated cable	RWU75 RL90, RWU90	75 90
	Aluminum-sheathed cable	RA75 RA90	75 90
For direct earth burial (with protection as required by inspection authority)	Mineral-insulated cable	MI, LWMI	90
	Thermoplastic-insulated cable	TWU TWU75	60 75
	Airport series lighting cable Tray cable	ASLC TC	90 90
For service entrance above ground	Armored cable	AC90 ACWU90 TECK90	90 90 90
	Aluminum-sheathed cable	RA75 RA90	75 90
	Mineral-insulated cable	MI	90
	Neutral supported cable	NS-75 NSF-90	75 90

Continued on next page >>

Table 15.1—CSA Cable Types (Continued)

Conditions of Use	Trade Designation	CSA Type Designation	Maximum Allowable Conductor Temperature °C
For service entrance below ground	Service-entrance cable	USEI75 USEI90 USEB90	75 90 90
	Thermoplastic-insulated wire	TWU TWU75	60 75
	Rubber- (thermoset-) insulated cable	RWU75 RWU90	75 90
	Armored cable	TECK90 ACWU90	90 90
	Aluminum-sheathed cable	RA75 RA90	75 90
For high-voltage wiring in luminous-tube signs	Luminous-tube sign cable	GTO, GTOL	60
For use in raceways in hoistways	Hoistway cable	—	60
For use in Class 2 circuits, in exposed or concealed wiring or use in raceways, in dry or damp locations	Extra-low-voltage control cable	LVT	60
For use in Class 2 circuits in dry locations in concealed wiring or exposed wiring where not subject to mechanical injury	Extra-low-voltage control cable	ELC	60
For use when concealed indoors under carpet squares in dry or damp locations	Flat conductor cable	FCC	60
For use in communication circuits when exposed, concealed or used in raceways, indoors in dry or damp locations, or in ceiling air handling plenums	Communication cable	CMP, CMR, CMG, CM, CMX, CMH	60

Continued on next page >>

Table 15.1–CSA Cable Types (Continued)

Conditions of Use	Trade Designation	CSA Type Designation	Maximum Allowable Conductor Temperature °C
For use in fire alarm, signal and voice communication circuits where exposed, concealed or used in raceways, indoors in dry or damp locations	Fire alarm and signal cable	FAS FAS 90 FAS 105 FAS 200	60 90 105 200
For use in raceways including ventilated, nonventilated and ladder type cable trays in wet locations and where exposed to weather	Tray cable	TC	As marked on cable
For use in cable trays in Class I Division 2 and Class II Division 2 hazardous locations	Tray cable	TC	As marked on cable
For use in buildings in dry or damp locations, where exposed, concealed or used in raceways including cable trays, or in plenums	Non-conductive optical fiber cable	OFNP, OFNR, OFNG, OFN, OFNH	As marked on cable
For use in buildings in dry or damp locations, where exposed, concealed or used in raceways including cable trays, or in plenums	Conductive optical fiber cable	OFCP, OFCR, OFCG, OFC, OFCH	As marked on cable
For use in buildings in dry or damp locations, where exposed or concealed	Hybrid conductor cable	NMDH90	90
For use in ventilated non-ventilated and ladder type trays, direct earth burial, in ceiling air handling plenums, for exposed or concealed wiring (or damp or dry) locations	Control and instrumentation Cable (without armor)	CIC	250 (dry or damp locations) 90 wet locations)
For use in ventilated non-ventilated and ladder type trays, direct earth burial, for exposed or concealed wiring in wet (or damp or dry) locations in ceiling air handling plenums	Armored control and instrumentation cable (other than steel wire armor)	ACIC	250 (dry or damp locations) 90 wet locations)

Source: Canadian Electrical Code (CSA C22.1, Table 19)

15.3 SUPPLY VOLTAGE AND PLUG CONFIGURATIONS

Table 15.2–Canadian Supply Voltage

Frequency (Hz)	Voltage
60	120/240

The neutral wire of the secondary distribution system is grounded. Three-phase, 4-wire systems such as 120/208 volts are available as well as 347/600 volts for commercial establishments.

Table 15.3–Canadian Plug Configurations

Jack	Plug	Description
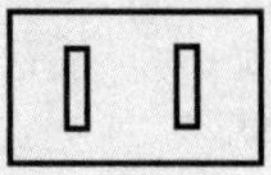	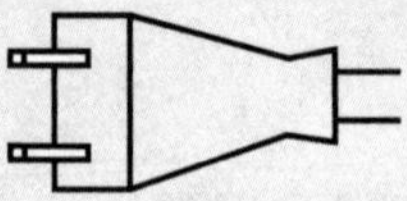	**North American Ungrounded**
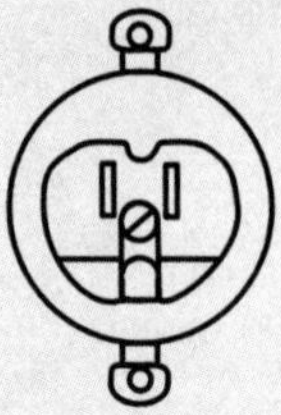		**North American** NEMA 5-15

15.4 FIRE RATINGS

The Canadian Electrical Code, published by the Canadian Standards Association, is the national safety code for electrical installations that is adopted into law by each province and territory with amendments or local rules. The Code includes references to a stringent series of tests developed for flame testing of wires and cables. Cables are marked from "FT1" to "FT6," depending on which of the specified flame test requirements they fulfill.

15.4.1 "FT1" Fire Test

The FT1 test procedure is known as the "Vertical Test" (published in CSA Standard C22.2 No. 0.3 Test Methods for Electrical Wires and Cables, para 4.11.1).

Cables are subjected to five, 15 second duration applications of a specified flame. For the cable to pass the test, burning must cease within 60 seconds after removal of the flame source, and not more than 25 percent of the extended portion of the indicator can be burned.

15.4.2 "FT4" Fire Test

The FT4 test procedure is known as the Vertical Flame Test – Cables in trays (published in CSA Standard C22.2 No. 0.3 Test Methods for Electrical Wires and Cables, para 4.11.4).

Cables are mounted on a vertical tray and exposed for 20 minutes to a 70,000 Btu/hr flame. For the cable to pass the test, the resulting char distance must not be greater than 1.5 meters from the point of flame application.

15.4.3 "FT6" Fire Test

The FT6 test procedure is known as the Plenum Flame Test (published in CSA Standard C22.2 No. 0.3 Test Methods for Electrical Wires and Cables, para 4.11.6).

15.4.4 Designations

The markings for wires and cables meeting the flame spread requirements of the National Building Code of Canada (without additional fire protection) are:

FT1 – Wires and cables that are suitable for installation in buildings of combustible construction; and

FT4 – Wires and cables that are suitable for installation in:
(a) buildings of noncombustible and combustible construction; and
(b) spaces between a ceiling and floor, or ceiling and roof, that may be used as a plenum in buildings of combustible or noncombustible construction.

Wires and cables with combustible outer jackets or sheaths that do not meet the above classifications should be located in noncombustible raceways, masonry walls or concrete slabs.

15.5 SINGLE CONDUCTOR TECK 90 TERMINATIONS

Single conductor high-current armored cables require special termination methods to prevent excessive heating caused by induced "circulating currents" in nearby conductors and "eddy currents" in nearby magnetic materials. One acceptable method is shown in Figure 15.1.

Figure 15.1–Termination of Single Phase Single-Conductor Teck 90 Cables for Circuit Ampacity Over 425 Amps

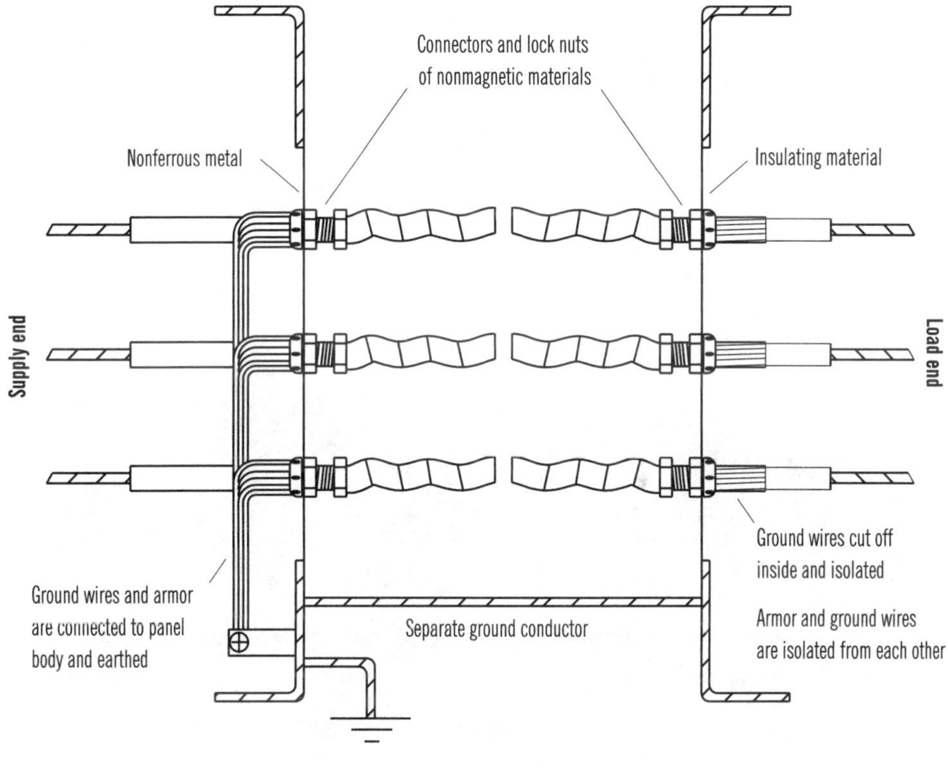

16. ASIA PACIFIC

16.1 AUSTRALIAN STANDARDS

16.1.1 SAA
Standards Australia
www.standards.org.au or www.saiglobal.com

Document No.	Title
AS/NZS 1026	Impregnated Paper Insulated Cables for Electricity Supply at Working Voltages Up To and Including 33 kV (Metric Units)
AS/NZS 1125	Conductors in Insulated Electric Cables and Flexible Cords
AS/NZS 1995	Welding Cables
AS/NZS 2380	Electrical Equipment for Explosive Atmospheres — Explosion Protection Techniques
AS/NZS 2381	Electrical Equipment for Explosive Atmospheres — Selection, Installation and Maintenance
AS/NZS 2430	Classification of Hazardous Areas
AS/NZS 3000	Australian/New Zealand Wiring Rules
AS/NZS 3008.1	Electrical Installations — Selection of Cables: Cables for Alternating Voltages Up To and Including 0.6/1 kV
AS/NZS 3112	Plugs and Socket-Outlets
AS/NZS 3123	Plugs, Socket-Outlets and Couplers for General Industrial Application
AS/NZS 3158	Fibrous-Insulated Electric Cables and Flexible Cables for Working Voltages of 0.6/1 kV
AS/NZS 3166	Cables for High-Voltage Luminous Discharge Tube Installations
AS/NZS 4961	Electric Cables — Polymeric Insulated — For Distribution and Service Applications
AS/NZS 3191	Electric Flexible Cords
AS/NZS 3560	Electric Cables — Aerial-Bundled Voltages Up To and Including 0.6/1 kV
AS/NZS 1972	Mining Cables Other Than Reeling and Trailing
AS/NZS 5000.1	Electric Cables — Polymeric Insulated — For Working Voltages Up To and Including 0.6/1 kV
AS/NZS 5000.2	Electric Cables — Polymeric Insulated — For working voltages Up To 450/750 V
AS/NZS 5000.3	Electric Cables — Polymeric Insulated — Multicore Control Cables
AS/NZS 60702	Mineral-Insulated Cables and Their Terminations Up To 750 V
AS/ACIF S008	Requirements for Authorized Cabling Products
AS/ACIF S009	Installation Requirements for Customer Cabling (Wiring Rules)

16.1.2 ACMA
Australian Communications and Media Authority
www.acma.gov.au

16.1.3 Supply Voltages and Plug Configuration

Table 16.1–Australian Supply Voltages

Frequency (Hz)	Voltage
50	240/415
50	250/440

The neutral wire of the secondary distribution system is grounded. A grounding conductor is required in the electrical cord attached to appliances that are not double insulated.

Table 16.2–Australian Plug Configuration

Jack	Plug	Description
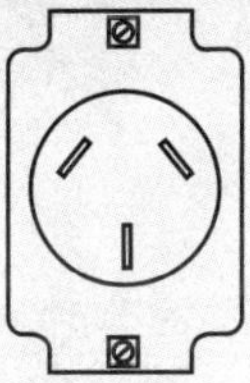		**Australian Standard** AS/NZS 3312

16.1.4 Limiting Temperatures for Insulated Cables

Table 16.3–Limiting Temperatures for Australian Insulated Cables

Type of Cable Insulation	Cable Operating Temperature, °C	
	Normal Use	Minimum
Elastomer compounds		
Type R-EP-90	90	-40
Type R-CSP-90	90	-20
Type R-CPE-90	90	-20
Type R-HF-90	90	-20
Type R-S-150	150	-50
Type R-S-200	200	-50
Thermoplastic compounds		
Type V-75	75	0
Type HFI-75-TP	75	-20
Type V-90	75	0
Type HFI-90-TP	75	-20
Type V-90-HT	75	0
Heat-resisting fibrous insulation		
Type 150	150	0
Type 200	200	0
Type 350	350	0
Cross-linked polyethylene (XLPE)		
Type X-90	90	-70
Type X-HF-90	90	-70

Source: Australian/New Zealand Wiring Rules (AS/NZS 3000)

16.2 SINGAPORE STANDARDS

16.2.1 SPRING Singapore (formerly PSB)
www.spring.gov.sg

Document No.	Title
SS254	Electrical Apparatus for Explosive Gas Atmospheres
SS299 Part 1	Fire Resistant Circuit Integrity Power Cables
SS300	Methods of Test for Gases Evolved During Combustion of Electric Cables
SS358	PVC Insulated Power Cables up to and Including 450/750 V

The following British and international cable standards are also used in Singapore: BS6004, BS6346, BS6500, IEC 60331 and IEC 60502.

16.2.2 Supply Voltage and Plug Configurations

Table 16.4–Singapore Supply Voltage

Frequency (Hz)	Voltage
50	230/400

A grounding conductor is required in the electrical cord attached to appliances.

Table 16.5–Singapore Plug Configurations

Jack	Plug	Description
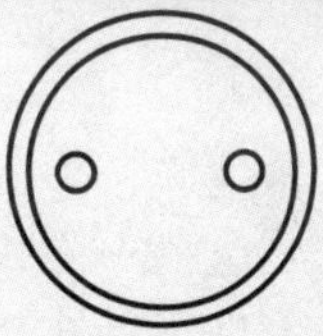		**Ungrounded Europlug** CEE 7/16
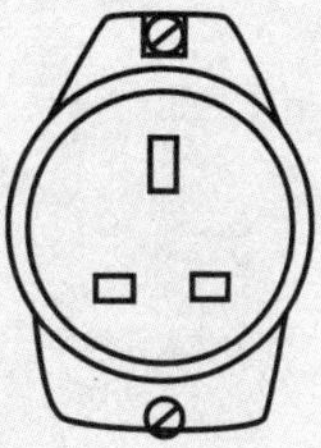		**British Standard** BS 1363

16.3 JAPANESE STANDARDS

16.3.1 JSA
Japanese Standards Association
www.jsa.or.jp

Document No.	Title
JIS C3307	600 V PVC Insulated Wires
JIS C3316	PVC Insulated Wire for Electrical Apparatus
JIS C3401	Control Cables
JIS C3406	Low Voltage Cables for Automobiles
JIS C3605	600 V Polyethylene Insulated Cables
JIS C3606	High-Voltage Cross Linked Polythylene Insulated Cables
JIS C3609	High-Voltage Drop Wires for Pole Transformers
JIS C3612	600 V Flame Retardant Polythylene Insulated Wires
JIS C3621	600 V Ethylene-Propylene Rubber Insulated Cables
JIS C3650	Installation Methods of Cables Embedded in Concrete
JIS C3653	Installation Methods of Cables Buried Underground
JIS C3660	Test Methods for Insulating and Sheathing Materials of Electric Cables
JIS C3662	PVC Insulated Cables Rated Up To 450/750 V
JIS C3663	Rubber Insulated Cables Rated Up To 450/750 V

16.3.2 Plug Configuration

Table 16.6–Japanese Plug Configuration

Jack	Plug	Description
	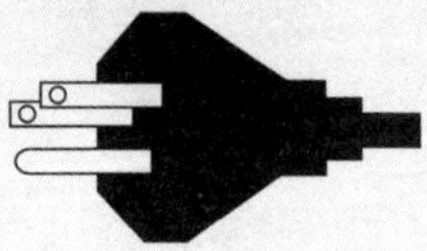	**Japanese Standard** JIS C 8303 (Same as NEMA 5-15 and CSA 22.2, No. 42)

16.4 CHINESE STANDARDS

16.4. CQC

Chinese Quality Certification Center
www.cqc.com.cn

16.4.2 CNIS

Chinese National Institute of Standardization
www.cnis.gov.cn

Document No.	Title
GB 5013	Rubber Insulated Cables of Rated Voltages Up To and Including 450/750 V
GB 5023	Polyvinyl Chloride Insulated Cables for Rated Voltages Up To and Including 450/750 V
GB 12528.1	Insulated Cables (Wires) for Railway Vehicles of Rated Voltage Up To and Including 3 kV
GB 12972	Flexible Rubber-Sheathed Cables for Mining Purposes
GB 15934	Cord Sets

17. CONVERSION TABLES

17.1 METRIC TO ENGLISH CONDUCTOR SIZE

Table 17.1–Conductor Size Conversion: Metric to English and English to Metric

Metric Conductor Area	Standard Metric Sizes	Standard English Sizes	English Conductor Area
(mm^2)	(mm^2)	(kcmil/AWG)	(cmil)
1,015	–	2,000	2,000,000
1,000	1,000	–	1,970,000
900	–	–	1,780,000
888	–	1,750	1,750,000
800	800	–	1,580,000
761	–	1,500	1,500,000
635	–	1,250	1,250,000
630	630	–	1,240,000
626	–	–	1,233,000
520	–	–	1,024,000
508	–	1,000	1,000,000
500	500	–	987,000
449	–	–	884,000
400	400	–	789,000
381	–	750	750,000
380	–	–	748,000
322	–	–	634,000
305	–	600	600,000
300	300	–	592,000
273	–	–	537,000
254	–	500	500,000
240	240	–	474,000
231	–	–	455,000
230	–	400	400,000
195	–	–	384,000
185	185	–	365,000
178	–	350	350,000
165	–	–	325,000
152	–	300	300,000
150	150	–	296,000

Continued on next page >>

ANIXTER

Table 17.1–Conductor Size Conversion: Metric to English and English to Metric (Continued)

Metric Conductor Area	Standard Metric Sizes	Standard English Sizes	English Conductor Area
(mm^2)	(mm^2)	(kcmil/AWG)	(cmil)
140	–	–	276,000
127	–	250	250,000
120	120	–	237,000
119	–	–	234,000
107	–	4/0	211,600
100	–	–	197,000
95	95	–	187,000
85	–	3/0	167,800
72	–	–	141,800
70	70	–	138,000
67	–	2/0	133,100
60.9	–	–	119,900
54	–	1/0	105,600
51.6	–	–	101,600
50	50	–	98,500
43.7	–	–	86,100
42	–	1	83,700
37	–	–	72,900
35	35	–	69,100
34	–	2	66,360
31.3	–	–	61,660
27	–	3	52,620
25	25	–	49,300
22.4	–	–	44,128
21	–	4	41,740
19	–	–	37,430
17	–	5	33,090
16.1	–	–	31,717
16	16	–	31,600
13.6	–	–	26,792
13.3	–	6	26,240
11.5	–	–	22,655
10.5	–	7	20,820

Continued on next page >>

Table 17.1–Conductor Size Conversion: Metric to English and English to Metric (Continued)

Metric Conductor Area	Standard Metric Sizes	Standard English Sizes	English Conductor Area
(mm^2)	(mm^2)	(kcmil/AWG)	(cmil)
10.00	10	–	19,700
9.77	–	–	19,246
8.38	–	8	16,510
8.27	–	–	16,292
7.00	–	–	13,790
6.64	–	9	13,090
6.00	6	–	11,800
5.93	–	–	11,682
5.26	–	10	10,380
5.02	–	–	9,889
4.25	–	–	8,372
4.00	4	–	7,890
3.60	–	–	7,092
3.31	–	12	6,530
2.58	–	–	5,082
2.50	2.5	–	4,930
2.18	–	–	4,294
2.08	–	14	4,110
1.85	–	–	3,644
1.57	–	–	3,093
1.50	1.5	–	2,960
1.33	–	–	2,620
1.31	–	16	2,580
1.12	–	–	2,206
1.00	1	–	1,970
0.95	–	–	1,871
0.82	–	18	1,620
0.75	0.75	–	1,480
0.68	–	–	1,339
0.58	–	–	1,142
0.52	–	20	1,020
0.50	0.50	–	987
0.49	–	–	965

Continued on next page >>

Table 17.1–Conductor Size Conversion: Metric to English and English to Metric (Continued)

Metric Conductor Area	Standard Metric Sizes	Standard English Sizes	English Conductor Area
(mm^2)	(mm^2)	(kcmil/AWG)	(cmil)
0.324	–	22	640
0.205	0.2	24	404
0.128	–	26	253
0.081	–	28	159
0.051	–	30	100
0.032	–	32	63.2
0.020	–	34	39.8
0.013	–	36	25.0
0.0080	–	38	15.7
0.0050	–	40	9.61
0.0032	–	42	6.25
0.0020	–	44	4.00
0.0013	–	46	2.56

Reference: IEC 60228

17.2 CIRCULAR MEASUREMENTS – DIAMETER, CIRCUMFERENCE AND AREA

Table 17.2–Circular Measurements – Diameter, Circumference and Area

Diameter		Circumference	Area
Fractional (in.)	Decimal (in.)	(in.)	(sq. in.)
1/64	0.015625	0.049087	0.00019
1/32	0.031250	0.098175	0.00019
3/64	0.046875	0.147262	0.00173
1/16	0.062500	0.196350	0.00307
3/32	0.093750	0.294524	0.00690
1/8	0.125000	0.392699	0.01227
5/32	0.156250	0.490874	0.01917
3/16	0.187500	0.589049	0.02761
7/32	0.218750	0.687223	0.03758
1/4	0.250000	0.785398	0.04909
9/32	0.281250	0.883573	0.06213
5/16	0.312500	0.981748	0.07670
11/32	0.343750	1.07992	0.09281
3/8	0.375000	1.17810	0.11045
13/32	0.406250	1.27627	0.12962
7/16	0.437500	1.37445	0.15033
15/32	0.468750	1.47262	0.17257
1/2	0.500000	1.57080	0.19635
17/32	0.531250	1.66897	0.22166
9/16	0.562500	1.76715	0.24850
19/32	0.593750	1.86532	0.27688
5/8	0.625000	1.96350	0.30680
21/32	0.656250	2.06167	0.33824
11/16	0.687500	2.15984	0.37122
23/32	0.718750	2.25802	0.40574
3/4	0.750000	2.35619	0.44179
25/32	0.781250	2.45437	0.47937
13/16	0.812500	2.55254	0.51849
27/32	0.843750	2.65072	0.55914
7/8	0.875000	2.74889	0.60132
29/32	0.906250	2.84707	0.64504
15/16	0.937500	2.94524	0.69029
31/32	0.968750	3.04342	0.73708

Continued on next page >>

Table 17.2–Circular Measurements – Diameter, Circumference and Area (Continued)

Diameter		Circumference	Area
Fractional (in.)	Decimal (in.)	(in.)	(sq. in.)
1	1.000000	3.14159	0.78540
1 1/16	1.062500	3.33794	0.88664
1 1/8	1.125000	3.53429	0.99402
1 3/16	1.187500	3.73064	1.1075
1 1/4	1.250000	3.92699	1.2272
1 5/16	1.312500	4.12334	1.3530
1 3/8	1.375000	4.31969	1.4849
1 7/16	1.437500	4.51604	1.6230
1 1/2	1.500000	4.71239	1.7671
1 9/16	1.562500	4.90874	1.9175
1 5/8	1.625000	5.10509	2.0739
1 11/16	1.687500	5.30144	2.2365
1 3/4	1.750000	5.49779	2.4053
1 13/16	1.812500	5.69414	2.5802
1 7/8	1.875000	5.89049	2.7612
1 15/16	1.937500	6.08684	2.9483
2	2.000000	6.28319	3.1416
2 1/16	2.062500	6.47953	3.3410
2 1/8	2.125000	6.67588	3.5466
2 3/16	2.187500	6.87223	3.7583
2 1/4	2.250000	7.06858	3.9761
2 5/16	2.312500	7.26493	4.2000
2 3/8	2.375000	7.46128	4.4301
2 7/16	2.437500	7.65763	4.6664
2 1/2	2.500000	7.85398	4.9087
2 9/16	2.562500	8.05033	5.1572
2 5/8	2.625000	8.24668	5.4119
2 11/16	2.687500	8.44303	5.6727
2 3/4	2.750000	8.63938	5.9396
2 13/16	2.812500	8.83573	6.2126
2 7/8	2.875000	9.03208	6.4978
2 15/16	2.937500	9.22843	6.7771
3	3.000000	9.42478	7.0686
3 1/16	3.062500	9.62113	7.3662
3 1/8	3.125000	9.81748	7.6699
3 3/16	3.187500	10.01380	7.9798

Continued on next page >>

Table 17.2–Circular Measurements – Diameter, Circumference and Area (Continued)

Diameter		Circumference	Area
Fractional (in.)	Decimal (in.)	(in.)	(sq. in.)
3 1/4	3.250000	10.2102	8.2958
3 5/16	3.312500	10.4065	8.6179
3 3/8	3.375000	10.6029	8.9462
3 7/16	3.437500	10.7992	9.2806
3 1/2	3.500000	10.9956	9.6211
3 9/16	3.562500	11.1919	9.9678
3 5/8	3.625000	11.3883	10.3206
3 11/16	3.687500	11.5846	10.6796
3 3/4	3.750000	11.7810	11.0447
3 13/16	3.812500	11.9773	11.4159
3 7/8	3.875000	12.1737	11.7932
3 15/16	3.937500	12.3700	12.1767
4	4.000000	12.5664	12.566
4 1/16	4.062500	12.7627	12.962
4 1/8	4.125000	12.9591	13.364
4 3/16	4.187500	13.1554	13.772
4 1/4	4.250000	13.3518	14.186
4 5/16	4.312500	13.5481	14.607
4 3/8	4.375000	13.7445	15.033
4 7/16	4.437500	13.9408	15.466
4 1/2	4.500000	14.1372	15.904
4 9/16	4.562500	14.3335	16.349
4 5/8	4.625000	14.5299	16.800
4 11/16	4.687500	14.7262	17.257
4 3/4	4.750000	14.9226	17.721
4 13/16	4.812500	15.1189	18.190
4 7/8	4.875000	15.3153	18.665
4 15/16	4.937500	15.5116	19.147
5	5.000000	15.7050	19.635
5 1/16	5.062500	15.9043	20.129
5 1/8	5.125000	16.1007	20.629
5 3/16	5.187500	16.2970	21.135
5 1/4	5.250000	16.4943	21.648
5 5/16	5.312500	16.6897	22.166
5 3/8	5.375000	16.8861	22.691
5 7/16	5.437500	17.0824	23.221

Continued on next page >>

Table 17.2–Circular Measurements – Diameter, Circumference and Area (Continued)

Diameter		Circumference	Area
Fractional (in.)	Decimal (in.)	(in.)	(sq. in.)
5 1/2	5.500000	17.2788	23.758
5 9/16	5.562500	17.4751	24.301
5 5/8	5.625000	17.6715	24.850
5 11/16	5.687500	17.8678	25.406
5 3/4	5.750000	18.0642	25.967
5 13/16	5.812500	18.2605	26.535
5 7/8	5.875000	18.4589	27.109
5 15/16	5.937500	18.6532	27.688
6	6.000000	18.8496	28.274
6 1/8	6.125000	19.2423	29.465
6 1/4	6.250000	19.6350	30.680
6 3/8	6.375000	20.0277	31.919
6 1/2	6.500000	20.4202	33.183
6 5/8	6.625000	20.8131	34.472
6 3/4	6.750000	21.2058	35.785
6 7/8	6.875000	21.5984	37.122
7	7.000000	21.9911	38.485
7 1/8	7.125000	22.3838	39.871
7 1/4	7.250000	22.7765	41.282
7 3/8	7.375000	23.1692	42.718
7 1/2	7.500000	23.5619	44.179
7 5/8	7.625000	23.9546	45.664
7 3/4	7.750000	24.3473	47.173
7 7/8	7.875000	24.7400	48.707
8	8.000000	25.1327	50.265
8 1/8	8.125000	25.5254	51.849
8 1/4	8.250000	25.9181	53.456
8 3/8	8.375000	26.3108	55.088
8 1/2	8.500000	26.7035	56.745
8 5/8	8.625000	27.0962	58.426
8 3/4	8.750000	27.4889	60.132
8 7/8	8.875000	27.8816	61.862
9	9.000000	28.2743	63.617
9 1/8	9.125000	28.6670	65.397
9 1/4	9.250000	29.0597	67.201
9 3/8	9.375000	29.4524	69.029

Continued on next page >>

Table 17.2–Circular Measurements – Diameter, Circumference and Area (Continued)

Diameter		Circumference	Area
Fractional (in.)	Decimal (in.)	(in.)	(sq. in.)
9½	9.500000	29.8451	70.882
9⅝	9.625000	30.2378	72.760
9¾	9.750000	30.6305	74.662
9⅞	9.875000	31.0232	76.589
10	10.000000	31.4159	78.540

Notes: 1. Multiply square inches by 645.16 to calculate mm^2.
2. Multiply inches by 25.4 to calculate mm.
3. Multiply inches by 1,000 to calculate mils.

17.3 LENGTH, WEIGHT, AREA, POWER AND ENERGY

Table 17.3–Conversion Factors

Note: Frequently used conversions are shown in **bold type**. To convert in the reverse direction, divide by the factor given in the table.

To Convert From	To	Multiply By
Area		
Circular mils	Square inches	0.0000007854
Circular mils	Square mils	0.7854
Circular mils	**Square millimeters**	**0.0005067**
Square centimeters	Square inches	0.155
Square feet	Square meters	0.0929
Square inches	Circular mils	1,273,240.00
Square inches	Square centimeters	6.4516
Square inches	Square millimeters	645.16
Square inches	Square mils	1,000,000.00
Square meters	Square feet	10.764
Square millimeters	Square inches	0.00155
Square millimeters	**Circular mils**	**1,973.53**
Square mils	Circular mils	1.2732
Square mils	Square inches	0.000001
Length		
Centimeters	Inches	0.3937
Centimeters	Feet	0.03281
Feet	Centimeters	30.48
Feet	**Meters**	**0.3048**
Inches	Centimeters	2.54
Inches	Meters	0.0254
Inches	**Millimeters**	**25.4**
Inches	Mils	1,000.00
Kilometers	Miles	0.6214
Meters	**Feet**	**3.2808**
Meters	Inches	39.3701
Meters	Yards	1.0936
Miles	Kilometers	1.6093
Millimeters	**Inches**	**0.03937**
Millimeters	Mils	39.3701
Mils	Inches	0.001
Mils	Millimeters	0.0254
Yards	Meters	0.9144

Continued on next page >>

Table 17.3–Conversion Factors (Continued)

To Convert From	To	Multiply By
Power		
Foot-pounds per minute	Horsepower	0.0000303
Foot-pounds per minute	Watts	0.0226
Foot-pounds per second	Horsepower	0.001818
Foot-pounds per second	Watts	1.356
Horsepower	Foot-pounds per minute	33,000.00
Horsepower	Foot-pounds per second	550.00
Horsepower	**Watts**	**746**
Kilogram-meters per sec.	Watts	9.807
Watts	Foot-pounds per minute	44.25
Watts	Foot-pounds per second	0.7375
Watts	Horsepower	0.001341
Watts	Kilogram-meters per sec.	0.1020
Energy		
British thermal units	Foot-pounds	778.00
British thermal units	Joules	1,055.00
British thermal units	Watt-hours	0.293
Foot-pounds	British thermal units	0.001285
Foot-pounds	Joules	1.356
Foot-pounds	Kilogram-meters	0.1383
Gram calories	Joules	4.186
Joules	British thermal units	0.000947
Joules	Ergs	107.00
Joules	Foot-pounds	0.7375
Joules	Gram-calories	0.2388
Joules	Kilogram-meters	0.10198
Kilogram-meters	Foot-pounds	7.233
Kilogram-meters	Joules	9.8117
Watt-hours	British thermal units	3.4126

Continued on next page >>

Table 17.3–Conversion Factors (Continued)

To Convert From	To	Multiply By
Weight		
Kilograms	Pounds	2.205
Kilograms per kilometer	**Pounds per 1,000 feet**	**0.6719**
Ohms per kilometer	Ohms per 1,000 feet	0.3048
Ohms per 1,000 feet	Ohms per kilometer	3.2808
Ohms per 1,000 yards	Ohms per kilometer	1.0936
Pounds	Kilograms	0.4536
Pounds per 1,000 feet	**Kilograms per kilometer**	**1.488**
Pounds per 1,000 yards	Kilograms per kilometer	0.4960
Pounds per 1,000 yards	Pounds per kilometer	1.0936
Newtons	Pound-force	0.2248
Pound-force	Newtons	4.4482
Miscellaneous		
Diameter circle	Circumference	3.1416
Diameter circle	Side of equal square	0.8862
Diameter sphere cubed	Volume of sphere	0.5236
U.S. gallons	Imperial gallons (British)	0.8327
U.S. gallons	Cubic feet	0.1337
U.S. gallons	Pounds of water (20°C)	8.33
Cubic feet	Pounds of water (4°C)	62.427
Feet of water (4°C)	Pounds per square inch	0.4336
Inches of mercury (0°C)	Pounds per square inch	0.4912
Knots	Miles per hour	1.1516

17.4 TEMPERATURE CONVERSION

Table 17.4–Degrees Celsius (Centigrade) vs. Degrees Fahrenheit

Conversion Formula: °F = 9/5°C + 32°
°C = 5/9 (°F – 32°)

°C	°F	°C	°F	°C	°F	°C	°F	°C	°F	°C	°F	°C	°F	°C	°F
-80	-112.0	-15	5.0	15	59.0	45	113.0	75	167.0	105	221.0	135	275.0	600	1,112
-70	-94.0	-14	6.8	16	60.8	46	114.8	76	168.8	106	222.8	136	276.8	700	1,292
-60	-76.0	-13	8.6	17	62.6	47	116.6	77	170.6	107	224.6	137	278.6	800	1,472
-50	-58.0	-12	10.4	18	64.4	48	118.4	78	172.4	108	226.4	138	280.4	900	1,652
-45	-49.0	-11	12.2	19	66.2	49	120.2	79	174.2	109	228.2	139	282.2	1,000	1,832
-40	-40.0	-10	14.0	20	68.0	50	122.0	80	176.0	110	230.0	140	284.0	1,000	2,012
-39	-38.2	-9	15.8	21	69.8	51	123.8	81	177.8	111	231.8	141	285.8	1,200	2,192
-38	-36.4	-8	17.6	22	71.6	52	125.6	82	179.6	112	233.6	142	287.6	1,300	2,372
-37	-34.6	-7	19.4	23	73.4	53	127.4	83	181.4	113	235.4	143	289.4	1,400	2,552
-36	-32.8	-6	21.2	24	75.2	54	129.2	84	183.2	114	237.2	144	291.2	1,500	2,732
-35	-31.0	-5	23.0	25	77.0	55	131.0	85	185.0	115	239.0	145	293.0	1,600	2,912
-34	-29.2	-4	24.8	26	78.8	56	132.8	86	186.8	116	240.8	146	294.8	1,700	3,092
-33	-27.4	-3	26.6	27	80.6	57	134.6	87	188.6	117	242.6	147	296.6	1,800	3,272
-32	-25.6	-2	28.4	28	82.4	58	136.4	88	190.4	118	244.4	148	298.4	1,900	3,452
-31	-23.8	-1	30.2	29	84.2	59	138.2	89	192.2	119	246.2	149	300.2	2,000	3,632
-30	-22.0	*0	*32.0	30	86.0	60	140.0	90	194.0	120	248.0	150	302.0	–	–
-29	-20.2	1	33.8	31	87.8	61	141.8	91	195.8	121	249.8	160	320.0	–	–
-28	-18.4	2	35.6	32	89.6	62	143.6	92	197.6	122	251.6	170	338.0	–	–
-27	-16.6	3	37.4	33	91.4	63	145.4	93	199.4	123	253.4	180	356.0	–	–
-26	-14.8	4	39.2	34	93.2	64	147.2	94	201.2	124	255.2	190	374.0	–	–
-25	-13.0	5	41.0	35	95.0	65	149.0	95	203.0	125	257.0	200	392	–	–
-24	-11.2	6	42.8	36	96.8	66	150.8	96	204.8	126	258.8	210	410	–	–
-23	-9.4	7	44.6	37	98.6	67	152.6	97	206.6	127	260.6	220	428	–	–
-22	-7.6	8	46.4	38	100.4	68	154.4	98	208.4	128	262.4	230	446		
-21	-5.8	9	48.2	39	102.2	69	156.2	99	210.2	129	264.2	240	464	–	–
-20	-4.0	10	50.0	40	104.0	70	158.0	†100	†212.0	130	266.0	250	482	–	–
-19	-2.2	11	51.8	41	105.8	71	159.8	101	213.8	131	267.8	300	572	–	–
-18	-0.4	12	53.6	42	107.6	72	161.6	102	215.6	132	269.6	350	662	–	–
-17	1.4	13	55.4	43	109.4	73	163.4	103	217.4	133	271.4	400	752	–	–
-16	3.2	14	57.2	44	111.2	74	165.2	104	219.2	134	273.2	500	932	–	–

*Freezing †Boiling

17.5 KVA TO AMPERES

Table 17.5–kVA to Amperes

Single Phase Circuits			
kVA Rating	Current in Amperes at: 120 V	240 V	480 V
1	8.33	4.16	2.08
1.5	12.5	6.24	3.12
2	16.66	8.33	4.16
3	25	12.5	6.1
5	41	21	10.4
7.5	62	31	15.6
10	83	42	21
15	124	62	31
25	208	104	52
37.5	312	156	78
50	416	208	104
75	624	312	156
100	830	415	207
167	1,390	695	348
200	1,660	833	416

Continued on next page >>

Table 17.5–kVA to Amperes (Continued)

Three Phase Circuits				
kVA Rating	Current in Amperes at: 208 V	240 V	480 V	600 V
3	8.3	7.2	3.6	2.9
6	16.6	14.4	7.2	5.8
9	25.0	21.6	10.8	8.7
15	41.6	36	18	14.4
30	83.0	72	36	28.8
45	125	108	54	43
75	208	180	90	72
112.5	312	270	135	108
150	415	360	180	144
200	554	480	240	192
225	625	540	270	216
300	830	720	360	288
400	1,110	960	480	384
500	1,380	1,200	600	480
750	2,080	1,800	900	720

17.6 HORSEPOWER TO AMPERES

Table 17.6–Horsepower to Amperes

Three Phase Motor			
Horsepower Rating	Current in Amperes at: 208 V	240 V	480 V
1	2.9	2.5	1.2
2	5.8	5.0	2.5
5	14.4	12.5	6.2
10	28.8	25.0	12.5
20	57.6	49.9	25.0
50	144	125	62.4
100	288	250	125
200	576	500	250
300	–	–	624

Note: Based on a motor efficiency of 90 percent and system power factor of 0.8. Above values are theoretical values based solely on numerical calculations. Check with applicable codes for specific ampere values to size overcurrent protective devices or conductors. For example, see Table 430.250 of the National Electrical Code.

18. FORMULAS AND CONSTANTS

18.1 ELECTRICAL PROPERTIES OF CIRCUITS

Table 18.1–Electrical Properties of Circuits

	Alternating Current		Direct Current
Desired Data	**Single Phase**	**Three Phase**	
Kilowatts (kw)	$\frac{I \times V \times \cos\theta}{1{,}000}$	$\frac{1.73 \times I \times V \times \cos\theta}{1{,}000}$	$\frac{I \times V}{1{,}000}$
Kilovolt-amperes (kVA)	$\frac{I \times V}{1{,}000}$	$\frac{1.73 \times I \times V}{1{,}000}$	$\frac{I \times V}{1{,}000}$
Horsepower output	$\frac{I \times V \times \cos\theta \times \text{Eff.}}{746}$	$\frac{1.73 \times I \times V \times \cos\theta \times \text{Eff.}}{746}$	$\frac{I \times V \times \text{Eff.}}{746}$
Amperes (I) when horsepower is known	$\frac{\text{hp} \times 746}{V \times \cos\theta \times \text{Eff.}}$	$\frac{\text{hp} \times 746}{1.73 \times V \times \cos\theta \times \text{Eff.}}$	$\frac{\text{hp} \times 746}{V \times \text{Eff.}}$
Amperes (I) when kilowatts are known	$\frac{\text{kw} \times 1{,}000}{V \times \cos\theta}$	$\frac{\text{kw} \times 1{,}000}{1.73 \times V \times \cos\theta}$	$\frac{\text{kw} \times 1{,}000}{V}$
Amperes (I) when kilovolt-amperes are known	$\frac{\text{kVA} \times 1{,}000}{V}$	$\frac{\text{kVA} \times 1{,}000}{1.73 \times V}$	$\frac{\text{kVA} \times 1{,}000}{V}$

$\cos\theta$ = Power factor of load (pf)
V = Volts between conductors
Eff. = Efficiency of motor
I = Current (amperes)
kw = Kilowatts
kVA = Kilovolt–amperes
hp = Horsepower

18.2 RESISTANCE AND WEIGHT OF CONDUCTORS

The resistance and weight of any uncoated copper wire at 20°C (68°F) having a conductivity of 100 percent IACS may be calculated from the following formulas:

$$\text{Ohms per 1,000 feet} = \frac{0.0081455}{\text{Cross-sectional area in sq. in.}} \text{ or } \frac{10371.176}{\text{Cross-sectional area in cmils}}$$

$$\text{Pounds per 1,000 feet} = \text{Area in sq. in.} \times 3{,}854.09 \text{ or area in cmils} \times 0.0030269$$

18.3 RESISTANCE, INDUCTANCE AND CAPACITANCE IN AC CIRCUITS

Table 18.2–Resistance, Inductance and Capacitance in AC Circuits

If Circuit Contains	Reactance	Impedance	"V" for Current "I"	Power Factor
Resistance (R) **Only**	0	R	IR	1
Inductance (L) **Only**	$2\pi fL$	$2\pi fL$	$I2\pi fL$	0
Capacitance (C) **Only**	$\frac{1}{2\pi fC}$	$\frac{1}{2\pi fC}$	$I\frac{1}{2\pi fC}$	0
Resistance and Inductance in Series (R and L)	$2\pi fL$	$\sqrt{R^2 + (2\pi fL)^2}$	$I\sqrt{R^2 + (2\pi fL)^2}$	$\frac{R}{\sqrt{R^2 + (2\pi fL)^2}}$
Resistance and Capacitance in Series (R and C)	$\frac{1}{2\pi fC}$	$\sqrt{R^2 + \left(\frac{1}{2\pi fC}\right)^2}$	$I\sqrt{R^2 + \left(\frac{1}{2\pi fC}\right)^2}$	$\frac{R}{\sqrt{R^2 + \left(\frac{1}{2\pi fC}\right)^2}}$
Resistance, Inductance and Capacitance in Series (R, L and C)	$2\pi fL - \frac{1}{2\pi fC}$	$\left(2\pi fL - \frac{1}{2\pi fC}\right)^2$	$\sqrt{R^2 + \left(2\pi fL - \frac{1}{2\pi fC}\right)^2}$	$\frac{R}{\sqrt{R^2 + \left(2\pi fL - \frac{1}{2\pi fC}\right)^2}}$

V = Voltage in volts
R = Resistance in ohms
I = Current in amperes
C = Capacitance in farads
L = Inductance in henries
π = 3.1416
f = Frequency in cycles per second

18.4 SERIES AND PARALLEL CONNECTIONS

Table 18.3–Series and Parallel Connections

	Resistance (R)	Inductance (L)	Capacitance (C)
Series	$R = R_1 + R_2 + R_3 + \ldots$	$L = L_1 + L_2 + L_3 + \ldots$	$\frac{1}{C} = \frac{1}{C_1} + \frac{1}{C_2} + \frac{1}{C_3} + \ldots$
Parallel	$\frac{1}{R} = \frac{1}{R_1} + \frac{1}{R_2} + \frac{1}{R_3} + \ldots$	$\frac{1}{L} = \frac{1}{L_1} + \frac{1}{L_2} + \frac{1}{L_3} + \ldots$	$C = C_1 + C_2 + C_3 + \ldots$

18.5 ENGINEERING NOTATION

Table 18.4–Engineering Notation

		Multiplying Factor	
Prefix	**Symbol**	**Scientific**	**Conventional**
tera	T	10^{12}	1,000,000,000,000
giga	G	10^{9}	1,000,000,000
mega	M	10^{6}	1,000,000
kilo	k	10^{3}	1,000
hecto	h	10^{2}	100
deca	da	10^{1}	10
deci	d	10^{-1}	0.1
centi	c	10^{-2}	0.01
milli	m	10^{-3}	0.001
micro	μ	10^{-6}	0.000001
nano	n	10^{-9}	0.000000001
pico	p	10^{-12}	0.000000000001
femto	f	10^{-15}	0.000000000000001
atto	a	10^{-18}	0.000000000000000001

Table 18.5–Engineering Notation

$e = 2.7183$	
$\pi = 3.1416$	
$\sqrt{2} = 1.4142$	
$\sqrt{3} = 1.7321$	
$\pi/4 = 0.7854$	
1/C =	one conductor
3/C =	three conductor
$>$	greater than
$\leq$	less than or equal to
$<$	less than
$\geq$	greater than or equal to

18.6 DIAMETER OF MULTICONDUCTOR CABLES

To calculate the overall diameter of a group of round conductors of uniform diameters twisted together, multipy the diameter of an individual conductor by the applicable factor below.

Table 18.6–Diameter of Multiconductor Cables

Number of Conductors	Factor
1	1.000
2	2.000
3	2.155
4	2.414
5	2.700
6	3.000
7	3.000
8	3.310
9	3.610

18.7 DETERMINATION OF LARGEST POSSIBLE CONDUCTOR IN CABLE INTERSTICES

The following factors permit the calculation of the maximum size conductor that will fit into the interstices (open spaces) of various conductor configurations, while keeping within a circumscribing circle. Multiply the diameter of one main conductor by the factor from the chart below to obtain the largest diameter that will fit into the interstices.

Table 18.7–Determination of Largest Possible Conductor in Cable Interstices

Number of Conductors	Factor
2	0.667
3	0.483
4	0.414
5	0.377
6	0.354

18.8 CONDUCTOR DIAMETER FROM WIRE DIAMETER

To calculate the nominal diameter of any concentric-lay-stranded conductor made from round wires of uniform diameters, multiply the diameter of an individual wire by the applicable factor below:

Table 18.8–Concentric Stranded Conductor Diameter from Wire Diameter

Number of Wires in Conductor	Factor to Calculate Conductor Diameter
3	2.155
7	3.000
12	4.155
19	5.000
37	7.000
61	9.000
91	11.00
127	13.00
169	15.00
217	17.00
271	19.00

For a greater number of wires use the formula: Conductor Diameter $=$ Wire Diameter $\times \sqrt{1.332 \times \text{No. of Wires}}$

18.9 COAXIAL CAPACITANCE

$$C = \frac{7.354e}{\text{Log}_{10}(1 + 2t/d)}$$

Where:

- C is capacitance in picofarads per foot
- e is the dielectric constant (also known as SIC)
- t is insulation thickness in mils
- d is diameter over the conductor (diameter under the insulation) in mils

Other forms of this equation include:

$$C_1 = \frac{0.0169e}{\text{Ln}\left(\frac{D}{d}\right)}$$

or

$$C_2 = \frac{0.0556e}{\text{Ln}\left(\frac{D}{d}\right)}$$

Where:

- C_1 is capacitance in microfarads per 1,000 ft.
- C_2 is capacitance in microfarads per kilometer
- e is the dielectric constant
- D is diameter over the insulation
- d is diameter under the insulation

18.10 INDUCTIVE REACTANCE

The inductive reactance of a shielded 3-conductor medium-voltage power cable at 60 Hz can be calculated with the following formulas:

$$X_L = 0.023 \text{ Ln}\left(\frac{GMD}{GMR}\right) \text{ ohms/1,000 ft.}$$

or

$$X_L = 0.0754 \text{ Ln}\left(\frac{GMD}{GMR}\right) \text{ ohms/km}$$

Where:

- GMD = geometric mean distance (equivalent conductor spacing)
- GMR = geometric mean radius of conductor

For conductors in a triplexed configuration, GMD is equal to the center-to-center spacing. For round, concentric stranded conductors, GMR ranges from 0.363d for a 7-wire strand up to 0.386d for a 61-wire strand where "d" is the diameter of the conductor.

GLOSSARY

0–10 V – A common analog process control signal voltage range.

4–20 mA – A common analog process control signal current range.

A

A – Common abbreviation for Ampere (see ampere)

AAR – American Association of Railroads.

ABRASION RESISTANCE – Ability to resist surface wear.

AB Switch – A coaxial cable switch capable of switching one cable to one of two branch cables, A or B.

AC – (1) Alternating current, (2) A UL cable type with flexible metal tape armor.

ACAR – Aluminum Conductor, Aluminum-Reinforced cable.

ACCELERATED LIFE TEST – A test in which a cable is subjected to extreme conditions to determine the life of a cable.

ACCR – Aluminum Conductor, Composite Reinforced aerial cable. Contains a ceramic strength member to reduce sag at high temperatures (up to 210°C).

ACRF – Attenuation to Crosstalk Ratio Far-end. See ELFEXT.

ACSR – Aluminum Conductor, Steel Reinforced. A bare composite of aluminum and steel wires, usually aluminum around steel.

ACSR/AW – Aluminum Conductor, Steel Reinforced, using Aluminum clad steel Wire.

ACSR/AZ – Aluminum Conductor, Steel Reinforced, using Aluminum coated steel wire.

ACSR/GA – Aluminum Conductor, Steel Reinforced, using Class A zinc-coated steel wire.

ACSR/GB – Aluminum Conductor, Steel Reinforced, using Class B zinc-coated steel wire.

ACSR/GC – Aluminum Conductor, Steel Reinforced, using Class C zinc-coated steel wire.

A/D – Analog/Digital. An integrated circuit device that converts analog signals to digital signals.

ADDRESS – The location of a terminal, a peripheral device, a node, or any other unit or component in a network, or process control system.

ADHESIVE-BONDED – Cables bonded by adding an adhesive coating to the surface of the cable components, then joining and curing the adhesive to form a cable. See Bonded Cables.

ADMITTANCE – A measure of how easily alternating current flows in a circuit. Admittance is the reciprocal of impedance. It is expressed in mhos.

AEIC – Association of Edison Illuminating Companies.

AERIAL CABLE – A cable suspended in the air on poles or other overhead structure.

AF – Audio Frequency.

AFCI – Arc Fault Circuit Interrupter. A protective device that detects arcing and then turns the circuit off.

AGC – Automatic Gain Control.

AGING – The irreversible change of material properties after exposure to an environment for an interval of time.

AIA – Aluminum Interlocked Armor. A type of cable sheath.

AIR CORE CABLE – A cable in which the interstices in the cable core are not filled with a moisture barrier.

AIRCRAFT WIRE – An electrical wire primarily designed for the extreme conditions (temperature, altitude, solvents, fuels, etc.) of airborne equipment.

AIR SPACED COAX – A coaxial cable in which air is basically the dielectric material. The conductor may be centered by means of a spirally wound synthetic filament, beads or braided filaments. This construction is also referred to as an air dielectric.

AL – Aluminum

ALLOY – A substance (usually metallic) composed of two or more individual substances.

ALS – A type of cable consisting of insulated conductors enclosed in a continuous, closely fitting aluminum tube.

ALTERNATING CURRENT – Electric current that periodically reverses direction. Alternating current is generally abbreviated AC.

AM – Amplitude Modulation. A method of adding information to an electronic signal where the height (amplitude) of the wave is changed to convey the added information.

AMBIENT – Conditions existing at a location prior to energizing of equipment (example: ambient temperature).

AMPACITY – The rms current which a device can carry within specified temperature limitations in a specified environment: dependent upon a) temperature rating, b) power loss, c) heat dissipation.

AMPERE – A standard unit of current. Designated as the amount of current that flows when one volt of emf is applied across one ohm of resistance. An ampere of current is flowing when one coulomb of charge is passing a point every second.

AMPERE-TURN – The product of amperes times the number of turns in a coil.

AMPLIFIER – A device used to boost the strength of an electronic signal.

AMPLITUDE – The maximum value of a varying wave form.

AMPLITUDE MODULATION (AM) – Transmission method in which variations in the voltage or current waveform of a signal carry encoded information.

ANALOG – Not digital. A continuously varying waveform.

ANNEAL – To soften and relieve strains in any solid material, such as metal or glass, by heating to just below its melting point and then slowly cooling it. This also generally lowers the tensile strength of the material, while improving its flex life.

ANNEALED WIRE – See Soft Wire.

ANNULAR CONDUCTOR – A number of wires stranded in reversed concentric layers around a core.

ANNUNCIATOR WIRE – Usually single solid copper, sometimes twisted pair or triplexed for open wiring of bell circuits and other low-voltage systems.

ANSI (American National Standards Institute) – An organization that publishes nationally recognized standards.

ANTENNA LEAD-IN WIRE – (Not coaxial) Parallel twin lead construction, plastic jacketed with fixed 300 ohm impedance for connecting a remote antenna to a receiver.

ANTENNA ROTOR CABLE – Multiconductor flat or round cable used to supply power to a motorized antenna and control wires for changing direction of rotation.

ANTIOXIDANT – Retards or prevents degradation of materials exposed to oxygen (air).

APPLIANCE WIRE AND CABLE – A classification covering insulated wire and cable for internal wiring of appliances and equipment. UL type AWM wire.

ARC RESISTANCE – The time required for an arc to establish a conductive path in a material.

ARMATURE – (1) Rotating machine: the member in which alternating voltage is generated; (2) electromagnet: the member which is moved by magnetic force.

ARMOR – Mechanical protector for cables; usually a helical winding of metal tape, formed so that each convolution locks mechanically upon the previous one (interlocked armor); may be a formed metal tube or a helical wrap of wires.

ARMOR-X – Southwire's trademark for CCW aluminum armor.

ARRHENIUS PLOT – A statistical method used to predict time-to-failure, based on a device's performance at different temperatures. One method is given in IEEE Standard 101.

ASCII – American National Standard Code for Information Interchange. A seven bit plus parity code established by the American National Standards Institute to achieve compatibility among data services and consisting of 96 displayed upper and lower case characters and 32 nondisplayed control codes.

ASKAREL – A synthetic insulating oil which is nonflammable but very toxic. It has been replaced by silicone oils.

ASTM – American Society for Testing Materials. An organization that sets standards on various material tests for industry.

ATTENUATION – The decrease in magnitude of a signal as it travels through any transmitting medium, such as a cable or circuitry. Attenuation is measured as a ratio or as the logarithm of a ratio (decibel).

ATTENUATION CONSTANT – A rating for a cable or other transmitting medium, which is the relative rate of amplitude decrease of voltage or current in the direction of travel. It is measured in decibels per unit length of cable.

AUDIO – A term used to describe sounds within the range of human hearing. Also used to describe devices which are designed to operate within this range.

AUDIO FREQUENCY – The range of frequencies audible to the human ear. Usually 20–20,000 Hz.

AUI – Attachment Unit Interface. The interface between the Ethernet/IEEE 802.3 controller and the baseband transceiver or broadband modem.

AWG – American Wire Gauge. A wire diameter specification. The lower the AWG number the larger the wire diameter.

AWM – Appliance Wiring Material.

B

BACKFILL – The materials placed to fill an excavation, such as sand in a trench.

BALANCED CIRCUIT – A circuit so arranged that the impressed voltages on each conductor of the pair are equal in magnitude but opposite in polarity with respect to ground.

BALANCED LINE – A cable having two identical conductors with the same electromagnetic characteristics in relation to other conductors and to ground.

BALLAST – A device designed to stabilize current flow.

BAND MARKING – A continuous circumferential band applied to a conductor at regular intervals for identification.

BANDWIDTH – The width of a communication channel, measured as frequency (in cycles per second, or hertz). A channel's bandwidth is a major factor in determining how much information it can carry.

BARE CONDUCTOR – A conductor having no insulation or jacket.

BARREL-PACKED – Method of coiling wire into a drum for shipment.

BASEBAND – A signalling technique in which the signal is transmitted in its original form and not changed by modulation.

BASEBAND LAN – A local area network employing baseband signalling.

BELDFOIL – Belden trademark for a highly effective electrostatic shield using reinforced metallic foil.

BELT – Layers of insulation on a conductor, or layers of jacket on a cable.

BELTED-TYPE CABLE – Multiple conductor cable having a layer of insulation over the assembled insulated conductors.

BER – Bit Error Rate. The ratio of received bits that are in error, relative to a specific number of bits received; usually expressed as a number referenced to a power of 10.

BIL – Basic Impulse Level. The crest value of a lightning impulse voltage of a specified wave shape which a high-voltage cable or termination is required to withstand under specified conditions.

BIMETALLIC WIRE – A wire formed of two different metals joined together (not alloyed). It can include wire with a steel core, plated or coated wire.

BINDER – A tape or thread used for holding assembled cable components in place.

BINDING POST – A device for clamping or holding electrical conductors in a rigid position.

BIRDCAGE – The undesirable unwinding of a stranded cable.

BIT – Abbreviation for binary digit. A unit of information equal to one binary decision or the designation of one of two possible and equally likely states (such as 1 and 0) of anything used to store or convey information.

BITS PER SECOND (bps) – The number of bits of data transmitted through a digital process control cable in one second.

BNC – Common connector for coax. BNC is said to be an abbreviation for Bayonet-Neill-Concelman.

BONDED CABLE – Cable consisting of preinsulated conductors or multiconductor components laid in parallel and bonded into a flat cable.

BONDED CONSTRUCTION – An insulation construction in which the glass braid and nylon jacket are bonded together.

BONDING – The method used to produce good electrical contact between metallic parts of any device. Used extensively in automobiles and aircraft to prevent static buildup. Also refers to the connectors and straps used to ground equipment.

BOOSTER – A device inserted into a line (or cable) to increase the voltage. Boosting generators are also used to raise the level of a DC line. Transformers are usually employed to boost AC voltages. The term booster is also applied to antenna preamplifiers.

BOOT – (1) Protective coating over a cable, wire or connector in addition to the normal jacketing or insulation. (2) A form placed around the wire termination of a multicontact connector to contain the liquid potting compound before it hardens.

BORDER LIGHT CABLE – Same as stage cable but more than 2 conductors. Type SO cable is often used.

BORE HOLE CABLE – Power and/or communication cable suspended down a vertically drilled hole to equipment undergound.

BRAID – Textile or metallic filaments interwoven to form a tubular structure that may be applied over one or more wires or flattened to form a strap.

BRAID ANGLE – The smaller of the angles formed by the shielding strand and the axis of the cable being shielded.

BRAID CARRIER – A spool or bobbin on a braiding machine which holds one group of strands or filaments consisting of a specific number of ends. The carrier revolves during braiding operations.

BRAID ENDS – The number of strands used to make up one carrier. The strands are wound side by side on the carrier bobbin and lie parallel in the finished braid.

BRAIDING MACHINE – Machine used to apply braids to wire and cable and to produce braided sleeving and braids for tying or lacing purposes. Braiding machines are identified by the number of carriers.

BRANCH JOINT – A cable joint used for connecting one or more cables to a main cable.

BRAZING – The joining of ends of two wires, rods, or groups of wires with nonferrous filler metal at temperatures above 800°F (427°C).

BREAKDOWN (PUNCTURE) – A disruptive discharge through the insulation.

BREAKDOWN VOLTAGE – The voltage at which the insulation between two conductors breaks down.

BREAKING STRENGTH – The maximum load that a conductor can withstand when tested in tension to rupture.

BREAKOUT – The point at which a conductor or group of conductors breaks out from a multiconductor cable to complete circuits at various points along the main cable.

BRIDGE – A circuit which measures by balancing four impedances through which the same current flows:

Wheatstone measures resistance
Kelvin measures low resistance
Schering measures capacitance, dissipation factor, dielectric constant
Wien measures capacitance, dissipation factor

BRIDGED TAP – The multiple appearances of the same cable pair at several distribution points.

BRITISH STANDARD WIRE GAUGE – A modification of the Birmingham Wire Gauge and the legal standard of Great Britain for all wires. Also known as Standard Wire Gauge (SWG), New British Standard (NBS), English Legal Standard and Imperial Wire Guide.

BROADBAND LAN – LAN which uses FDM (frequency division multiplexing) to divide a single physical channel into a number of smaller independent frequency channels. The different channels created by FDM can be used to transfer different forms of information – voice, data and video.

BROADCAST – The act of sending a signal from one station on a LAN to all other stations.

B and S – Brown and Sharpe wire gauge – same as AWG.

BSL (Basic Switching impulse insulation Level) – The crest value of a switching impulse voltage of a specified wave shape which a high-voltage cable termination is required to withstand under specified conditions.

BUFFER – A protective coating in intimate contact with an optical fiber.

BUFFER TUBE – A loose, crush-resistant polymer tube applied over optical fibers to provide mechanical protection.

BUILDING WIRE – Commercial wires used in the building trades such as: Types RHH, RHW, THW and THHN wire.

BUNA – A synthetic rubber insulation of styrenebutadiene; was known as GR-S, now as SBR.

BUNCH STRAND – A conductor in which all individual wires are twisted in the same direction without regard for geometrical arrangement.

BUNCHER – A machine that twists wires together in a random arrangement.

BUOYANT CABLE – Originally military type MIL-C-2401 with built-in floatation ability. Many applications have been developed using buoyancy to advantage – numerous types and sizes for power, communications, telecommunications have resulted.

BURIED CABLE – A cable installed directly in the earth without use of underground conduit. Also called "direct burial cable."

BUS – A network topology which functions like a signal line which is shared by a number of nodes.

BUS-BAR WIRE – Uninsulated tinned copper wire used as a common lead.

BUSHING – A mechanical device used as a lining for an opening to prevent abrasion to wire and cable.

BUTT SPLICE – A splice wherein two wires from opposite ends butt against each other, or against a stop, in the center of a splice.

BUTT WRAP – Tape wrapped around an object or conductor in an edge-to-edge condition.

BUTYL RUBBER – Synthetic rubber formerly used for electrical insulating purposes.

BX – A common type of armored building wire rated 600 volts.

BYTE – Generally, an 8-bit quantity of information, used mainly in referring to parallel data transfer, semiconductor capacity and data storage; also generally referred to in data communications as an octet or character.

C

C – Symbol for capacitance and Celsius.

CABLE – A cable may be a small number of large conductors or a large number of small conductors, cabled together, usually color coded and with a protective jacket overall.

CABLE ASSEMBLY – A cable assembly is a cable with plugs or connectors on each end for a specific purpose. It may be formed in various configurations.

CABLE, BELTED – A multiconductor cable having a layer of insulation over the assembled insulated conductors.

CABLE CLAMP – A device used to give mechanical support to the wire bundle or cable at the rear of a plug or receptacle.

CABLE CLAMP ADAPTER – A mechanical adapter that attaches to the rear of a plug or receptacle to allow the attachment of a cable clamp.

CABLE CORE – The portion of an insulated cable lying under a protective covering.

CABLE CORE BINDER – A wrapping of tapes or cords around the conductors of a multiple-conductor cable used to hold them together.

CABLE FILLER – The material used in multiple-conductor cables to occupy the interstices formed by the assembly of the insulated conductors, thus forming a cable core.

CABLE JOINT – A complete insulated splice, or group of insulated splices, contained within a single protective covering or housing. In some designs, the insulating material may also serve as the protective covering.

CABLE LOSS – The amount of RF (radio frequency) signal attenuated by coaxial cable transmission. The cable attenuation is a function of frequency, media type and cable length. For coaxial cable, higher frequencies have greater loss than lower frequencies and follow a logarithmic function. Cable losses are usually calculated for the highest frequency carried on the cable.

CABLE, PRESSURIZED – A cable having a pressurized fluid (gas or oil) as part of the insulation; nitrogen and oil are the most common fluids.

CABLE SHEATH – The protective covering applied to cables.

CABLE SUPPORT – A device to mount a cable on a supporting member.

CABLING – The method by which a group of insulated conductors is mechanically assembled (or twisted together).

CAD – Computer-Aided Design.

CAM – Computer-Aided Manufacture.

CAPACITANCE – Capacitance is that property of a system of conductors and dielectrics which permits the storage of electricity when potential differences exist between the conductors.

CAPACITANCE COUPLING – Electrical interaction between two conductors caused by the potential difference between them.

CAPACITANCE, DIRECT – The capacitance measured from one conductor to another conductor through a single insulating layer.

CAPACITANCE, MUTUAL – The capacitance between two conductors (typically of a pair) with all other conductors, including shield, short circuited to ground.

CAPACITANCE, UNBALANCE – An inequality of capacitance between the wires of two or more pairs which result in a transfer of unwanted signal from one pair to others.

CAPACITANCE, UNBALANCE-TO-GROUND – An inequality of capacitance between the ground capacitance of the conductors of a pair which results in a pickup of external noise energy, usually from power transmission lines.

CAPACITIVE REACTANCE – The opposition to alternating current due to the capacitance of a capacitor, cable or circuit. It is measured in ohms and is equal to 1/(6.28 fC) where f is the frequency in Hz and C is the capacitance in farads.

CAPACITOR – Two conducting surfaces separated by a dielectric material. The capacitance is determined by the area of the surface, type of dielectric and spacing between the conducting surfaces.

CAPILLARY ACTION – The travelling of liquid along a small interstice due to surface tension.

CARRIER – (1) An AC electrical signal that is used to carry information, (2) The woven element of a braid consisting of one or more ends (strands) which creates the interlaced effect. (3) A spindle, spool, tube, or bobbin (on a braiding machine) containing yarn or wire, employed as a braid.

CATHODE – (1) The negative electrode through which current leaves a nonmetallic conductor, such as an electrolytic cell, (2) the positive pole of a storage battery.

CATHODIC PROTECTION – Reduction or prevention of corrosion by making the metal to be protected the cathode in a direct current circuit.

CATV – Community Antenna TeleVision. Refers to the use of a coaxial or fiber cable to transmit television or other signals to subscribers from a single head-end location.

CATV CABLE – General term for all cables used for community antenna TV service and feeders, distribution and house drops.

CB – Citizens Band. One type of two-way radio communication.

C CONDITIONING – A type of line conditioning that controls attenuation, distortion and delay distortion so they lie within specific limits.

C CONNECTOR – A bayonet-locking connector for coax; C is named after Carl Concelman.

CCTV – Closed-Circuit TeleVision. One of the many services often found on broadband networks.

CCW – Continuously Corrugated and Welded. A type of cable sheath.

CD – Carrier Detect. An RS-232 control signal (on Pin 8) which indicates that the local modem is receiving a signal from the remote modem. Also called Received Line Signal Detector (RLSD) and Data Carrier Detect (DCD).

CENELEC – Comité Européen de Normalisation ELECtrotechnique. One of the European Union's key electrical standards-writing bodies.

CELLULAR POLYETHYLENE – Expanded or "foam" polyethylene, consisting of individual closed cells of inert gas suspended in a polyethylene medium, resulting in a desirable reduction of dielectric constant.

CERTIFICATE OF COMPLIANCE – A written statement; normally generated by a Quality Control Department, which states that the product being shipped meets customer's specifications.

CERTIFIED TEST REPORT (CTR) – A report reflecting actual test data on the cable shipped. Tests are normally conducted by the Quality Control Department and shows that the product being shipped meets the required test specifications.

CHANNEL – (1) A path for electrical transmission. Also called a circuit facility, line, link, or path. (2) A specific and discrete bandwidth allocation in the radio frequency spectrum (for example, in a broadband LAN) utilized to transmit one information signal at a time.

CHANNEL TRANSLATOR – Device used in broadband LANs to increase carrier frequency, converting upstream (toward the head-end) signals into downstream signals (away from the head-end).

CHARACTERISTIC IMPEDANCE – An electrical characteristic of transmission lines. When terminated in its characteristic impedance, reflections from the end of a line are minimized.

CHEMICAL STRIPPING – Removal of insulation by chemical means.

CHLOROSULFONATED POLYETHYLENE (CSP) – A rubbery polymer used for insulations and jackets. Manufactured by E.I. DuPont under the trade name of Hypalon.

CIGARETTE WRAP – Tape insulation wrapped longitudinally instead of spirally over a conductor.

CIRCUIT SWITCHING – A switching technique in which an information path (i.e., circuit) between calling and called stations is established on demand for exclusive use by the connected parties until the connection is released.

CIRCUIT TRACING – Locating or identifying a specific conductive path.

CIRCULAR MIL (CM) – A term universally used to define cross-sectional areas of conductors. It is an area equal to the area of a circle 1/1000 of an inch in diameter. As the number of circular mils increase, the size of a wire increases.

CLAD WIRE – Different from coated wire, is any metal covered with a relatively heavy coating of different metal, such as copperweld (copper over steel) or alum-o-weld (aluminum over steel). See Coated Wire.

COATED WIRE – Any metal covered by a relatively thin coating of a different metal such as tin, zinc or other alloy by a dip bath and wipe process, often at high speeds in line with insulating equipment. See Tinned Wire.

COAXIAL CABLE – A cylindrical transmission line comprised of a conductor centered inside a metallic tube or shield, separated by a dielectric material and usually covered by an insulating jacket.

COHERENT SOURCE – A fiber optic light source which emits a very narrow, unidirectional beam of light of one wavelength (monochromatic).

COIL EFFECT – The inductive effect exhibited by a spiral wrapped shield, especially above audio frequencies.

COLD BEND – Generally refers to a test to determine cable or wire characteristics at low temperatures. The test specimen is cooled in a low temperature box to a specified temperature. The wire specimen is then wound around a mandrel after which it is examined for cracks or other defects caused by bending at low temperatures.

COLD-DRAWING – Reducing the cross section by pulling through a die or dies, at a temperature lower than the recrystallization temperature.

COLD FLOW – Permanent deformation of the insulation due to mechanical pressure (not due to heat softening).

COLOR CODE – A color system for wire or circuit identification by use of solid colors, tracers, braids, surface printing, etc.

COMBINATION STRANDED CONDUCTOR – A conventional concentric conductor in which the wires in the outer layer are larger in diameter than the wires in the inner layer or layers and the diameters of all wires are within plus and minus 5 percent of the nominal wire diameter for the same size noncombination stranded conductor.

COMMON AXIS CABLING – In multiconductor constructions, a twisting of all conductors about a "common axis" to result in smaller diameter constructions. Tends to result in greater susceptibility to electromagnetic and electrostatic interference.

COMMON MODE NOISE – Noise caused by a difference in "ground potential." By grounding at either end rather than both ends (usually grounded at source) one can reduce this interference.

COMPACT STRANDED CONDUCTOR – A unidirectional or conventional concentric conductor manufactured to a specified diameter, approximately 8 to 10 percent below the nominal diameter of a noncompact conductor of the same cross-sectional area.

COMPOSITE CABLE – A cable containing more than one gauge size or a variety of circuit types, e.g., pairs, triples, quads, coaxials, etc.

COMPOSITE (CLAD) WIRE – A wire having a core of one metal with a fused outer shell of a different metal.

COMPOSITE CONDUCTOR – A conductor consisting of two or more types of wire, each type of wire being plain, clad, or coated-stranded together to operate mechanically and electrically as a single conductor.

COMPRESSED STRANDED CONDUCTOR – A conventional concentric conductor manufactured to a diameter not more than 3 percent below the nominal diameter of a noncompressed conductor of the same cross-sectional area.

COMPRESSION LUG OR SPLICE – A connection installed by compressing the connector onto the strand, hopefully into a cold weld.

CONCENTRICITY – The measurement of the location of the center of the conductor with respect to the geometric center of the circular insulation.

CONCENTRIC-LAY CONDUCTOR – A layer of uninsulated wires twisted around a central wire with subsequent layers spirally wrapped around the inner layers to form a single conductor.

CONDUCTANCE – The ability of a conductor to carry an electric charge. The ratio of the current flow to the potential difference causing the flow. The reciprocal of resistance.

CONDUCTIVITY – Capacity of a material to carry electrical current – usually expressed as a percentage of copper conductivity (copper being 100 percent).

CONDUCTOR – A material suitable for carrying an electric current. Several types are as follows:

> **COMPACT ROUND CONDUCTOR** – A conductor constructed with a central wire surrounded by one or more preshaped (nonround) helically laid wires and formed into final shape by rolling, drawing, or other means.
>
> **CONCENTRIC-LAY CONDUCTOR** – A conductor constructed with a central wire surrounded by one or more layers of helically laid wires.
>
> **CONVENTIONAL CONCENTRIC CONDUCTOR** – A conductor constructed with a central wire surrounded by one or more layers of helically laid wires. The direction of lay is reversed in successive layers and generally with an increase in length of lay for successive layers.

EQUILAY CONDUCTOR – A conductor constructed with a central wire surrounded by more than one layer of helically laid wires, all layers having a common length of lay, direction of lay being reversed in successive layers.

PARALLEL CORE CONDUCTOR – A conductor constructed with a central core of parallel-laid wires surrounded by one layer of helically laid wires.

ROPE-LAY CONDUCTOR – A conductor constructed of a bunch-stranded or a concentric-stranded member or members, as a central wire, around which are laid one or more helical layers of such members.

UNIDIRECTIONAL CONDUCTOR – A conductor constructed with a central wire surrounded by more than one layer of helically laid wires, all layers having a common direction of lay, with increase in length of lay for each successive layer.

UNILAY CONDUCTOR – A conductor constructed with a central wire surrounded by more than one layer of helically laid wires, all layers having a common length and direction of lay.

CONDUCTOR CORE – The center strand or member about which one or more layers of wires or members are laid helically to form a concentric-lay or rope-lay conductor.

CONDUCTOR SHIELD – A conducting layer applied to make the conductor a smooth surface in intimate contact with the insulation; sometimes called Extruded Strand Shield (ESS).

CONDUIT – A tube or trough for protecting electrical wires or cables.

CONNECTION, DELTA – Interconnection of 3 electrical equipment windings in a delta (triangular) configuration.

CONNECTION, WYE – Interconnection of 3 electrical equipment windings in wye (star) configuration.

CONNECTOR – A metallic device of suitable electric conductance and mechanical strength, used to splice the ends of two or more cable conductors, or as a terminal connector on a single conductor. Connectors usually fall into one of the following types:

- solder
- welded
- mechanical
- compression or indent

Conductors are sometimes spliced without connectors, by soldering, brazing, or welding.

CONTACT – The part of a connector which carries the electrical current.

CONTACT SIZE – The largest size wire which can be used with the specific contact. Also, the diameter of the engagement end of the pin.

CONTINUITY CHECK – A test performed on a length of finished wire or cable to determine if the electrical current flows continuously throughout the length.

CONTINUOUS VULCANIZATION – Simultaneous extrusion and vulcanization (cross-linking) of wire insulating and jacketing materials.

CONTRAHELICAL – Cable spiralling in an opposite direction than the preceding layer within a wire or cable.

CONTROL CABLE – A cable used for remote control operation of any type of electrical power equipment.

CONTROLLED IMPEDANCE CABLE – A package of two or more insulated conductors where impedance measurements between respective conductors are kept essentially constant throughout the entire length.

COPOLYMER – A compound resulting from the polymerization of two different monomers.

COPPER-CLAD STEEL – Steel with a coating of copper welded to it before drawing as opposed to copper-plated. Synonymous with Copperweld.

COPPERWELD – Trademark of Copperweld Steel Co. for copper-clad steel conductor.

CORD – A flexible insulated cable.

CORD SET – Portable cords fitted with a connector at one or both ends.

CORE – (1) In cables, a component or assembly of components over which other materials are applied, such as additional components, shield, sheath, or armor. (2) In fiber optics, the transparent glass or plastic section with a high refractive index through which the light travels by internal reflections.

CORONA – A discharge due to ionization of the air around a conductor due to a potential gradient exceeding a certain critical value. See Partial Discharge.

CORONA RESISTANCE – The time that the insulation will withstand a specified level of ionization that does not result in the complete breakdown of the insulation.

CORROSION – The destruction of the surface of a metal by chemical reaction.

COULOMB – The derived SI unit for quantity of electricity or electrical charge: One coulomb equals one ampere-second.

COUNTER EMF – The voltage opposing the applied voltage and the current in a coil; caused by a flow of current in the coil; also known as back emf.

COUNTERPOISE WIRE – Bare copper wire used to offset the impact of lightning surges along high-voltage overhead lines and around the base of towers. Buried counterpoise wire is connected to overhead ground wires and towers. Numerous methods of application are used, dependent upon resistance of the soil at the tower base.

COUPLING – The transfer of energy between two or more cables or components of a circuit.

COUPLING LOSS – Signal losses in an optical fiber due to small differences in numerical aperture, core diameter, core concentricity and tolerances in connectors when two fibers are spliced together. Also known as Splicing loss and Transfer loss.

COVERAGE – The calculated percentage which defines the completeness with which a metal braid covers the underlying surface. The higher percentage of coverage, the greater the protection against external interference.

CPE – Chlorinated PolyEthylene. A jacketing compound sold by Dow Chemical under the trademark Tyrin.

CROSS-LINKED – Inter-molecular bonds created between long chain thermoplastic polymers by chemical or electron bombardment means. The properties of the resulting thermosetting material are usually improved.

CROSS-LINKED POLYETHYLENE – A dielectric material used for insulating and jacketing. Also referred to as "XLP" or "XLPE."

CROSS TALK – A type of interference caused by audio frequencies from one circuit being coupled into an adjacent circuit. The term is loosely used to also include coupling at higher frequencies.

CRT – Cathode-Ray Tube. A television-like picture tube used in terminals; CRT is commonly used as a synonym for the CRT terminal.

CRT WIRE – High-voltage lead wire for energizing cathode ray tubes.

CSA (Canadian Standards Association) – Similar to UL in the United States.

CSA Certified – A product that has been tested and found to comply with applicable Canadian standards.

CSPE – A jacketing compound based on DuPont's chlorosulfonated polyethylene (Hypalon). Sometimes abbreviated CSP.

CT – Cable Tray, NEC Article 392. A cable marking which indicates a cable is suitable for use in a cable tray.

CURE – To change the properties of a polymeric material into a more stable, usable condition by the use of heat, radiation, or reaction with chemical additives. To cross-link.

CURING CYCLE – The time, temperature and pressure required for curing.

CURL – The degree to which a wire tends to form a circle after removal from a spool.

CURRENT – The rate of transfer of electricity. The unit of current is the ampere, a rate of one coulomb/second.

CURRENT, ALTERNATING (AC) – An electric current that periodically reverses direction of electron flow. The number of cycles in a given unit of time (generally a second) is called the frequency of the current.

CURRENT CARRYING CAPACITY – The maximum current an insulated conductor can safely carry without exceeding its insulation and jacket temperature limitations. Same as Ampacity.

CURRENT, CHARGING – The current needed to bring the cable up to voltage; determined by the capacitance of the cable. The charging current will be 90° out of phase with the voltage.

CURRENT DENSITY – The current per cross sectional area. Usually in units of amperes/square meter.

CURRENT, DIRECT (DC) – Electrical current whose electrons flow in one direction only. It may be constant or pulsating as long as their movement is in the same direction.

CUT-THROUGH RESISTANCE – The ability of a material to withstand mechanical pressure without damage.

CV – Continuous Vulcanization. An insulation and jacketing curing process.

CYCLE – The complete sequence including reversal of the flow of an alternating electric current.

D

D/A – Digital to Analog.

DAC – Digital to Analog Converter. A device that converts a digital input signal to an analog output signal carrying equivalent information.

DATA – Digitally represented information, which includes voice, text, facsimile and video.

dB – DeciBel. The standard unit used to express the relative strength of two signals. When referring to a single signal measured at two places in a transmission system, it expresses either a gain or loss in power between the input and output devices.

dBmV – (DeciBel MilliVolt) The level at any point in a system expressed in dBs above or below a 1 millivolt/75 ohm standard is said to be the level in decibel-millivolts or dBmV. Zero dBmV is equal to 1 millivolt across an impedance of 75 ohms.

DC – Direct Current. (see Current, Direct.)

DCE – Data Communications Equipment. In common usage, synonymous with modem; the equipment that provides the functions required to establish, maintain and terminate a connection as well as the signal conversion required for communications between the DTE and the telephone line or data circuit.

DCL – Data Carrier Level.

DCS – Distributed Control System. A type of industrial automation system.

DC RESISTANCE – See Resistance.

DecaBDE – DecaBromoDiphenyl Ether. A type of brominated flame retardant sometimes used in wire and cable and other products. A type of PolyBrominated Diphenyl Ether (PBDE).

DELAY SKEW – The difference between the propagation delay of any two pairs within the same cable sheath. Delay skew is caused primarily because twisted pairs are designed to have different twists per foot (lay lengths). Delay skew could cause data transmitted over one channel to arrive out of sync with data over another channel.

DEMAND – (1) The measure of the maximum load of a utility's customer over a short period of time, (2) The load integrated over a specified time interval.

DERATING FACTOR – A factor used to reduce the current carrying capacity of a wire when used in environments other than that for which the value was established.

DETECTOR – A fiber optic device that picks up light from the fiber and converts the information into an electrical signal.

DIELECTRIC – An insulating (nonconducting) medium.

DIELECTRIC BREAKDOWN – Any change in the properties of a dielectric that causes it to become conductive. Normally the failure of an insulation because of excessive voltage.

DIELECTRIC CONSTANT – The property of an insulation which determines the electrostatic energy stored per unit volume for unit potential gradient. It is expressed as a ratio. "K" for air is 1.0, while that for polyethylene is 2.3. Therefore, the capacitance of polyethylene is 2.3 times that of air. It is also referred to as Specific Inductive Capacity or Permitivity.

DIELECTRIC DISPERSION – The change in relative capacitance due to a change in frequency.

DIELECTRIC HEATING – The heating of an insulating material when placed in a radio-frequency field, caused by internal losses during the rapid polarization reversal of molecules in the material.

DIELECTRIC LOSS – The power dissipated in a dielectric as the result of the friction produced by molecular motion when an alternating electric field is applied.

DIELECTRIC STRENGTH – The maximum voltage which an insulation can withstand without breaking down; usually expressed as a gradient in V/mil (volts per mil). Polyethylene for example has a dielectric strength of about 800 V/mil.

DIELECTRIC STRENGTH TESTING – A common test performed on electrical products which is often called hi-pot testing. A voltage higher than normal operating voltage is applied across the insulation. This test can increase product reliability by detecting faulty workmanship.

DIGITAL – Refers to communications procedures, techniques and equipment by which information is encoded as either a binary "1" or "0"; the representation of information in discrete binary form, discontinuous in time, as opposed to the analog representation of information in variable, but continuous, waveforms.

DIN – Deutsches Institut für Normung (DIN). The German Standard for many products.

DIP COATING – An insulating coating applied to the conductor by passing the conductor through an applicator containing liquid insulating medium.

DIRECT BURIAL CABLE – A cable installed directly in the earth.

DIRECT CAPACITANCE – The capacitance measured directly from conductor to conductor through a single insulating layer.

DIRECTIONAL COUPLER – A passive device used in a cable system to divide or combine unidirectional RF power sources.

DIRECTION OF LAY – The lateral direction, designated as left-hand or right-hand, in which the wires of a conductor run over the top of the conductor as they recede from an observer looking along the axis of the conductor.

DISPERSION – The variation of the refractive index of an optical fiber with wavelength, causing light of different wavelengths to travel at different velocities in the fiber.

DISSIPATION FACTOR – Energy lost when voltage is applied across an insulation. The cotangent of the phase angle between voltage and current in a reactive component. Dissipation factor is quite sensitive to contamination and deterioration of insulation. Also known as Power Factor.

DISTORTION FACTOR – An undesirable change in waveform as the signal passes through a device.

DISTRIBUTION CABLE – (1) In a CATV system, the transmission cable from the distribution amplifier to the drop cable. (2) In an electric power system, provides low voltage service to the customer.

DISTURBED CONDUCTOR – A conductor that receives energy generated by the field of another conductor or an external source such as a transformer.

DISTURBING CONDUCTOR – A conductor carrying energy whose field(s) create spurious energy in another conductor.

DOWNLOAD – The process of loading software into the nodes of a network from one node or device over the network media.

DRAIN WIRE – An uninsulated wire in contact with a shield throughout its length, used for terminating the shield.

DRAWING – In wire manufacture, pulling the metal through a die or series of dies to reduce diameter to a specified size.

DROP CABLE – In a CATV system, the transmission cable from the distribution cable to a dwelling.

DSR – Data Set Ready. One of the control signals on a standard RS-232-C connector. It indicates whether the data communications equipment is connected and ready to start handshaking control signals so that transmission can start.

DTE – Data Terminal Equipment.

DTR – Data Terminal Ready. An RS-232 modem interface control signal (sent from the DTE to the modem on pin 20) which indicates that the DTE is ready for data transmission and which requests that the modem be connected to the telephone circuit.

DUAL CABLE – A two-cable system in broadband LANs in which coaxial cables provides two physical paths for transmission, one for transmit and one for receive, instead of dividing the capacity of a single cable.

DUCT – An underground or overhead tube for carrying electrical conductors.

DUOFOIL – Belden trademark for a shield in which metallic foil is applied to both sides of a supporting plastic film.

DUPLEX – Two way data transmission on a four-wire transmission cable.

DUPLEX CABLE – A cable composed of two insulated single conductor cables twisted together.

E

E – (1) Symbol for voltage. Usually used to represent direct voltage or the effective (root-mean-square) value of an alternating voltage. (2) A UL cable type. Elevator lighting and control cable.

EARTH – British terminology for zero-reference ground.

ECCENTRICITY – Like concentricity, a measure of the center of a conductor's location with respect to the circular cross section of the insulation. Expressed as a percentage of displacement of one circle within the other.

ECTFE – Ethylene ChloroTriFluoroEthylene. Halar is a Solvay Solexis trademark for this material. Used as an insulation or jacketing material.

EDDY CURRENT – Circulating currents induced in conducting materials by varying magnetic fields.

EIA – Electronic Industries Alliance. The U.S. national organization of electronic manufacturers. It is responsible for the development and maintenance of industry standards for the interface between data processing machines and data communications equipment.

ELASTOMER – Any material that will return to its original dimensions after being stretched or distorted.

ELECTROMAGNET – A device consisting of a ferromagnetic core and a coil that produces appreciable magnetic effects only when an electric current exists in the coil.

ELECTROMAGNETIC – Referring to the combined electric and magnetic fields caused by electron motion through conductors.

ELECTROMAGNETIC COUPLING – The transfer of energy by means of a varying magnetic field. Inductive coupling.

ELECTRO-MECHANICAL CABLES – Dual purpose composite cables made up of support strands capable of supporting predetermined loads together with communication, coaxial, or power as integral members of a finished cable.

ELECTROMOTIVE FORCE (E.M.F.) – Pressure or voltage. The force which causes current to flow in a circuit.

ELECTRON – An elementary particle containing the smallest negative electric charge; Charge = 0.16 attocoulomb. Diameter = 1 femtometer.

ELECTRON VOLT – A measure of the energy gained by an electron passing through an electric field produced by one volt.

ELECTRONIC WIRE AND CABLE – Wire or cable used in electronic applications.

ELECTRO-OSMOSIS – The movement of fluids through dielectrics because of electric current.

ELECTROSTATIC – Pertaining to static electricity, or electricity at rest. An electric charge, for example.

ELECTROSTATIC COUPLING – The transfer of energy by means of a varying electrostatic field. Capacitive coupling.

ELECTROSTATIC DISCHARGE – (ESD) An instantaneous flow of an electrical charge on a nonconductor through a conductor to ground.

ELECTRO-TINNED – Electrolytic process of tinning wire using pure tin.

ELEXAR – Shell trademark for a thermoplastic elastomer (TPE).

ELFEXT – Equal-level Far-end Crosstalk. A measure of the unwanted signal coupling from a transmitter at the near-end into a neighboring pair measured at the far-end, relative to the received signal level measured on that same pair. Referred to as ACR-F (insertion loss to crosstalk ratio far-end) in the TIA/EIA-568-B.2-Addendum 10 draft. (ELFEXT is FEXT adjusted to discount insertion loss.)

ELONGATION – The fractional increase in the length of a material stressed in tension.

EMA – (Electrical Moisture Absorption) A water tank test during which sample cables are subjected to voltage and water maintained at rated temperature; the immersion time is long, with the object being to accelerate failure due to moisture in the insulation; simulates buried cable.

EMBOSSING – Identification by means of thermal indentation which leaves raised lettering on the sheath material of cable.

EMERGENCY OVERLOAD – A situation in which larger than normal currents are carried through a cable or wire for a limited period of time.

EMI – Electromagnetic Interference. External signals that disrupt the data being transmitted on the local area network or electronic device being operated. Typically, these external signals emanate from universal motors with brushes, fluorescent lights, personal computers, printers or other devices including copy machines, etc. The Federal Communications Commission (FCC) regulates this emission area.

ENDOSMOSIS – The penetration of water into a cable by osmosis; aggravated and accelerated by DC voltage on the cable.

ENDS – In braiding, the number of essentially parallel wires or threads on a carrier.

ENERGIZE – To apply rated voltage to a circuit or device in order to activate it.

EO – A UL cable type. Elevator lighting and control cable with thermoset insulation.

EOT – End of Transmission Character. A transmission control character used to indicate the end of transmission, which may include one or more texts and any associated message headings.

EP, EPR, EPM, EPDM – Designations for a synthetic rubber based upon the hydrocarbon ethylene propylene.

EPA – Environmental Protection Agency. The federal regulatory agency responsible for keeping and improving the quality of our living environment – mainly air and water.

EPDM – Ethylene Propylene Diene Monomer.

EPROM – Erasable Programmable Read Only Memory.

EPR – Ethylene Propylene Rubber.

EQUILAY CONDUCTOR – See Concentric-lay Conductor.

ET – A UL cable type. Elevator lighting and control cable with thermoplastic insulation, three braids, flame-retardant and moisture-retardant finish. May have steel supporting strand in the center, 300 V.

ETCHED WIRE – A process applied to Teflon wire in which the wire is passed through a sodium bath to create a rough surface to allow epoxy resin to bond to the Teflon.

ETFE – Ethylene TetraFluoroEthylene. Tefzel™ is DuPont's trademark for this material.

ETHERNET – A baseband local area network specification developed jointly by Xerox Corporation, Intel Corporation and Digital Equipment Corporation to interconnect computer equipment using coaxial cable and "Transceivers."

ETL – Electrical Testing Laboratories, Inc.

ETPC – Abbreviation for electrolytic tough pitch copper. It has a minimum conductivity of 99.9 percent.

EVA – Ethylene Vinyl Acetate. A polymer often used in low-smoke, non-halogenated cables.

EXIT ANGLE – The angle between the output radiation vectors and the axis of the fiber or fiber bundle.

EXPANDED DIAMETER – Diameter of shrink tubing as supplied. When heated the tubing will shrink to its extruded diameter.

EXTERNAL WIRING – Electronic wiring that interconnects subsystems within the system.

EXTRUDED CABLE – Cable that is insulated by applying insulation material in a continuous extrusion process.

EXTRUSION – A method of applying insulation to a conductor or jacketing to a cable. The process is continuous and utilizes rubber, neoprene or a variety of plastic compounds.

F

FACSIMILE – The remote reproduction of graphic material; an exact copy.

FARAD – A unit of capacitance when a difference of potential of 1 volt produces a displacement of one coulomb in a capacitor. The farad is a very large unit and a much smaller unit, the microfarad (μf), is more commonly used.

FATIGUE RESISTANCE – Resistance to metal crystallization which leads to conductors or wires breaking from flexing.

FAULT, GROUND – A fault to ground.

FCC – Federal Communications Commission.

FDDI (Fiber Distributed Data Interface) – An ANSI defined token-passing ring using fiber optic media to attain a 100 Mbps transmission rate.

FDX – Full Duplex. Transmission in two directions simultaneously, or, more technically, bidirectional simultaneous two-way communications.

FEP – Fluorinated Ethylene Propylene. Teflon is DuPont's trademark for this material.

FEPB – A UL cable type. Fluorinated ethylene propylene insulated wire with glass braid.

FEXT – Far-end crosstalk. A measure of the unwanted signal coupling from a transmitter at the near-end into a neighboring pair measured at the far-end.

FFH-2 – A UL type of fixture wire with a 600 V rating.

FIBER DISPERSION – Pulse spreading in an optical fiber caused by differing transit times of various modes.

FIBER OPTICS – Transmission of energy by light through glass fibers. A technology that uses light as an information carrier. Fiber optic cables (light guides) are a direct replacement for conventional cable and wire pairs. The glass-based transmission cable occupies far less physical volume for an equivalent transmission capacity; the fibers are immune to electrical interference.

FIBER TUBING – A loose, crush-resistant cylinder applied over individual fibers to provide mechanical protection. Also called a buffer tube.

FIELD COIL – A suitable insulated winding mounted on a field pole to magnetize it.

FIELD MOLDED SPLICE – A joint in which the solid dielectric joint insulation is fused and cured thermally at the job site.

FIELD TESTS – Tests that may be made on a cable system after installation as an acceptance or proof test.

FIGURE 8 CABLE – An aerial cable configuration in which the conductors and the steel strand which supports the cable are integrally jacketed. A cross section of the finished cable approximates the figure "eight."

FILLED CABLE – A cable construction in which the cable core is filled with a material that will prevent moisture from entering or passing through the cable.

FILLER – Fillers are used in multiconductor cables which occupy the interstices formed by the assembled conductors. This is done so that the finished cable will be round.

FILLING COMPOUND – A dielectric material poured or injected into a splice housing or cable to prevent the entry of water. Filling compounds may require heating or mixing prior to filling. Some filling compounds may also serve as the insulation.

FINE STRANDED WIRE – Stranded wire with component strands of 36 AWG or smaller.

FLAME RESISTANCE – The ability of a material to not propagate flame once the heat source is removed.

FLAMMABILITY – The measure of a material's ability to support combustion.

FLASHOVER – A disruptive discharge around or over the surface of a solid or liquid insulator.

FLAT BRAID – A woven braid of tinned copper strands rolled flat at the time of manufacture to a specified width.

FLAT CABLE – A cable with two essentially flat surfaces.

FLAT CONDUCTOR – A wire having a rectangular cross section as opposed to a round or square conductor.

FLEX-LIFE – The measurement of the ability of a conductor or cable to withstand repeated bending.

FLEXIBILITY – The ease with which a cable may be bent.

FLEXIBLE – That quality of a cable or cable component that allows for bending under the influence of an outside force, as opposed to limpness that is bending due to the cable's own weight.

FLOATING – Refers to a circuit that has no electrical connection to ground.

FLUOROPOLYMER – A class of polymers used as insulating and jacketing materials. Common ones include Teflon, Tefzel, Kynar and Halar.

FLUX – (1) The lines of force which make up an electrostatic field. (2) The rate of flow of energy across or through a surface. (3) A substance used to promote or facilitate fusion.

FM – Frequency modulation. A modulation technique in which the carrier frequency is shifted by an amount proportional to the value of the modulating signal. The deviation of the carrier frequency determines the signal content of the message.

FOAMED INSULATION – Insulations having a cellular structure.

FOIL – A thin, continuous sheet of metal.

FREQUENCY – The number of cycles per second at which an analog signal occurs, expressed in Hertz (Hz). One Hertz is one cycle per second.

FREQUENCY ANALYZER – An instrument to measure the intensity of various component frequencies from a transmitting source.

FREQUENCY COUNTER – An electronic measuring instrument that counts the number of cycles of a periodic electrical signal during a given time interval.

FREQUENCY MODULATION (FM) – Method of encoding a carrier wave by varying the frequency of the transmitted signal.

FREQUENCY PLAN – Specification of how the various frequencies of a broadband cable system are allocated for use.

"F" TYPE CONNECTOR – A low cost connector used by the TV industry to connect coaxial cable to equipment.

FULL DUPLEX – Two-way communications in which each modem simultaneously sends and receives data at the same rate.

FUSE WIRE – Wire made from an alloy that melts at a relatively low temperature.

FUSED COATING – A metallic coating which has been melted and solidified, forming a metallurgical bond to the base material.

FUSED CONDUCTORS – Individual strands of heavily tinned copper wire stranded together and then bonded together by induction heating.

FUSED SPIRAL TAPE – A PTFE insulation often used on hookup wire. The spiral wrapped tape is passed through a sintering oven where the overlaps are fused together.

G

G – A UL cable type. Rubber insulated, neoprene, Hypalon or CPE jacketed portable power cable with two to five #8 AWG or larger conductors with ground wires.

GALVANIZED STEEL WIRE – Steel wire coated with zinc.

GANG STRIP – Stripping all or several conductors simultaneously.

GAS FILLED CABLE – A self-contained pressurized cable in which the pressure medium is an inert gas having access to the insulation.

GAUGE – A term used to denote the physical size of a wire.

GAUSS – A unit of magnetic induction (flux density) equal to 1 maxwell per square centimeter.

GENERAL PURPOSE INSTRUMENTATION BUS (GPIB) – A protocol standard defined by the IEEE.

GFCI – Ground fault circuit interrupter. A protective device that detects abnormal current flowing to ground and then interrupts the circuit.

G-GC – A UL cable type. A portable power cable similar to Type G, but also having a ground check conductor to monitor the continuity of the grounding circuit.

GHz – Gigahertz; 1,000,000,000 cycles per second.

GIGA – A numerical prefix denoting one billion (10^9).

GND – Ground.

GROUND – A voltage reference point that is the same as earth or chassis ground.

GROUND CONDUCTOR – A conductor in a transmission cable or line that is grounded.

GROUND FAULT – A type of electrical failure in which current flows to ground.

GROUND LOOP – The generation of undesirable current flow within a ground conductor, owing to the circulation currents which originate from a second source of voltage.

GROUND PLANE – Expanded copper mesh which is laminated into some flat cable constructions as a shield.

GROUND POTENTIAL – Zero potential with respect to the ground or earth.

GROUNDED NEUTRAL – The neutral wire that is metallically connected to ground.

GTO – Gas tube sign cable, 5 kV–15 kV.

GUY – A tension wire connected to a tall structure and another fixed object to add strength to the structure.

H

HALAR – Solvay Solexis trademark for Ethylene ChloroTriFluoroEthylene (ECTFE).

HALF DUPLEX – Two-way communications in which data are sent in only one direction at a time.

HALOGENS – Chemical elements such as chlorine and bromine that when present in a cable are released when burned. These materials can cause additional damage to human respiratory systems and to electrical equipment.

HARD-DRAWN WIRE – As applied to aluminum and copper, wire that has been cold drawn to final size so as to approach the maximum strength attainable.

HARNESS – An arrangement of wires and cables, usually with many breakouts, which have been tied together or pulled into a rubber or plastic sheath, used to interconnect an electric circuit.

HASH MARK STRIPE – A noncontinuous helical stripe applied to a conductor for identification.

HAZARDOUS LOCATION – An area of ignitable vapors, dust, or fibers that may cause fire or explosion as defined in Article 500 of the NEC. UL Listed cable types PLTC, ITC, TC, MC and MV can be used in hazardous locations when installed in accordance with the NEC.

HDPE – High-Density PolyEthylene.

HDTV – High-Definition TeleVision.

HDX – Half-Duplex Transmission. Transmission in either direction but not in both directions simultaneously. Compare with full-duplex transmission.

HEAD-END – A central point in broadband networks that receives signals on one set of frequency bands and retransmits them on another set of frequencies.

HEAT DISTORTION – Distortion or flow of a material or configuration due to the application of heat.

HEAT SEAL – Method of sealing a tape wrap jacket by means of thermal fusion.

HEAT SHOCK – A test to determine stability of material by sudden exposure to a high temperature for a short period of time.

HEAT SINK – A device that absorbs heat.

HEATER CORD – A group of cable types defined in Article 400 of the NEC such as types HPD, HPN, HS, HSJ, HSJO and HSO.

HELICAL STRIPE – A continuous, colored, spiral stripe applied to a conductor for circuit identification.

HELIX – Spiral winding.

HENRY – A unit of inductance equal to the inductance of a current changing at the rate of 1 ampere per second inducing a counter electromotive force of 1 volt.

HERTZ (Hz) – Cycles per second. A cycle that occurs once every second has a frequency of 1 Hertz. The bandwidth of the average phone line is between 300 and 3,000 cycles per second.

HF – High Frequency.

HID – High Intensity Discharge as in mercury metal halide and sodium lamps.

HIGH BOND INSULATION – Insulation exhibiting great bond strength to the conductors.

HIGH-SPLIT – A broadband cable system in which the bandwidth utilized to send toward the head-end (reverse direction) is approximately 6 MHz to 180 MHz and the bandwidth utilized to send from the head-end (forward direction) is approximately 220 MHz to 400 MHz. The guard band between the forward and reverse directions (180 MHz to 220 MHz) provides isolation from interference.

HIGH TEMPERATURE WIRE AND CABLE – Electrical wire and cables having thermal operating characteristics of 150°C and higher.

HIGH TENSION CABLES – Generally the high voltage ignition wires for combustion engines, gas and oil ignitors, or neon signs, etc. (Unshielded.) Usually Type GTO.

HIGH-VOLTAGE CABLE TERMINATION – A device used for terminating alternating current power cables having laminated or extruded insulation rated 2.5 kV and above.

HIGH-VOLTAGE POWER (system voltage ratings) – A class of system voltages equal to or greater than 69,000 volts or less than 230,000 volts.

HINGE CABLE – A cable connected between a hinged or swinging device and a stationary object.

HIPOT – A DC high potential test used on medium and high voltage cables. See Dielectric Strength Testing.

HMWPE – High Molecular Weight Polyethylene.

HOLDING STRENGTH – Ability of a connector to remain assembled to a cable when under tension.

HOOKUP WIRE – Small wires used to hook up instruments or electrical parts, usually 12 AWG and smaller.

HOT MODULUS – Stress at 100 percent elongation after 5 minutes of conditioning at a given temperature (normally 130°C).

HOT STAMPING – Method of alphanumeric coding. Identification markings are made by pressing heated tape and marking foil into softened insulation surfaces.

HOT STICK – A long insulated stick having a hook at one end which is used to open energized switches, etc.

HOT TIN DIP – A process of passing bare wire through a bath of molten tin to provide a coating.

HOUSING – A metallic or other enclosure for an insulated splice.

HPD – A UL cable type. Two, three or four conductor heater cord with thermoset insulation. For use in dry locations.

HPN – A UL cable type. Two or three conductor, thermosetting-insulated heater cord. Parallel construction. For use in damp locations.

HSO – A UL cable type. Thermoset jacketed heater cord.

HV – High Voltage.

HYBRID CABLE – Multiconductor cable containing two or more types of components.

HYDROSCOPIC – Readily absorbing and retaining moisture.

HYPALON – DuPont's trademark for ChloroSulphonated Polyethylene (CSP).

HYPOT – (see Hipot) Registered trade name of Associated Research, Inc. for its high-voltage tester.

HYSTERESIS – The time lag exhibited by a body in reacting to changes in forces affecting it; an internal friction.

Hz – Hertz. A measure of frequency or bandwidth equal to one cycle per second. Named after experimenter Heinrich Hertz.

I

I – Symbol used to designate current.

IACS – International Annealed Copper Standard for copper used in electrical conductors. 100 percent conductivity at 20°C is 0.017241 ohm-mm^2/m.

ICEA – Insulated Cable Engineers Association. The association of cable manufacturing engineers who make nationally recognized specifications for cables. Formerly IPCEA.

IEC – International Electrotechnical Commission.

IEEE – Institute of Electrical and Electronic Engineers. An international professional society that issues its own standards and is a member of ANSI and ISO.

IEEE 10BASE2 Network – A network conforming to the IEEE 802.3 local area network standard. The network is capable of carrying information at rates up to 10 Mbps over distances up to 2,800 meters (9,184 feet).

IEEE 10BROAD36 – 10 million bits per second over broadband coaxial cable with node-to-node coverage of 3,600 meters. The IEEE 802.3 specification for running Ethernet on broadband.

IEEE-488 – An IEEE standard for a parallel interface bus consisting of eight bidirectional data lines, eight control lines and eight signal grounds, which provides for connection to an IEEE-488 device.

IEEE-802 – Standards for the interconnection of local networking computer equipment. The IEEE-802 standard deals with the Physical Link Layers of the ISO Reference Model for OSL.

IEEE 802.3 – An IEEE standard describing the physical and data link layers of a local area network based on bus topology and CSMA/CD. Ethernet.

IEEE 802.5 – A physical layer standard specifying a LAN with a token-passing access method on a ring topology. Token ring.

IEEE 802.11 – Wireless LAN. (Wi-Fi)

IEEE 802.15 – Wireless PAN. (Bluetooth)

IF – Intermediate-frequency.

IMPACT TESTS – Tests designed to reveal the behavior of material of a finished part if it were subjected to impact or shock loading.

IMPEDANCE – The total opposition a circuit, cable, or component offers to alternating current. It includes both resistance and reactance and is generally expressed in ohms.

IMPEDANCE MATCH – A condition whereby the impedance of a particular cable or component is the same as the impedance of the circuit, cable or device to which it is connected.

IMPEDANCE MATCHING STUB – A section of transmission line or a pair of conductors cut to match the impedance of a load. Also called matching stub.

IMPEDANCE MATCHING TRANSFORMER – A transformer designed to match the impedance of one circuit to that of another.

IMSA – International Municipal Signal Association.

IN-BAND SIGNALING – The transmission of signalling information at some frequency or frequencies that lie within a carrier channel normally used for information transmission.

INCOHERENT SOURCE – A fiber optic light source which emits wide, diffuse beams of light of many wave lengths.

INDEX EDGE – Edge of flat (ribbon) cable from which measurements are made, normally indicated by the location of the printing which is near the index edge. Sometimes indicated by a thread or other identification stripe.

INDOOR TERMINATION – A cable termination intended for use where it is protected from direct exposure to both solar radiation and precipitation.

INDUCTANCE – A property of a conductor or circuit which resists a change in current. It causes current changes to lag behind voltage changes and is measured in henrys.

INDUCTION – The phenomenon of a voltage, magnetic field, or electrostatic charge being produced in an object by lines of force from the source of such fields.

INDUCTION HEATING – Heating a conducting material by placing it in a rapidly changing magnetic field. The changing field induces electric currents in the material and I^2R losses account for the resultant heat.

INDUCTIVE COUPLING – Cross talk resulting from the action of the electromagnetic field of one conductor on the other.

INPUT – (1) A signal (or power) which is applied to a piece of electric apparatus, (2) The terminals on the apparatus to which a signal or power is applied.

INSERTION LOSS – A measure of the attenuation of a device by determining the output of a system before and after the device is inserted into the system.

INSERTION TOOL – A small, hand-held tool used to insert contacts into a connector.

INSULATED RADIANT HEATING WIRE – Similar to blanket wire but heavier construction for applications such as in ceiling panels, buried in ground or driveway and concrete walks.

INSULATED SPLICE – A splice with a dielectric medium applied over the connected conductors and adjacent cable insulation.

INSULATING (ISOLATING) JOINT – A cable joint which mechanically couples and electrically separates the sheath, shield and armor on contiguous lengths of cable.

INSULATION – A material having good dielectric properties which is used to separate close electrical components, such as cable conductors and circuit components.

INSULATION LEVEL – A thickness rating for power cable insulation. Circuits having fault detectors which interrupt fault currents within 1 minute are rated 100 percent level, within 1 hour are rated 133 percent level and over 1 hour are rated 173 percent level.

INSULATION TEMPERATURE RATING – A maximum temperature assigned to insulations based on laboratory tests.

INSULATION RESISTANCE – The electrical resistance of an insulating material at a specific time and condition as measured between two conductors.

INSULATION STRESS – The potential difference across an insulator. The stress on insulation is expressed in volts per mil (V/m) or kilovolts per meter (kV/m).

INSULATION THICKNESS – The wall thickness of wire insulation.

INSULATION VOLTAGE RATING – The nominal phase-to- phase operating voltage of a three-phase cable system.

INTERAXIAL SPACING – Center-to-center conductor spacing in paired wire or center-to-center spacing between conductors in a flat cable.

INTERCALATED TAPES – Two or more tapes helically wound and overlapped on a cable.

INTERCONNECTING CABLE – The wiring between modules, between units, or the larger portions of a system.

INTERCONNECTION – Mechanically joining devices together to complete an electrical circuit.

INTERFACE – The two surfaces on the contact side of both halves of a multiple-contact connector which face each other when the connector is assembled.

INTERFERENCE – Disturbances of an electrical or electromagnetic nature that introduce undesirable responses into other electronic equipment.

INTERMEDIATE FREQUENCY – A frequency to which a signal is converted for ease of handling. Receives its name from the fact that it is an intermediate step between the initial and final conversion or detection stages.

INTERMEDIATE TEMPER – As applied to aluminum, any temper between soft and hard drawn.

INTERNAL WIRING – Electronic wiring which interconnects components, usually within a sealed subsystem.

INTERSTICE – The space or void between assembled conductors and within the overall circumference of the assembly.

INTRINSICALLY SAFE – Incapable of releasing sufficient electrical or thermal energy under normal or abnormal conditions to cause ignition of a specific hazardous atmospheric mixture in its most ignitable concentration. See Article 504 of the NEC.

I/O – Input/Output. The process of transmitting data to and from the processor and its peripherals.

IONIZATION – (1) The creation of ions when polar compounds are dissolved in a solvent, (2) when a liquid, gas or solid is caused to lose or gain electrons due to the passage of an electric current.

IONIZATION FACTOR – This is the difference between percent dissipation factors at two specified values of electrical stress; the lower of the two stresses is usually so selected that the effect of the ionization on dissipation factor at this stress is negligible.

IONIZATION VOLTAGE – The potential at which a material ionizes. The potential at which an atom gives up an electron.

IR DROP – A method of designating a voltage drop in terms of both current and resistance.

IRRADIATION – In insulations, the exposure of the material to high-energy emissions for the purpose of favorably altering the molecular structure.

ISDN – Integrated Services Digital Network. A standard which covers a wide range of data communication issues but primarily the total integration of voice and data.

ISO – International Standards Organization.

ISO 9000 – A set of quality standards widely used around the world.

ISOLATION – The ability of a circuit or component to reject interference, usually expressed in dB.

ITU – International Telecommunications Union.

I^2R – Formula for power in watts, where I = current in amperes, R = resistance in ohms. See Watt.

J

JACK – A plug-in type terminal widely used in electronic apparatus for temporary connections.

JACKET – Pertaining to wire and cable, the outer sheath that protects against the environment and may also provide additional insulation.

JAN SPECIFICATION – Joint Army-Navy specification (replaced by current Military Specifications).

JET STARTER CABLE – Single conductor 600 V cable used for external aircraft power.

JITTER – The slight movement of a transmission signal in time or phase that can introduce errors and loss of synchronization in high-speed synchronous communications.

JOINT – That portion of the conductor where the ends of two wires, rods or groups of wires are joined by brazing, soldering, welding or by mechanical means.

JOULE'S LAW – When electricity flows through a material the rate of heating in watts will equal the resistance of the material in ohms times the square of the current in amperes. $W = I^2R$.

JUMPER CABLE – Extra flexible cables with high-voltage insulation for use as temporary connections. Usually has red jacket.

K

KAPTON – DuPont's trademark for polyimide.

kB – K-byte. 1,024 bytes. Transmission speeds are measured in kB/second.

kbps – Thousands of bits per second (bps).

kcmil – One thousand circular mils, replaced "MCM" in the 1990 NEC. Sometimes shortened to "kcm."

KEVLAR – A high-strength DuPont polymer used as a cable messenger or strength member.

K-FIBER – A polyaramid-based material used for jacketing high temperature cables.

KILO – Prefix meaning thousand.

kV – Kilovolt (1,000 volts).

kVA – Kilovolt ampere.

kW – Kilowatt. 1,000 watts of power.

KYNAR – Arkema Inc. trademark for PolyVinyliDene Fluoride (PVDF).

L

L – Symbol for inductance.

LACING AND HARNESSING – A method of grouping wires by securing them in bundles of designated patterns.

LACQUER – A liquid resin or compound applied to textile braid to prevent fraying, moisture absorption, etc.

LAMINATED TAPE – A tape consisting of two or more layers of different materials bonded together.

LAN – Local area network. A user-owned, user-operated, high-volume data transmission facility connecting a number of communicating devices within a single building or campus of buildings.

LASER DIODE – A semiconductor diode that emits coherent light.

LAUNCH ANGLE – The angle between the radiation vector and the axis of an optical fiber.

LAY – Pertaining to wire and cable, the axial distance required for one cabled conductor or conductor strand to complete one revolution about the axis around which it is cabled.

LAY DIRECTION – The twist in the cable as indicated by the top strands while looking along the axis of the cable away from the observer. Described as "right hand" or "left hand."

LAYER – Consecutive turns of a coil lying in a single plane.

L BAND – The band of frequencies between 390 and 1,550 megahertz.

LEACHING AND NONLEACHING – In a leaching wire the plasticizer will migrate when exposed to heat. A nonleaching wire will retain its plasticizer under extreme temperature conditions and remain flexible after baking.

LEAD – A wire, with or without terminals, that connects two points in a circuit.

LEAD CURED – A cable that is cured or vulcanized in a metallic lead mold.

LEAD-IN – The conductor or conductors that connect the antenna proper to electronic equipment.

LEAKAGE CURRENT – An undesirable flow of current through or over the surface of an insulating material.

LEAKAGE DISTANCE – The shortest distance along an insulation surface between conductors.

LED – Light-emitting diode; device that accepts electrical signals and converts the energy to a light signal; with lasers, the main light source for optical-fiber transmission, used mainly with multimode fiber.

LENGTH OF LAY – The axial length of one turn of the helix of a wire or member. See Lay.

LEVEL – A measure of the difference between a quantity or value and an established reference.

LF – Low frequency. A band of frequencies extending from 30 to 300 kHz in the radio spectrum, designated by the Federal Communications Commission.

LIFE CYCLE TESTING – A test to determine the length of time before failure in a controlled, usually accelerated environment.

LIGHTNING GROUND CABLE – A specially stranded single conductor cable used to connect lightning rods (air terminals) to grounding rods.

LIGHT SOURCE – An object capable of emitting light. In fiber optics, the light source is normally a LED or a laser.

LIMITS OF ERROR – The maximum deviation (in degrees or percent) of the indicated temperature of a thermocouple from the actual temperature.

LIMPNESS – The ability of a cable to lay flat or conform to a surface.

LINE BALANCE – The degree to which the conductors of a cable are alike in their electrical characteristics with respect to each other, to other conductors and to ground.

LINE DROP – A voltage loss occurring between any two points in a power transmission line. Such loss, or drop, is due to the resistance, or leakage of the line.

LINE EQUALIZER – A reactance (inductance and/or capacitance) connected in series with a transmission line to alter the frequency-response characteristics of the line.

LINE FAULT – A fault such as an open circuit, short circuit or ground in an electrical line or circuit.

LINE LEVEL – The level of a signal at a certain point on a transmission line. Usually expressed in decibels.

LINE LOSS – A total of the various energy losses occurring in a transmission line.

LINE VOLTAGE – The value of the potential existing on a supply or power line.

LITZ WIRE – Very fine, e.g. #44 AWG bare copper, each strand is enamel insulated. Used for low inductance coil windings.

LOAD – A device that consumes or converts the power delivered by another device.

LOAD CELL CABLE – Small multiconductor shielded cables for connecting load cells with instruments in electronic strain gauges. Also used for weighing and force measurement applications.

LOADED LINE – A transmission line that has lumped elements (inductance or capacitance) added at uniformly spaced intervals. Loading is used to provide a given set of characteristics to a transmission line.

LOC-TRAC – Alpha's registered trademark for a zipper tubing closure track which does not require any sealants to keep it closed, even during extreme flexing.

LOCAL AREA NETWORK (LAN) – A network that is located in a localized geographical area (e.g., an office, building, complex of buildings, or campus), and whose communications technology provides a high-bandwidth, low-cost medium to which many nodes can be connected.

LOGGING CABLE – Usually FEP/Tefzel self-supporting instrumentation cable. Generally dropped through borings in subsurface mining or well applications.

LONGITUDINAL SHIELD – A tape shield, flat or corrugated, applied longitudinally with the axis of the cable.

LONGITUDINAL SHRINKAGE – A term generally applied to shrink products denoting the axial length lost through heating in order to obtain the recovered diameter.

LONGITUDINAL WRAP – Tape applied longitudinally with the axis of the core being covered.

LONGWALL MACHINE – A mining machine used to undercut coal.

LOOP RESISTANCE – The total resistance of two conductors measured round trip from one end. Commonly used term in the thermocouple industry.

LOOP TEST – A long line test where a good line is connected to a faulty line to form a loop in which measurements will locate the fault.

LOSS – The portion of energy applied to a system that is dissipated and performs no useful work.

LOSS FACTOR – The power factor times the dielectric constant.

LOW BOND INSULATION – An insulation that exhibits a small bond strength to the conductors.

LOW FREQUENCY – A band of frequencies extending from 30 to 300 kHz in the radio spectrum, designated by the Federal Communications Commission.

LOW LOSS DIELECTRIC – An insulating material that has a relatively low dielectric loss, such as polyethylene or Teflon.

LOW NOISE CABLE – A cable specially constructed to eliminate spurious electrical disturbances caused by capacitance changes or self-generated noise induced by either physical movement or adjacent circuitry.

LOW TENSION – Low voltage, as applied to ignition cable.

LOW VOLTAGE – (1) As defined in the National Electrical Code, a system rated nominal 24 volts or less, supplied from a transformer, converter, or battery, (2) A power system voltage rating of 1,000 volts or less.

LPF – Low pass filter. A filter which greatly attenuates signals of higher than a specified frequency, but passes with minimal attenuation all signals lower in frequency.

LUMEN – A unit of measurement for light output.

LV – Low voltage.

M

mA – Milliampere (one-thousandth of an ampere).

MAGNET WIRE – Insulated wire used in the windings of motors, transformers and other electromagnetic devices.

MAGNETIC FIELD – The field created when current flows through a conductor, especially a coiled conductor.

MAP – Manufacturing automation protocol. The OSI profile championed by General Motors Corporation to provide interconnectivity between plant hosts, area managers and cell controllers over a broadband token-passing bus network.

MARKER TAPE – A tape laid parallel to the conductors under the sheath in a cable, imprinted with the manufacturer's name and the specification to which the cable is made.

MARKER THREAD – A colored thread laid parallel and adjacent to the strand in an insulated conductor which identifies the manufacturer and sometimes the specification to which the wire is made.

MASTIC – A meltable coating used on the inside of some shrink products which when heated flows to help create a waterproof seal.

MATV – Master Antenna Television System. A combination of components providing multiple television receiver operations from one antenna or group of antennas.

MAXIMUM CABLE DIAMETER – The largest cable diameter that a high-voltage cable termination is designed to accommodate.

MINIMUM CABLE DIAMETER – The smallest cable diameter that a high-voltage cable termination is designed to accommodate.

MAXIMUM DESIGN VOLTAGE – The maximum voltage at which a high-voltage cable termination is designed to operate continuously under normal conditions.

MC – (1) Main cross-connect, (2) A UL cable type (metal clad).

MECHANICAL WATER ABSORPTION – A check of how much water will be absorbed by material in warm water for seven days (mg/sq. in. surface).

MEDIUM FREQUENCY – The band of frequencies between 300 and 3,000 kilohertz.

MEDIUM-HARD DRAWN WIRE – As applied to copper wire, having tensile strength less than the minimum for hard-drawn wire, but greater than the maximum for soft wire.

MEDIUM VOLTAGE – A class of nominal power system voltage ratings between 2.4 and 46 kV.

MEGA – Prefix meaning million.

MEGAHERTZ (MHz) – One million cycles per second.

MEGGER – A special ohmmeter for measuring very high resistance. Primarily used for checking the insulation resistance of cables; however, it is also useful for equipment leakage tests.

MELT INDEX – The extrusion rate of a material through a specified orifice under specified conditions.

MEMBER – A group of wires stranded together that is in turn stranded into a multiple-membered conductor.

MESSENGER WIRE – A metallic supporting member either solid or stranded that may also perform the function of a conductor.

MFD – Microfarad (one-millionth of a farad). Obsolete abbreviation.

MFT – Abbreviation for 1,000 feet.

MG – Glass reinforced mica tape insulated cable with an overall sheath of woven glass yarn impregnated with a flame, heat and moisture resistant finish. 450°C, 600 V appliance wire.

MHO – The unit of conductivity. The reciprocal of an ohm.

MHz – MegaHertz (one million cycles per second).

MI – A UL cable type. One or more conductors insulated with highly compressed refractory minerals and enclosed in a liquid-tight and gas-tight metallic tube sheathing.

MICA – A transparent silicate which separates into layers and has high insulation resistance, high dielectric strength and high heat resistance.

MICRO – Prefix meaning one-millionth.

MICROBENDING LOSS – A signal loss due to small geometrical irregularities along the core-cladding interface of optical fibers.

MICROFARAD – One-millionth of a farad (abbreviated μf).

MICROMICROFARAD – One-millionth of a microfarad (abbreviated $\mu\mu f$). Also, a picofarad (pf).

MICROPHONE CABLE – A very flexible, usually shielded cable used for audio signals.

MICROPHONICS – Noise caused by mechanical movement of a system component. In a single conductor microphone cable, for example, microphonics can be caused by the shield rubbing against the dielectric as the cable is flexed.

MICROWAVE – A short (usually less than 30 cm) electrical wave.

MID-SPLIT – A broadband cable system in which the cable bandwidth is divided between transmit and receive frequencies. The bandwidth utilized to send toward the headend (reverse direction) is approximately 5 MHz to 100 MHz and the bandwidth utilized to send away from the headend (forward direction) is approximately 160 MHz to 300 Mhz.

MIL – A unit of length equal to one-thousandth of an inch.

MIL-SPEC – Military specification.

MIL-DTL-17 – A military specification covering many coaxial cables. Formerly MIL-C-17.

MIL-DTL-16878 – A military specification covering various wires intended for internal wiring of electric and electronic equipment. Formerly MIL-C-16878.

MIL-W-22759 – A military specification for fluorocarbon insulated copper and copper alloy wire. Replaced by SAE AS22759.

MILLI – Prefix meaning one-thousandth.

MIPS – Millions of instructions per second. One measure of processing power.

MODULATION – Systematic changing of properties, e.g., amplification, frequency, phase of an analog signal to encode and convey (typically digital) information.

MODULUS OF ELASTICITY – The ratio of stress (force) to strain (deformation) in a material that is elastically deformed.

MOISTURE ABSORPTION – The amount of moisture, in percentage, that a material will absorb under specified conditions.

MOISTURE RESISTANCE – The ability of a material to resist absorbing moisture from the air or when immersed in water.

MOLDED PLUG – A connector molded on either end of a cord or cable.

MONO FILAMENT – A single strand filament as opposed to a braided or twisted filament.

MONOMER – The basic chemical unit used in building a polymer.

MOTOR LEAD WIRE – Wire that connects to the fragile magnet wire found in coils, transformers and stator or field windings.

MPF – Mine power feeder cables. Usually rated 5, 8, or 15 kV.

MSHA – Mine Safety and Health Administration. The Federal enforcement agency for employee safety in mines and mills. Formerly known as MESA, Bureau of mines. MSHA regulations appear in CFR Title 30, Chapter 1.

MTW – Machine tool wire. Thermoplastic insulated, 90°C to 105°C, 600 V. A UL cable type.

MULTICAST – The ability to broadcast messages to one node or a select group of nodes.

MULTIMODE – Optical fiber which allows more than one mode of light to propagate.

MULTIPLE-CONDUCTOR CABLE – A combination of two or more conductors cabled together and insulated from one another and from sheath or armor where used.

MULTIPLE-CONDUCTOR CONCENTRIC CABLE – An insulated central conductor with one or more tubular stranded conductors laid over it concentrically and insulated from one another.

MULTIPLEX – The use of a common physical channel in order to make two or more logical channels, either by splitting of the frequency band (frequency-division multiplex), or by utilizing this common channel at different points in time (time-division multiplex).

MULTIPLEXER – Equipment that permits simultaneous transmission of multiple signals over one physical circuit.

MULTIPOINT CIRCUIT – A single line connecting three or more stations.

MURRAY LOOP TEST – A method used to localize cable faults.

MUTUAL CAPACITANCE – Capacitance between two conductors in a cable.

MUX – Multiplex. To transmit two or more signals over a single channel.

mV – Millivolt (one-thousandth of a volt).

MV – Medium voltage cables. Usually rated 5–35 kV.

mW – Milliwatt (one-thousandth of a watt).

MYLAR – DuPont's trademark for polyethylene terephthalate (polyester) film.

N

NBR – Butadiene-acrylonitrile copolymer rubber, a material with good oil and chemical resistance.

NBR/PVC – A blend of acrylonitrile-butadiene rubber and polyvinyl chloride (PVC). Used for jacketing.

NBS – National Bureau of Standards. Now called NIST (National Institute of Standards and Technology).

N CONNECTOR – A threaded connector for coax; N is named after Paul Neill.

NEC – National Electrical Code.

NEMA – National Electrical Manufacturers Association.

NEOPRENE – A synthetic rubber with good resistance to oil, chemicals and flame. Also called polychloroprene.

NETWORK – A series of nodes connected by communications channels.

NEWTON – The derived SI unit for force; the force which will give one kilogram mass an acceleration of one meter per second. Equals 0.2248 pounds force.

NEXT – Near-end crosstalk. A measure of the unwanted signal coupling from a transmitter at the near-end into a neighboring (non-energized) pair measured at the near-end.

NFPA – National Fire Protection Association.

NICKEL CLAD COPPER WIRE – A wire with a layer of nickel on a copper core where the area of the nickel is approximately 30 percent of the conductor area.

NIST – National Institute of Standards and Technology. Formerly the National Bureau of Standards.

NM-B – A UL cable type rated 600 volts and intended for use per Article 334 of the NEC. Nonmetallic sheathed cable. For dry use, 90°C conductor rating.

NMC-B – A UL cable type rated 600 volts and intended for use per Article 334 of the NEC. Nonmetallic sheathed cable. Wet or dry use, 90°C conductor rating.

NODE – A station.

NOISE – In a cable or circuit any extraneous sounds or signal which tends to interfere with the sound or signal normally present in or passing through the system.

NOMEX – DuPont's trademark for a heat resistant, flame retardant nylon.

NOMINAL – Name or identifying value of a measurable property by which a conductor or component or property of a conductor is identified and to which tolerances are applied.

NOMINAL VOLTAGE (NATIONAL ELECTRICAL CODE) – A nominal value assigned to a circuit or system for the purpose of conveniently designating its voltage class (as 120/240, 480Y/277, 600 volts, etc.). The actual voltage at which a circuit operates can vary from the nominal within a range that permits satisfactory operation of equipment.

NOMOGRAPH – A chart or diagram with which equations can be solved graphically by placing a straight edge on two known values and reading the answer where the straight edge crosses the scale of the unknown value.

NONCONTAMINATING – A type of PVC jacket material whose plasticizer will not migrate into the dielectric of a coaxial cable and thus avoid contaminating and destroying the dielectric.

NONCONTAMINATING PVC – A polyvinyl chloride formulation, which does not produce electrical contamination through plasticizer migration.

NONFLAMMABLE – The property of a material that is not capable of being easily ignited.

NONMIGRATING PVC – Polyvinyl chloride compound formulated to inhibit plasticizer migration.

NRZI – Non-Return to Zero Inverted. A binary encoding technique in which a change in state represents a binary 0 and no change in state represents a binary 1.

N-SERIES CONNECTOR – A coaxial connector (RG-8/U) used in standard Ethernet networks.

NTSC – National Television Standard Committee. The U.S. color TV standard.

NUMERICAL APERTURE – The acceptance angle of an optical fiber which determines the angle at which light can enter the fiber; expressed as a number which is equivalent to the sine of the angle.

NYLON – An abrasion-resistant thermoplastic with good chemical resistance. Polyamide.

O

O.D. – Outside diameter.

OEM – Original equipment manufacturer.

OFHC – Oxygen-Free High-conductivity Copper.

OHM – The electrical unit of resistance. The value of resistance through which a potential difference of one volt will maintain a current of one ampere.

OHM'S LAW – Stated $V = IR$, $I = V/R$, or $R = V/I$ where V is voltage, I is current in amperes and R is resistance in ohms.

OIL AGING – Cable aged in an accelerated manner by placement in an oil bath and heated to a preset temperature for a stated time.

OPEN CELL – Foamed or cellular material with cells which are generally interconnected.

OPEN CIRCUIT – A break in an electrical circuit so that there can be no current flow.

OPTICAL CONDUCTOR – Materials which offer a low optical attenuation to transmission of light energy.

OPTICAL CROSS-CONNECT – A cross-connect unit used for circuit administration. It provides for the connection of individual optical fibers with optical fiber patch cords.

OPTICAL ENCODER – A device whose position is determined by a photoelectric device and converted to an electrical data output.

OPTICAL FIBER – Any filament or fiber, made of dielectric materials, that is used to transmit light signals; optical fiber usually consists of a core, which carries the signal and cladding, a substance with a higher refractive index than the core, which surrounds the core and serves to reflect the light signal. See also Fiber Optics.

OPTICAL WAVEGUIDE – A fiber used for optical communications. Analogous to a waveguide used for microwave communications.

OSCILLATORY SURGE – A surge which includes both positive and negative polarity values.

OSCILLOSCOPE – Test instrument for showing visually the changes in a varying voltage by means of the wavy line made on a fluorescent screen by the deflection of a beam of cathode rays.

OSHA – Abbreviation for Occupational Safety and Health Act. Specifically the Williams-Steiger laws passed in 1970 covering all factors relating to safety in places of employment.

OSMOSIS – The diffusion of fluids through membranes.

OTDR – Optical Time Domain Reflectometer. A device used for testing fiber optic cables.

OUTDOOR TERMINATION – A cable termination intended for use where it is not protected from direct exposure to either solar radiation or precipitation.

OUTGASSING – Dissipation of gas from a material.

OUTPUT – The useful power or signal delivered by a circuit or device.

OVERALL DIAMETER – Finished diameter over wire or cable.

OVERCOAT CONDUCTOR – A stranded conductor made from individual strands of tin-coated wire stranded together and then given an overall tin coat.

OVERLAP – The amount the trailing edge laps over the leading edge of a spiral tape wrap.

OXYGEN INDEX – A test to rate flammability of materials in a mixture of oxygen and nitrogen. More formally referred to as Limiting Oxygen Index (LOI).

OZONE – An extremely reactive form of oxygen, normally occurring around electrical discharges and present in the atmosphere in small but active quantities. In sufficient concentrations it can break down certain insulations.

P

PAIR – Two insulated wires of a single circuit associated together; also known as a "balanced" transmission line.

PARALLEL CIRCUIT – A circuit in which identical voltage is presented to all components and the current divides among the components according to the resistances or the impedances of the components.

PARALLEL STRIPE – A stripe applied longitudinally on a wire or cable parallel to the axis of the conductor.

PARALLEL TRANSMISSION – A type of data transfer in which all bits of a character, or multiple-bit data blocks, are sent simultaneously, either over separate communications lines or circuits, over a single channel using multiple frequencies, or over a multiple-conductor cable.

PARTIAL DISCHARGE (CORONA) EXTINCTION VOLTAGE – The voltage at which partial discharge (corona) is no longer detectable on instrumentation adjusted to a specific sensitivity, following the application of a specified higher voltage.

PATCH CABLE – A cable with plugs or terminals on each end of the conductors to connect circuits of equipment together.

PAYOFF – The process of feeding a cable or wire from a bobbin, reel, or other package.

PBDE – Polybrominated Diphenyl Ether. A class of flame retardants sometimes used in wire and cable products. PBDEs are generally banned by European Union RoHS regulations.

PCB – Printed Circuit Board.

PCP – Polychloroprene (Neoprene).

PDN – Public Data Network. A packet switched or circuit switched network available for use by many customers. PDNs may offer value-added services at a reduced cost because of communications resource sharing and usually provide increased reliability due to built-in redundancy.

PE – PolyEthylene. A widely used thermoplastic insulation and jacket compound.

PEAK – The maximum instantaneous value of a varying current or voltage. Also called Crest.

PEEK – Poly Ether Ether Ketone.

PEEL STRENGTH – The force necessary to peel a flexible member from another member that may be either flexible or rigid.

PERCENT CONDUCTIVITY – The ratio of the resistivity of the International Annealed Copper Standard (IACS) at 20°C to the resistivity of a material at 20°C, expressed in percent. Results are calculated on a weight basis or volume basis and so specified.

PERCENT PLATING – Quantity of plating on a conductor expressed as a percentage by weight.

PERCENTAGE CONDUCTIVITY – Conductivity of a material expressed as a percentage of that of copper.

PFA – PerFluoroAlkoxy. Teflon is DuPont's trademark for this material.

PHASE – The location of a position on a waveform of an alternating current, in relation to the start of a cycle. Measured in degrees, with 360 corresponding to one complete cycle.

PHASE SEQUENCE – The order in which successive members of a periodic wave set reach their positive maximum values: a) zero phase sequence – no phase shift, b) plus/minus phase sequence – normal phase shift.

PHASE SHIFT – A change in the phase relationship between two alternating quantities. The phase angle between the input and output signals of a system.

PICK – Distance between two adjacent crossover points of braid filaments. The measurement in picks per inch indicates the degree of coverage.

PICO – Prefix meaning one-millionth of one-millionth (10^{-12}).

PICOFARAD – One-millionth of one-millionth of a farad. A micromicrofarad, or picofarad (abbreviation pf).

PIGTAIL WIRE – Fine stranded, extra flexible, rope lay lead wire attached to a shield for terminating purposes.

PILC CABLE – Paper Insulated, Lead Covered.

PIN ASSIGNMENT – A predetermined relationship between the terminals in a connector and the conductors in a cable that specifies the terminals to which each conductor is to be terminated.

PITCH – In flat cable, the nominal distance between the index edges of two adjacent conductors.

PITCH DIAMETER – Diameter of a circle passing through the center of the conductors in any layer of a multiconductor cable.

PLANETARY TWISTER – A cabling machine whose payoff spools are mounted in rotating cradles that hold the axis of the spool in a fixed direction as the spools are revolved so the wire will not kink as it is twisted.

PLASTICIZER – A chemical added to plastics to make them softer and more flexible.

PLATED HOLE – A hole with walls that have been plated with conductive material to provide an electrical connection between the conductive patterns on both sides of a printed circuit or an anchor for soldering an inserted wire.

PLC – Programmable Logic Controller. A type of industrial control system.

PLENUM – The air return path of a central air handling system, either ductwork or open space over a suspended ceiling.

PLENUM CABLE – Cable approved by a recognized agency such as UL for installation in plenums without the need for conduit.

PLTC – Power Limited Tray Cable, rated 300 volts.

PLUG – The part of the two mating halves of a connector that is movable when not fastened to the other mating half.

PLY – The number of individual strands or filaments twisted together to form a single thread.

POINT-TO-POINT WIRING – An interconnecting technique wherein the connections between components are made by wires routed between connecting points.

POLARIZATION – The orientation of a flat cable or a rectangular connector.

POLISHING – Act of smoothing ends of optical fibers to an "optically smooth" finish, generally using abrasives.

POLYAMIDE – The chemical name for Nylon.

POLYBUTADIENE – A type of synthetic rubber often blended with other synthetic rubbers to improve their properties.

POLYCHLOROPRENE – See Neoprene.

POLYESTER – Polyethylene terephthalate, used extensively as a moisture resistant cable core wrap. Mylar is DuPont's trademark for polyester.

POLYETHYLENE – A thermoplastic material having excellent electrical properties.

POLYHALOCARBON – A general name for polymers containing halogen atoms. The halogens are flourine, chlorine, bromine and iodine.

POLYIMIDE – A relatively high temperature plastic developed for use as a dielectric or jacketing material. Kapton is DuPont's trademark for polyimide.

POLYMER – A substance made of many repeating chemical units or molecules. The term polymer is often used in place of plastic, rubber or elastomer.

POLYOLEFINS – A family of plastics including cross-linked polyethylene and various ethylene copolymers.

POLYPROPYLENE – A thermoplastic similar to polyethylene but stiffer and having a higher temperature softening point.

POLYURETHANE – Broad class of polymers noted for good abrasion and solvent resistance. Can be in solid or cellular form.

POLYVINYL CHLORIDE (PVC) – A general purpose thermoplastic used for wire and cable insulations and jackets.

POROSITY – Multiple voids in an insulation cross section.

PORT – A point of access into a computer, a network or other electronic device; the physical or electrical interface through which one gains access; the interface between a process and a communications or transmission facility.

P.O.S. – Abbreviation for point-of-sale.

POSITION CODING – Identification of conductors by their location, possible only when conductors are located in assigned positions with relation to each other throughout the entire length of a cable.

POSITIVE SEQUENCE IMPEDANCE – See Phase Sequence. Sometimes referred to as Plus Sequence Impedance.

POTTING – Sealing by filling with a substance to exclude moisture.

POWER – The amount or work per unit of time. Usually expressed in watts and equal to I^2R.

POWER CABLES – Cables of various sizes, constructions and insulations, single or multiconductor, designed to distribute primary power to various types of equipment.

POWER FACTOR – The cosine of the phase difference between current and applied voltage.

POWER LOSS – The difference between the total power delivered to a circuit, cable, or device and power delivered by that device to a load.

POWER RATIO – The ratio of the power appearing at the load to the input power. Expressed in dB, it is equal to 10 $\log_{10} (P_2/P_1)$ where P_1 is input power and P_2 is the power at the load.

PPE – PolyPropylene Ethylene.

PREBOND – Stranded wire that has been fused, topcoat tinned or overcoat tinned.

PREMOLDED SPLICE – A joint made of premolded components assembled in the field.

PRIMARY – The transformer winding that receives the energy from a supply current.

PRIMARY INSULATION – The first layer of nonconductive material applied over a conductor, whose prime function is to act as electrical insulation.

PRINTING WIRING – A printed circuit intended to provide point-to-point electrical connections.

PRODUCTION TESTS – Tests made on components or subassemblies during production for the purpose of quality control.

PROPAGATION DELAY – The time it takes a signal, composed of electromagnetic energy to travel from one point to another over a transmission channel; usually most noticeable in communicating with satellites; normally, the speed-of-light delay.

PROPAGATION TIME – Time required for a wave or pulse to travel between two points on a transmission line.

PROPAGATION VELOCITY – The velocity of the propagation of a wave or pulse along a transmission path.

PROTECTIVE COVERING – A field-applied material to provide environmental protection over a splice or housing, or both.

PROXIMITY EFFECT – Nonuniform current distribution over the crosssection of a conductor caused by the variation of the current in a neighboring conductor.

PSAACRF – Powersum insertion loss to alien crosstalk ratio far-end. A computation of signal coupling from multiple pairs of disturbing channels, to a disturbed pair in another channel measured at the far-end and relative to the received signal level in the disturbed pair at the far-end. Also referred to as powersum alien equal-level far-end crosstalk (PSAELFEXT).

PSANEXT – Powersum alien near-end crosstalk. A computation of signal coupling from multiple near-end disturbing channel pairs into a disturbed pair of a neighboring channel or part thereof, measured at the near-end.

PSAFEXT – Powersum alien far-end crosstalk. A computation of signal coupling from multiple near-end disturbing channel pairs into a disturbed pair of a neighboring channel or part thereof, measured at the far-end.

PSNEXT – Powersum near-end crosstalk. A computation of the unwanted signal coupling from multiple transmitters at the near-end into a neighboring (non-energized) pair measured at the near-end.

PT – Thermostat cable with solid conductor, individual insulation, twisted together.

PTFE – PolyTetraFluoroEthylene. One type of Teflon. Sometimes abbreviated TFE.

PULLING EYE – A device used to pull cable into or from a duct.

PULSE – A current or voltage that changes abruptly from one value to another and back to the original value in a finite length of time.

PULSE CABLE – A type of coaxial cable constructed to transmit repeated high-voltage pulses without degradation.

PVC – PolyVinyl Chloride. A common insulating and jacketing material used on cables.

PVC-I – A MIL-DTL-17 coax jacket type. A black polyvinyl chloride with excellent weathering and abrasion properties, but is a contaminating type and will cause cable attenuation to increase with age. Can be used for direct burial.

PVC-II – A MIL-DTL-17 coax jacket type. A grey polyvinyl chloride material which is semi-noncontaminating.

PVC-IIA – A MIL-DTL-17 coax jacket type. A black or grey polyvinyl chloride material which is noncontaminating. It has good weathering and abrasion-resistant properties and can be used for direct burial.

PVDF – PolyVinyliDene Fluoride. Arkema Inc.'s trademark for this material is Kynar.

PYROMETER – See Thermocouple.

Q

Q band – The band of frequencies between 36 and 46 gigahertz.

QPL – A Qualified Products List issued by the U.S. government.

QUAD – A series of four separately insulated conductors, generally twisted together in pairs.

R

R – Symbol for electrical resistance.

RADIO FREQUENCY – The frequencies in the electromagnetic spectrum that are used for radio communications. A band of frequencies between 10 kilohertz and 100 gigahertz.

RANDOM WINDING – A winding in rotating equipment wherein wires do not lie in an even pattern.

REA (RURAL ELECTRIFICATION ADMINISTRATION) – A federally supported program to provide electrical service to rural areas. See RUS.

REACH – A European Community Regulation on the safe use of Chemicals through Registration, Evaluation and Authorization.

REACTANCE – The opposition to an alternating electron flow by a capacitance or inductance. The amount of such opposition varies with the frequency of the current. The reactance of a capacitor decreases with an increase in frequency; the opposite occurs with an inductance.

RECOVERED DIAMETER – Diameter of shrinkable products after heating has caused it to return to its extruded diameter.

RED PLAGUE – A powdery, brown-red growth sometimes found on silver-coated copper conductors and shield braids.

REDRAW – The consecutive drawing of wire through a series of dies to reach a desired wire size.

REEL – A revolving flanged device made of plastic, wood or metal, used for winding flexible cable. Also called a spool.

REFERENCE EDGE – See preferred term Index Edge.

REFERENCE JUNCTION – The junction of a thermocouple that is at a known reference temperature. Also known as the "cold" junction, it is usually located at the emf measuring device.

REFLECTION – (1) The change in direction (or return) of waves or pulses striking a surface. For example, electromagnetic energy reflections can occur at an impedance mismatch in a transmission line, causing standing waves, (2) Change in direction of a light wave or ray in an optical fiber.

REFLECTION LOSS – The part of a signal that is lost to reflection of power at a line discontinuity.

REFLOW SOLDERING – The process of connecting two solder-coated conductive surfaces by remelting of the solder to cause fusion.

REFRACTION – The bending of lightwaves or rays as they go from one material to another due to the difference in velocities in the materials.

REINFORCED SHEATH – The outer covering of a cable that has a reinforcing material, usually a braided fiber, molded in place between layers.

RELIABILITY – The probability that a device will function without failure over a specified time period or amount of usage.

RESIN – A solid or semisolid organic substance, originally of plant origin but largely synthesized now. Resins are broadly classified as thermoplastic or thermoset according to whether they soften or harden with the application of heat.

RESISTANCE – In DC circuits, the opposition a material offers to current, measured in ohms. In AC circuits, resistance is the real component of impedance and may be higher than the value measured at DC.

RESISTIVE CONDUCTOR – A conductor with high electrical resistance.

RESISTIVITY – A material characteristic that opposes the flow of electrical energy through the material. It is affected by temper, temperature, contamination, alloying, etc. The unit of volume resistivity is the ohm-cm. The unit of surface resistivity is ohms/m^2.

RESISTOR – An electronic component designed to have a specific value of resistance.

RESISTOR COLOR CODE – A method of indicating resistance value and tolerance. The first color represents the first significant figure of the value. A second color represents the second significant figure and the third is the multiplier or the number of zeros that follow two significant figures. When there is a fourth color band, it indicates the tolerance.

RESONANCE – An AC circuit condition in which inductive and capacitive reactances interact to cause a minimum or maximum circuit impedance.

RETRACTILE CORD – A cord having specially treated insulation or jacket so that it will retract like a spring. Retractability may be added to all or part of a cord's length.

RETURN LOSS – A measure of the degree of impedance mismatch between two impedances. It is the ratio, expressed in decibels, of the amplitude of a reflected wave echo to the amplitude of the main wave at the junction of a transmission line and a terminating impedance.

RETURN WIRE – A ground wire or the negative wire in a direct-current circuit.

REW – A 600-volt Canadian wire type covered by CSA standards. Made with thermoset insulation.

RFI – Radio frequency interference. The disruption of radio signal reception caused by any source that generates radio waves at the same frequency and along the same path as the desired wave. Similar to EMI.

RF MODEM – Radio frequency modem. A device used to convert digital data signals to analog signals (and from analog to digital), then modulate/demodulate them to/from their assigned frequencies.

RG/U – "RG" is the military designation for coaxial cable and "U" stands for "general utility."

RHH – Rubber-insulated, heat-resistant wire, rated 90°C. A UL cable type.

RHW – Rubber-insulated building wire, heat and moisture-resistant, 75°C dry or wet. A UL cable type.

RHW-2 – Rubber-insulated building wire, heat and moisture-resistant, 90°C dry or wet. A UL cable type.

RIBBON CABLE – A flat cable of individually insulated conductors lying parallel and held together by means of adhesive or an extruded polymer web.

RIDGE MARKER – One or more ridges running laterally along the outer surface of a plastic insulated wire for purposes of identification.

RIGID COAXIAL CABLE – Nonflexible coaxial cable, usually a metal tube armored coaxial cable. Sometimes called "hardline."

RINGING OUT – Locating or identifying specific conductive paths by passing current through selected conductors.

RING TONGUE – A solderless terminal that connects wire to a stud.

RIP CORD – Two or more insulated conductors in a parallel configuration which may be separated to leave the insulation of each conductor intact.

RISE TIME – The time it takes the voltage to rise from 0.1 to 0.9 of its final value.

RIV – Radio influence voltage. The radio noise appearing on conductors of electric equipment or circuits.

RMS – See root mean square.

ROCKWELL HARDNESS – A measure of hardness determined by resistance to indention by a small diamond or steel ball under pressure.

RoHS – Restriction of Hazardous Substances. An environmental regulation that limits the amount of six substances in products including lead, cadmium, mercury, Hexavalent Chromium, PBB and PBDE.

ROMEX – A type of nonmetallic sheathed (Type NM) cable. A trademark of the Southwire Company.

ROOT MEAN SQUARE (RMS) – The effective value of an alternating current or voltages.

ROPE CONCENTRIC – A group of stranded conductors assembled in a concentric manner.

ROPE-LAY CONDUCTOR – See concentric-lay conductor.

ROPE STRAND – A conductor composed of a center group of twisted strands surrounded by layers of twisted strands.

ROPE UNILAY – A group of stranded conductors assembled in a unilay manner.

ROTATING CABLE – A coil of cable whose inner end is attached to a member that rotates in relation to a member to which the outer end of the cable is fastened.

ROUND CONDUCTOR FLAT CABLE – A cable made with parallel round conductors in the same plane.

ROUND WIRE SHIELDS – Shields constructed from bare, tinned or silver-plated copper wire that include braided, spiral and reverse spiral.

ROUTINE TESTS – Tests made on each high-voltage cable or upon a representative number of devices, or parts, during production for the purposes of quality control.

RS-232 – An EIA recommended standard (RS); a common standard for connecting data processing devices. RS-232 defines the electrical characteristics of the signals in the cable that connect DTE (Data Terminal Equipment) with DCE (Data Communications Equipment); it specifies a 25-pin connector (the DB-25 connector is almost universally used in RS-232 applications). It is similar to ITU-T V.24/V.28.

RS-232-C SERIAL I/O PORT – A standard connection interface for computer peripheral equipment.

RS-422 – An ANSI/TIA/EIA standard for cable lengths that exceed the RS-232 50-foot limit. Although introduced as a companion standard with RS-449, RS-422 is most frequently implemented on unused pins of DB-25 (RS-232) connectors. Similar to ITU-T V.11.

RS-423 – A TIA/EIA standard for cable lengths that exceed the RS-232 50-foot limit. Although introduced as a companion standard with RS-422, RS-423 is not widely used. Similar to ITU-T V.10.

RTD – Resistance temperature sensing device. Converts temperature to an electrical signal.

RTS – Request-To-Send. An RS-232 modem interface signal (sent from the DTE to the modem on pin 4) which indicates that the DTE has data to transmit.

RUBBER, ETHYLENE PROPYLENE (EPR) – A synthetic rubber insulation having excellent electrical properties.

RUBBER INSULATION – A general term used to describe wire insulations made of elastomers such as natural or synthetic rubbers, neoprene, Hypalon, EPR, CPE and others.

RUS – Rural Utilities Service. A federal agency formerly known as the REA.

S

S – A UL cable type. Hard service flexible cord with thermoset insulation and jacket. Also see SO and SJ.

SAE – Society of Automotive Engineers.

S BAND – A band of frequencies between 1,550 and 5,200 megahertz.

SBR – A copolymer of styrene and butadiene. Also GRS or Buna-S.

SCHERING BRIDGE – See Bridge.

SDN – A small diameter multiconductor control cable with neoprene jacket and nylon sheath over polyethylene insulation.

SECONDARY INSULATION – A nonconductive material that protects the conductor against abrasion and provides a second barrier.

SEGMENTAL CONDUCTOR – A stranded conductor consisting of three or more stranded conducting elements, each element having approximately the shape of the sector of a circle, assembled to give a substantially circular cross section.

SELF-EXTINGUISHING – Characteristic of a material whose flame is extinguished after the igniting flame source is removed.

SEMICONDUCTOR – In wire industry terminology, a material possessing electrical conduction properties that fall somewhere between conductors and insulators. Usually made by adding carbon particles to an insulator. Not the same as semiconductor materials such as silicon, germanium, etc., used for making transistors and diodes.

SEMICONDUCTING JACKET – A jacket having a sufficiently low-electrical resistance so that its outer surface can be kept at substantially ground potential.

SEMIRIGID CABLE – Generally refers to Type MI or Type ALS that can be bent or shaped into a required configuration from coils or reels.

SEMIRIGID PVC – A hard semiflexible polyvinylchoride compound with low plasticizer content.

SEMISOLID – An insulation crosssection having a partially open space between the conductor and the insulation perimeter.

SENSITIVE CONDUCTOR – A conductor terminated to a circuit that is adversely affected by spurious signals.

SEPARABLE INSULATED CONNECTOR – An insulated device to facilitate cable connections and separations.

SEPARATOR – Pertaining to wire and cable, a layer of insulating material such as textile, paper, Mylar, etc., which is placed between a conductor and its dielectric, between a cable jacket and the components it covers, or between various components of a multiple conductor cable. It can be utilized to improve stripping qualities and/or flexibility, or can offer additional mechanical or electrical protection to the components it separates.

SERIAL INTERFACE – An interface that requires serial transmission, or the transfer of information in which the bits composing a character are sent sequentially. Implies only a single transmission channel.

SERIES CIRCUIT – A circuit in which the components are arranged end to end to form a single path for current.

SERVE – A filament or group of filaments such as fibers or wires, wound around a central core.

SERVED WIRE ARMOR – Spiral wrap of soft galvanized steel wires wrapped around a cable to afford mechanical protection and increase the cable-pulling tension characteristic.

SERVING – A wrapping applied over the core of a cable or over a wire.

SE – A UL cable type. Service Entrance Cable, 600 volts.

SEW, SEWF – A CSA cable type. Silicone rubber-insulated equipment wire.

SF – A CSA cable type. Silicone rubber insulated fixture wire, solid or 7 strand conductor, 200°C.

SFF – A CSA cable type. Same as SF, except flexible stranding 150°C.

SG – A CSA cable type. Same as SW except with ground wires.

SGO – A CSA cable type. Same as SWO except with ground wires.

SHD – Portable mine power cable, three or four individually shielded conductors, with grounding conductors, 5 through 25 kV.

SHEATH – The outer covering or jacket over the insulated conductors to provide mechanical protection for the conductors.

SHIELD – A sheet, screen, or braid of metal, usually copper, aluminum, or other conducting material placed around or between electric circuits or cables or their components, to contain any unwanted radiation, or to keep out any unwanted interference.

SHIELD COVERAGE – See Shield Percentage.

SHIELDED INSULATED SPLICE – An insulated splice in which a conducting material is employed over the full length of the insulation for electric stress control.

SHIELDED LINE – A transmission line whose elements confine radio waves to an essentially finite space inside a tubular conducting surface called the sheath, thus preventing the line from radiating radio waves.

SHIELD EFFECTIVENESS – The relative ability of a shield to screen out undesirable radiation. Frequently confused with the term shield percentage, which it is not.

SHIELDING, POWER CABLE – A conducting layer applied to increase safety, control dielectric stresses and prevent partial discharges.

SHIELD PERCENTAGE – The physical area of a circuit or cable actually covered by shielding material, expressed in percent.

SHORT – A low-resistance path that results in excessive current flow and often in damage.

SHOVEL CABLE – Normally an SHD-GC type that supplies high-voltage (2 to 25 kV) power to mobile equipment.

SHRINKING RATIO – The ratio between the expanded diameter and recovered diameter of shrinkable products.

SHRINK TEMPERATURE – That temperature that effects complete recovery of a heat shrinkable product from the expanded state.

SHRINK TUBING – Tubing that has been extruded, cross-linked and mechanically expanded which when reheated or released will return to its original diameter.

SHUNT – A very low resistance component used to divert a portion of the electric current.

SHUNT WIRE – A conductor joining two parts of an electric circuit to divert part of the current.

SI – An international system of standardized units of measurement.

SIC (SPECIFIC INDUCTIVE CAPACITANCE) – See Dielectric Constant.

SIGNAL – Any visible or audio indication which can convey information. Also, the information conveyed through a communication system.

SIGNAL CABLE – A cable designed to carry current of usually less than one ampere per conductor.

SIGNAL-TO-NOISE RATIO – A ratio of the amplitude in a desired signal to the amplitude of noise, usually expressed in dB.

SILICONE – A material made from silicon and oxygen. Can be in thermosetting elastomer or liquid form. The thermosetting elastomer form is noted for high heat resistance.

SINGLE CABLE – A one-cable system in broadband LANs in which a portion of the bandwidth is allocated for send signals and a portion for receive signals, with a guard band in between to provide isolation from interference.

SINGLE-MODE – Optical fiber in which only one mode of light can propagate.

SINTERING – Fusion of an extruded paste or a spirally applied tape wrap insulation or jacket by the use of high heat to a homogenous continuum. Usually employed for fluorocarbon, nonextrudable materials.

SIS – Switchboard wiring made with cross-linked polyethylene insulation.

SJ – A UL or CSA cable type. Junior hard service, rubber-insulated pendant or portable cord. Same construction as type S, but 300 V.

SJO – Same as SJ, but with oil-resistant jacket.

SJOO – Same as SJO but with oil-resistant insulation as well as an oil-resistant jacket.

SJT – A UL or CSA cable type. Junior hard service thermoplastic or rubber insulated conductors with overall thermoplastic jacket. 300 V.

SJTO – Same as SJT but oil-resistant thermoplastic outer jacket.

SJTOO – Same as SJTO but with oil-resistant insulation.

SKIN EFFECT – The tendency of alternating current, as its frequency increases, to travel only on the surface of a conductor.

S METER – An instrument to measure signal strength.

S/N – See Signal-to-Noise Ratio.

SNM – Shielded nonmetallic sheathed cable.

SO – A UL or CSA cable type. Hard service cord, same construction as type S except oil-resistant thermoset jacket, 600 V.

SOFT WIRE – Wire that has been drawn or rolled to final size and then heated (annealed) to remove the effects of cold working.

SOLEF – Solvay's trademark for its PVDF polymer.

SOLID CONDUCTOR – A conductor consisting of a single wire.

SOO – Same as SO but with oil-resistant insulation.

SOOW – A UL or CSA cable type. Portable cord and control cable. 600 V. Same as SOO but UL Listed or CSA Certified for outdoor use.

SOURCE COUPLING LOSS – Loss of light intensity as the light from a source passes into an optical fiber.

SPACER CABLE – A type of overhead power distribution cable. Spacing is accomplished by ceramic or plastic hangers suspended from a support messenger.

SPAN – In flat conductors, distance between the reference edge of the first and the last conductor. In round conductors, distance between centers of the first and last conductors.

SPC – Statistical Process Control.

SPECIFIC INDUCTIVE CAPACITY (SIC) – Dielectric constant of an insulating material.

SPIRAL SHIELD – A metallic shield of fine stranded wires applied spirally rather than braided.

SPIRAL STRIPE – A color coding stripe applied helically to the surface of an insulated wire or cable.

SPIRAL WRAP – The helical wrap of a tape or thread over a core.

SPLICE – A connection of two or more conductors or cables to provide good mechanical strength as well as good electrical conductivity.

SPLITTER – A passive device used in a cable system to divide the power of a single input into two or more outputs of lesser power. Can also be used as a combiner when two or more inputs are combined into a single output.

SP-1 – A UL cable type. All thermoset, parallel-jacketed, two-conductor light duty cord for pendant or portable use in damp locations, 300 V.

SP-2 – Same as SP-1, but heavier construction, with or without a third conductor for grounding purposes, 300 V.

SP-3 – Same as SP-2, but heavier construction for refrigerators or room air-conditioners, 300 V.

SPT – A UL type of thermoplastic-insulated, 2 or 3 conductor parallel cord. Frequently called "zip cord" or "lamp cord."

SQUIRREL CAGE MOTOR – An induction motor having the primary winding (usually the stator) connected to the power and a current is induced in the secondary cage winding (usually the rotor).

SR – Silicone Rubber cable 600 V, 125°C.

SRG – A cable with ozone resistant silicone rubber insulation with an overall jacket of braided glass yarn impregnated with flame, heat and moisture resistant finish. 150/200°C 600 V appliance and motor lead wire.

SRGK – A cable with ozone resistant silicone rubber insulation with braided glass yarn conductor jacket. Cable core of insulated conductors shielded or unshielded and an overall jacket of braided K-fiber impregnated with flame, heat and moisture resistant finish. 150/200°C 600 V multiconductor cable.

SRK – A cable with ozone resistant silicone rubber insulation with an overall jacket of braided K-fiber impregnated with flame, heat and moisture resistant finish. 200°C 600 V fixture wire and power cable.

ST – A UL cable type. Hard service cord, jacketed, same as type S except thermoplastic construction. 600 V, 60°C to 105°C.

ST1 – An optional smoke test rating for wires in UL Standard 44.

STABILITY FACTOR – The difference between the percentage power factor at 80 volts/mil and at 40 volts/mil measured on wire immersed in water at 75°C for a specified time.

STANDARD – A set of rules or protocols that describe how a device should be manufactured so it will be reliable and interoperability (compatibility) with others of the same type from different manufacturers will be maintained.

STANDING WAVE – The stationary pattern of waves produced by two waves of the same frequency traveling in opposite directions on the same transmission line. The existence of voltage and current maxima and minima along a transmission line is a result of reflected energy from an impedance mismatch.

STANDING WAVE RATIO (SWR) – In a transmission line, waveguide or analogous system, a figure of merit used to express the efficiency of the system in transmitting power.

STATIC CHARGE – An electrical charge that is bound to an object. An unmoving electrical charge.

STAY CORD – A component of a cable, usually a high-tensile textile, used to anchor the cable ends at their points of termination and to keep any pull on the cable from being transferred to the electrical connections.

STEP INDEX FIBER – A multimode optical fiber consisting of a core of uniform refractive index, surrounded by cladding of slightly lower refractive index.

STIFFNESS – As applied to copper, the property of a conductor that causes it to resist permanent deformation by bending.

STO – A North American flexible cord type. Same as ST but with an oil-resistant thermoplastic outer jacket, 600 V, 60°C.

STOO – Same as STO but with oil-resistant insulation.

STOOW – Same as STOO but suitable for use in wet locations.

STOP JOINT – A splice that is designed to prevent any transfer of dielectric fluid between the cables being joined.

STP – Shielded twisted pair. Two wires, wound around each other to help cancel out any induced noise in balanced circuits. Multiple pairs of wires are contained in one sheath and each wire pair is shielded.

STRAIGHT JOINT – A cable splice used for connecting two lengths of cable, each of which consists of one or more conductors.

STRAIN GAUGE – A device for determining the amount of strain (change in dimension) when a stress is applied.

STRAIN HARDENING – An increase in hardness and strength caused by plastic deformation at temperatures lower than the recrystallization range.

STRAND – One of the wires of any stranded conductor.

STRANDED CONDUCTOR – A conductor composed of a group of wires, usually twisted, or of any combination of such groups of wires.

STRAND LAY – The distance of advance of one strand of a spirally stranded conductor, in one turn, measured longitudinally.

STRESS-RELIEF CABLE – Cable used to relieve stresses in the process of welding pipe joints by inducing heat in pipe sections to be welded, flexible copper strand.

STRESS-RELIEF CONE (TERMINATION) – A device used to relieve the electrical stress at a shielded cable termination; generally used above 2,400 volts.

STRIP – To remove insulation from a wire or cable.

STRUCTURAL RETURN LOSS – Backward reflected energies from uneven parts of the cable structure.

SUBSTRATE – Insulating material of a printed circuit.

SUGGESTED WORKING VOLTAGE – AC voltage that can be applied between adjacent conductors.

SUPERCONDUCTORS – Materials whose resistance and magnetic permeability are virtually zero at very low temperatures.

SUPPRESSOR – A device used to reduce or eliminate unwanted voltages in electric or electronic circuits. For example, a resistance conductor in, or a resistor in series with, a sparkplug cable, to suppress interference which would otherwise affect radio reception in and near the vehicle.

SURFACE RESISTIVITY – The resistance of a material between two opposite sides of a unit square of its surface. It is usually expressed in ohms.

SURGE – A temporary and relatively large increase in the voltage or current in an electric circuit or cable. Also called Transient.

SV – A North American cable type. Vacuum cleaner cord, two or three conductor, rubber insulated. Overall rubber jacket. For light duty in damp locations, 300 V 60°C.

SVO – A North American cable type. Same as SV except oil-resistant thermoset jacket, 300 V 60°C or 90°C.

SVT – A North American cable type. Same as SV except thermoplastic jacket. 300 V, 60°C or 90°C.

SVTO – A North American cable type. Same as SVT, except with oil-resistant thermoplastic jacket, 60°C.

SW – A CSA cable type. Rubber jacketed power supply cable (8 AWG to 2 AWG) 600 V.

SWEEP TEST – A test given to check attenuation at various frequencies, as in coaxial cable.

T

TAKE-UP – The process of accumulating wire or cable onto a reel, bobbin or some other type of pack. Also, the device for pulling wire or cable through a piece of equipment or machine.

TANK TEST – A dielectric strength test in which the test sample is submerged in water and voltage is applied between the conductor and water as ground.

TAP – (1) Baseband – The component of a connector that attaches a transceiver to a cable. (2) Broadband – (Also called a directional tap or multitap) a passive device used to remove a portion of the signal power from the distribution line and deliver it onto the drop line.

TAPED INSULATION – Insulation of helically wound tapes applied over a conductor or over an assembled group of insulated conductors.

TAPED SPLICE – A joint with hand-applied tape insulation.

TAPE WRAP – A spirally applied tape over an insulated or uninsulated wire.

TC – A UL cable type. See Tray Cable, NEC Article 336.

TCLP – Toxicity Characteristic Leaching Procedure. A test created by the EPA to determine whether an item can be safely discarded in an ordinary (nonhazardous) landfill.

T CONNECTOR – A cable adapter that attaches a PC with a network interface module to the network.

TEAR STRENGTH – The force required to initiate or continue a tear in a material under specified conditions.

TEFLON – Trademark of the DuPont Co. for FEP, PTFE and PFA polymers.

TEMPERATURE RATING – The maximum temperature at which an insulating material may be used in continuous operation without loss of its basic properties.

TENSILE STRENGTH – The maximum load per unit of original cross-sectional area that a conductor attains when tested in tension to rupture.

TERMINALS – Metal wire termination devices designed to handle one or more conductors, and to be attached to a board, bus or block with mechanical fasteners or clipped on.

TERMINATOR – A resistive device used to terminate the end of cable or an unused tap into its characteristic impedance. The terminator prevents interference-causing signal reflections.

TEST LEAD – A flexible, insulated lead wire used for making tests, connecting instruments to a circuit temporarily, or for making temporary electrical connections.

TEW – Canadian Standards Association type appliance wires. Solid or stranded single conductor, plastic insulated, 105°C, 600 V.

TEXTILE BRAID – Any braid made from threads of cotton, silk or synthetic fibers.

TF – A UL cable type. Fixture wire, thermoplastic-covered solid or 7 strands, 60°C.

TFE – One of three types of Teflon. Also known as PTFE (PolyTetraFluoroEthylene).

TFF – Same as TF but flexible stranding, 60°C.

TFFN – Same as TFF but with nylon outer jacket.

TFN – Same as TF but with nylon outer jacket.

TG – Flexible nickel or nickel-clad copper conductor, Teflon tape, glass braid, 200°C.

TGGT – PTFE Teflon tape insulation with an insulation covering of wrapped glass yarn and an overall sheath of braided glass yarn impregnated with a moisture, heat, flame and fraying resistant compound. 600 V, 250°C appliance wire.

TGS – Solid or flexible copper, nickel-clad iron or copper, or nickel conductor. Teflon tape, silicone glass braid, 600 V 250°C.

THERMAL AGING – Exposure to a thermal condition or programmed series of conditions for predescribed periods of time.

THERMOCOUPLE – A device consisting of two dissimilar metals in physical contact, which when heated will develop an emf output.

THERMOCOUPLE ELEMENT – A thermocouple designed to be used as part of an assembly, but without associated parts such as terminal block, connecting head or protecting tube.

THERMOCOUPLE EXTENSION CABLE – A cable comprised of one or more twisted thermocouple extension wires under a common sheath.

THERMOCOUPLE EXTENSION WIRE – A pair of wires of dissimilar alloys having emf temperature characteristics complementing the thermocouple with which it is intended to be used, such that when properly connected allows the emf to be faithfully transmitted to the reference junction.

THERMOCOUPLE LEAD WIRE – An insulated pair of wires used from the thermocouple to a junction box.

THERMOPLASTIC – A material that softens when heated and becomes firm on cooling.

THERMOSET – A material that has been hardened or set by the application of heat or radiation, and which, once set, cannot be resoftened by heating. The application of heat or radiation is called "curing."

THHN – A UL cable type. 600 V, 90°C nylon-jacketed building wire for use in dry locations.

THREE-PHASE CURRENT – Current delivered through three wires, with each wire serving as a return for the other two.

THREE-PHASE THREE-WIRE SYSTEM – An alternating current supply system comprising three conductors over which three-phase power is sent.

THREE-QUARTER-HARD WIRE – As applied to aluminum, wire that has been processed to produce a strength approximately midway between that of half-hard wire and that of hard-drawn wire.

THREE-WIRE SYSTEM – A DC or single-phase AC system comprising three conductors, one of which is maintained at a potential midway between the potential of the other two.

THW – A UL cable type. Thermoplastic vinyl-insulated building wire. Flame-retardant, moisture and heat resistant. 75°C rated in dry and wet locations.

THWN – A UL cable type. Same as THW but with nylon jacket overall. Rated 75°C in wet and dry locations.

TIA – Telecommunications Industries Association.

TINNED WIRE – See Coated Wire, Overcoat and Topcoated.

TIN OVERCOAT (TOC) – Tinned copper wire, stranded, then coated with tin.

TINSEL WIRE – A low voltage stranded wire, with each strand a very thin conductor ribbon spirally wrapped around a textile yarn.

TKGT – PTFE Teflon tape insulation with an insulating covering of felted K-fiber yarn and an overall sheath of braided glass yarn impregnated with a moisture, heat, flame and fraying resistant compound. 250°C 600 V apparatus and Motor Lead wire.

TL-9000 – A quality system for the telecommunications industry based on ISO 9000.

TNC – A threaded connector for miniature coax; TNC is said to be an abbreviation for threaded-Neill-Concelman. Contrast with BNC.

TOPCOATED – Bare (untinned) copper wire, stranded then coated with pure tin.

TPE – Thermoplastic Elastomer.

TRACER – A means of identifying polarity.

TRANSCEIVER – A device required in baseband networks which takes the digital signal from a computer or terminal and imposes it on the baseband medium.

TRANSCEIVER CABLE – Cable connecting the transceiver to the network interface controller allowing nodes to be placed away from the baseband medium.

TRANSITION SPLICE – A cable splice which connects two different types of cable.

TRANSMISSION – The dispatching of a signal, message, or other form of intelligence by wire, radio, telegraphy, telephony, facsimile, or other means (ISO); a series of characters, messages, or blocks, including control information and use data; the signaling of data over communications channels.

TRANSMISSION CABLE – Two or more transmission lines. See Transmission Line.

TRANSMISSION LINE – A signal-carrying circuit with controlled electrical characteristics used to transmit high-frequency or narrow-pulse signals.

TRANSMISSION LOSS – The decrease or loss in power during transmission of energy from one point to another. Usually expressed in decibels.

TRANSPOSITION – Interchanging the relative positions of wires to neutralize the effects of induction to or from other circuits or, to minimize interference pickup by the lead-in during reception.

TRAY – A cable tray system is an assembly of units or sections and ancillary fittings, made of noncombustible materials used to support cables. Cable tray systems include ladders, troughs, channels, solid bottom trays and similar structures.

TRAY CABLE – A factory-assembled multiconductor or multipair control cable approved under the National Electrical Code for installation in trays.

TREEING – Microscopic tree-like channels in medium voltage, e.g., 15 kV, cable insulation that can lead to cable failure.

TRIAXIAL – A three conductor cable with one conductor in the center, a second circular conductor concentric with and insulated from the first, and a third circular conductor insulated from and concentric with the second, and an impervious sheath overall.

TRIBOELECTRIC NOISE – Noise generated in a shielded cable due to variations in capacitance between shielding and conductor as the cable is flexed.

TROLLEY WIRE – A round or shaped solid, bare, hard conductor ordinarily used to supply current to motors through traveling current collectors.

TRUNK CABLE – A main cable used for distribution of signals over long distances throughout a cable system.

TRUE CONCENTRIC – A cable conductor in which each successive strand layer has a reversed direction of lay from the preceding layer.

TR-XLP – Water tree retardant cross-linked polyethylene.

TUBING – A tube of extruded nonsupported plastic material.

TURNKEY SYSTEM – Any system that is completely assembled and tested and that will be completely operational by turning it "on."

TV CAMERA CABLE – Multiconductor (often composite) to carry power for camera, lights, maneuvering motors, intercom signals to operators, video, etc. Usually heavy duty jacketed.

TW – A UL wire type. Thermoplastic vinyl-jacketed building wire, moisture resistant and rated 60°C.

TWINAXIAL CABLE – A shielded coaxial cable with two central insulated conductors.

TWIN COAXIAL – A configuration containing two separate, complete coaxial cables laid parallel or twisted around each other in one unit.

TWIN-LEAD – A transmission line having two parallel conductors separated by insulating material. Line impedance is determined by the diameter and spacing of the conductors.

TWINNER – A device for twisting together two conductors.

TWINNING – Synonymous with pairing.

TWISTED PAIR – A pair of insulated copper conductors that are twisted around each other, mainly to cancel the effects of electrical noise; typical of telephone and LAN wiring.

U

U-BEND TEST – A cable test in which the insulation is tested for resistance to corona and ozone.

UF – A UL cable type. Thermoplastic underground feeder or branch circuit cable.

UHF – Ultrahigh frequency, the band extending from 300 to 3,000 MHz as designated by the Federal Communications Commission.

UL – Underwriters Laboratories, Inc.

UL LISTED – A product that has been tested and found to comply with applicable standards.

ULTRASONIC CLEANING – Immersion cleaning aided by ultrasonic waves which cause microagitation.

ULTRASONIC DETECTOR – A device that detects ultrasonic noise such as that produced by corona or leaking gas.

ULTRAVIOLET – Radiant energy within the wavelength range 10 to 380 nanometers. It is invisible, filtered out by glass and causes suntan.

UNBALANCED LINE – A transmission line in which voltages on the two conductors are unequal with respect to ground, e.g., coaxial cable.

UNBALANCED-TO-GROUND – Describing a two-wire circuit, where the impedance-to-ground on one wire is measurably different from that on the other, compare with balanced-to-ground.

UNIDIRECTIONAL CONDUCTOR – See Concentric-lay Conductor.

UNIDIRECTIONAL STRANDING – A term denoting that in a stranded conductor all layers have the same direction of lay.

UNILAY – More than one layer of helically laid wires with the direction of lay and length of lay the same for all layers. See Concentric-lay Conductor.

USE – A UL cable type. Underground service entrance cable, XLP or rubber-insulated, Hypalon or XLP jacketed.

UTP – Unshielded Twisted Pair. Two wires, usually twisted around each other to help cancel out induced noise in adjacent circuits. An unshielded twisted-pair cable usually contains four pairs in a single cable jacket.

V

V – Volts. The SI unit of electrical potential difference. One volt is the difference in potential between two points of a conducting wire carrying a constant current of one ampere when the power dissipated between these two points is equal to one watt.

VA – Volt-ampere. A designation of power in terms of volts and amperes.

VAR – A unit of reactive power that means volt-amperes, reactive.

VAR METER – An instrument used by power companies to measure the kvar consumption.

V BAND – A band of frequencies between 46 and 56 gigahertz.

VC – Varnished-cambric insulation.

VCSEL – Vertical-cavity surface-emitting laser is a type of semiconductor laser diode operating in the 850 nm wavelength window that is commonly used in Ethernet-based networks.

VDE – Association of German Electrical Engineers.

VELOCITY OF PROPAGATION – The transmission speed of an electrical signal down a length of cable compared to its speed in free space. Usually expressed as a percentage.

VG – Varnished-glass or nylon braid, 600 V or 3,000 V, 130°C.

VHF – Very high frequency, the band extending from 30 to 300 MHz (television channels 2 to 13 and most FM radio) as designated by the Federal Communications Commission.

VIDEO PAIR CABLE – A transmission cable containing low-loss pairs with an impedance of 125 ohms. Used for TV pick ups, closed-circuit TV, telephone carrier circuits, etc.

VISCOSITY – Internal friction or resistance to flow of a liquid: the constant ratio of shearing stress to rate of shear.

VLF – Very low frequencies, the band extending from 10 to 30 kHz, as designated by the Federal Communications Commission.

VOICE FREQUENCY (VF) – Describes an analog signal within the range of transmitted speech, typically supported by an analog telecommunications circuit.

VOLT – A unit of electrical "pressure." One volt is the amount of pressure that will cause one ampere of current to flow through one ohm of resistance.

VOLTAGE – Electrical potential or electromotive force expressed in volts.

VOLTAGE BREAKDOWN – A test to determine the maximum voltage insulated wire can withstand before failure.

VOLTAGE, CORONA EXTINCTION – The minimum voltage that sustains corona (partial discharge), determined by applying a corona producing voltage, then decreasing the voltage until corona is extinct.

VOLTAGE DIVIDER – A network consisting of impedance elements connected in series to which a voltage is applied and from which one or more voltages can be obtained across any portion of the network.

VOLTAGE DROP – The voltage developed across a conductor by the current and the resistance or impedance of the conductor.

VOLTAGE, INDUCED – A voltage produced in a conductor by a change in magnetic flux from an outside source.

VOLTAGE RATING – The highest voltage that may be continuously applied to a wire or cable in conformance with standards or specifications.

VOLTAGE STANDING WAVE RATIO (VSWR) – The ratio of the maximum effective voltage to the minimum effective voltage measured along the length of a mismatched radio frequency transmission line.

VOLTAGE TO GROUND – The voltage between an energized conductor and earth.

VOLUME RESISTIVITY – The resistance in ohms of a body of unit length and unit cross-sectional area.

VULCANIZATION – A chemical reaction in which the physical properties of a polymer are changed by reacting it with cross-linking agents.

VW-1 – Vertical wire flame test. Formerly designated as FR1. A UL fire rating for single conductor cables. The test is described in UL Standard 1581.

W

W – (1) Symbol for watt or wattage, (2) A UL cable type. Heavy duty portable power cable, one to six conductors. 600 V, without grounds.

WALL THICKNESS – The thickness of the applied insulation or jacket.

WATER ABSORPTION – A test to determine the amount of water absorbed by a material after a given immersion period.

WATER BLOCKED CABLE – A multiconductor cable having interstices filled with a water-blocking compound to prevent water flow or wicking.

WATER COOLED LEADS – Furnace Cables. High Energy Cables. Usually welding cable strands cabled with a hose core for carrying coolant – used in heavy duty welding equipment, electric furnace applications, plating and various chemical processes.

WATER TREEING – A type of insulation deterioration that can occur after long term immersion in water with an electrical stress applied.

WATERFALL – The point at which cables installed horizontally in a tray transition to a vertical section of tray.

WATT – A unit of electrical power. One watt is equivalent to the power represented by one ampere of current under a pressure of one volt in a DC circuit.

WAVEFORM – A graphical representation of a varying quantity. Usually, time is represented on the horizontal axis and the current or voltage value is represented on the vertical axis.

WAVE FRONT – (1) That portion of an impulse (in time or distance) between the 10 percent point and the point at which the impulse reaches 90 percent of crest value, (2) the rising part of an impulse wave.

WAVELENGTH – The distance between the nodes of a wave. The ratio of the velocity of the wave to the frequency of the wave.

WAVESHAPE REPRESENTATION – The designation of current or voltage by a combination of two numbers. For other than rectangular impulses: (a) virtual duration of the wave front in microseconds, (b) time in microseconds from virtual zero to the instant at which one-half of the crest value is reached on the tail. For rectangular impulses: (a) minimum value of current or voltage, (b) duration in microseconds.

WEEE – Waste Electrical and Electronic Equipment. A European requirement that promotes the recycling of obsolete electrical equipment. Also see RoHS.

WEIGHT RESISTIVITY – The resistance in ohms at a specified temperature of a copper wire of uniform cross section and of unit weight and unit length.

WELDING – Joining the ends of two wires, rods, or groups of wires (a) by fusing, using the application of heat or pressure or both, by means of a flame torch, electric arc, or electric current or (b) by cold pressure.

WHEATSTONE BRIDGE – A device used to measure DC resistance. See Bridge.

WICKING – The longitudinal flow of a liquid in a wire or cable due to capillary action.

WIRE – A rod or filament of drawn or rolled metal whose length is great in comparison with the major axis of its cross section.

WIRE BRAID – Flexible wire constructed of small size strands in tubular form. Used for shielding or connections where constant flexing is required.

WIRE GAUGE (AWG) – The American Wire Gauge, originally called Brown & Sharpe Gauge. A system of numerical wire sizes starting with the lowest numbers for the largest sizes. Gauge sizes are each 20.6 percent apart based on the cross-sectional area.

WIRE NUT – A closed-end splice that is screwed on instead of crimped.

WIRE-WRAPPED CONNECTION – A solderless connection made by wrapping bare wire around a square or rectangular terminal with a power or hand tool.

WIRE WRAPPING TOOLS – Portable electric tools and automatic stationary machines used to make solderless wrapped connections of wires to terminals.

WITHSTAND TEST VOLTAGE – The voltage that the device must withstand without flashover, disruptive discharge, puncture, or other electric failure when voltage is applied under specified conditions.

WP – Weatherproof construction for overhead wires.

WORKSTATION – (1) Input/Output equipment at which an operator works; (2) a station at which a user can send data to, or receive data from, a computer or other workstation for the purpose of performing a job.

WRAPPER – An insulating barrier applied as a sheet of tape wrapped around a coil periphery.

X

X – Symbol for reactance.

X BAND – A band of frequencies between 5,200 and 10,000 megahertz.

XHHW – A UL cable type. Cross-linked polyethylene insulated small diameter building wire rated 75°C wet and 90°C dry.

XHHW-2 – A UL cable type. Cross-linked polyethylene insulated small diameter building wire rated 90°C wet and dry.

XLP – Cross-linked polyethylene. Also written XLPE.

Y

YIELD STRENGTH – The point at which a substance changes from elastic to viscous.

Z

Z – Symbol for impedance.

ZETABON – Dow's trade name for an acrylic acid copolymer coated aluminum tape.

ZIPPER TUBING – Alpha's trade name for harnessing/jacketing material containing a zipper-track type closure. The zipper arrangement allows installation with no need to disconnect installed wire. See Loc-Trac.

ZYTEL – DuPont's trademark for nylon resins.

GENERAL INDEX

ANIXTER

D

F

ANIXTER

R

S

T

V

W

X

Z

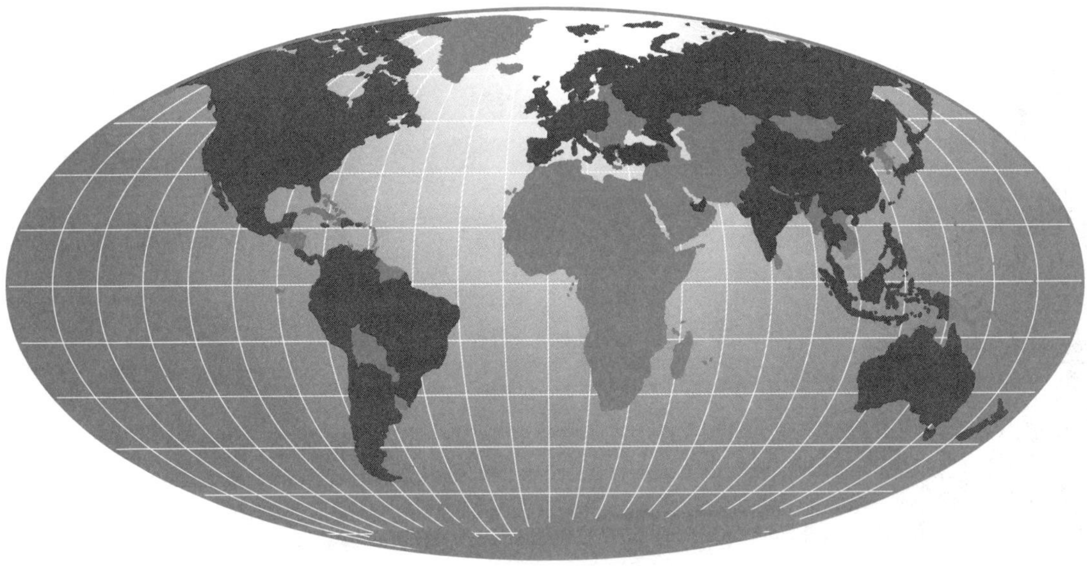

Anixter has over 249 sales and warehouse locations around the world in 49 different countries.

North America – USA

Worldwide HQ
Glenview, IL
224-521-8000

Alaska
Anchorage

Arizona
Phoenix

Arkansas
Little Rock

California
Anaheim
Burbank
Los Angeles
Pleasanton
Sacramento
San Diego
Union City

Colorado
Denver

Connecticut
Cheshire
Wallingford

District of Columbia
Washington

Florida
Fort Lauderdale
Jacksonville
Orlando
Tampa

Georgia
Atlanta

Idaho
Boise

Illinois
Bloomington
Mount Prospect

Indiana
Indianapolis

Iowa
Des Moines

Kansas
Kansas City

Kentucky
Louisville

Louisiana
New Orleans

Maryland
Baltimore

Massachusetts
Boston

Michigan
Detroit
Grand Rapids

Minnesota
Minneapolis

Missouri
St. Louis

Nebraska
Omaha

Nevada
Las Vegas

New Jersey
Cranbury
Florham Park

New Mexico
Albuquerque

New York
Albany
New York
Rochester

North Carolina
Charlotte
Raleigh

Ohio
Cincinnati
Cleveland
Columbus

Oklahoma
Tulsa

Oregon
Portland

Pennsylvania
Philadelphia
Pittsburgh

Tennessee
Memphis
Nashville

Texas
Austin
Dallas
El Paso
Houston

Utah
Salt Lake City

Virginia
Richmond

Washington
Seattle

Wisconsin
Appleton
Milwaukee

Canada

Canadian HQ
Toronto, Ontario
905-568-8999

Calgary
Edmonton
Halifax
London
Montreal
Ottawa
Quebec City
Rouyn-Noranda
St. John's
Saskatoon
Sudbury
Thunder Bay
Vancouver
Winnipeg

Latin America

Latin American HQ
Glenview, IL
224-521-8000

Argentina
Buenos Aires

Barbados

Brazil
São Paulo

Carribean (English)
Miami, FL

Chile
Santiago

Colombia
Bogotá

Costa Rica
San Jose

Dominican Republic
Santo Domingo

Ecuador
Quito

Jamaica
Kingston

Mexico
Ciudad Juarez
Guadalajara
Mexico City
Monterrey

Panama
Panama City

Peru
Lima

Puerto Rico
San Juan

Venezuela
Caracas

Europe/Middle East/Africa

European HQ
Uxbridge, England
44-1895-818181

United Kingdom
Aberdeen
Belfast
Birmingham
Glasgow
Uxbridge
Warrington

Austria
Vienna

Belgium
Willebroek

Czech Republic
Prague

Denmark
Copenhagen

Finland
Helsinki

France
Paris

Germany
Korntal-Münchingen

Greece
Athens

Hungary
Budapest

Ireland
Dublin

Italy
Milan
Rome

Netherlands
LE Capelle aan den IJssel

Norway
Oslo

Poland
Warsaw

Portugal
Lisbon

Russia
Moscow

Slovakia
Bratislava

Spain
Barcelona
Madrid

Sweden
Gothenburg
Stockholm

Switzerland
Montreux
Zurich

Turkey
Istanbul

United Arab Emirates
Dubai

Asia/Pacific

Asia/Pacific HQ
Singapore
+65-6756-7011

South Pacific HQ
Sydney, Australia
612-9-333-0800

Australia
Brisbane
Melbourne
Sydney

China
Shanghai

Hong Kong

India
Bangalore
Chennai
Mumbai
New Dehli

Indonesia
Jakarta

Japan

Malaysia
Kuala Lumpur
Penang

New Zealand
Auckland

Philippines
Manila

Singapore
Singapore

Taiwan
Taipei

Thailand
Bangkok

Vietnam

International Services

Glenview, IL
224-521-8000

After Hours

USA
800-323-8167

Canada
800-361-0250